Hillslope Hydrology

LANDSCAPE SYSTEMS

A Series in Geomorphology

Editor

M. J. Kirkby, *School of Geography, University of Leeds*

Hillslope Hydrology:
Edited by
M. J. Kirkby, *School of Geography, University of Leeds*

Hillslope Hydrology

Edited by

M. J. Kirkby

School of Geography
University of Leeds

A Wiley–Interscience Publication

JOHN WILEY & SONS

CHICHESTER · NEW YORK · BRISBANE · TORONTO

Library of Congress Cataloging in Publication Data:

Main entry under title:

Hillslope hydrology.

  'A Wiley–Interscience publication.'
  Includes bibliographical references.
  1. Hydrology. 2. Slopes (Geomorphology)
I. Kirkby, M. J.
GB665.H54   551.4'8'09144   77-2669
ISBN 0 471 99510 X

Set in Monophoto Times and printed in Gt Britain
by Page Bros (Norwich) Ltd, Norwich.

Acknowledgements

Acknowledgement is gratefully made for permission to reproduce the following:

Figure 1.1	The Geological Society of America.
Figure 1.3	The American Geophysical Union.
Figure 1.4	The United States Department of the Interior, Geological Survey.
Figure 1.5	The International Association of Hydrological Sciences.
Figure 1.6 and extract pp. 24/25	John D. Hewlett and the School of Forest Resources, University of Georgia.
Figures 1.8, 1.10	American Society of Civil Engineers.
Figure 1.9	Pergamon Press Limited.
Figure 2.8	H. J. Morel Seytoux and J. Khanji, *Water Resources Research*, **10**, 795–800, 1974, copyright by American Geophysical Union.
Figure 2.9	S. I. Bhuiyan, E. A. Hiler, C. H. M. van Bareland, and A. R. Aston, *Water Resources Research*, **7**, 1597–1606, 1971, copyright by American Geophysical Union.
Figure 2.10	K. K. Watson, *Journal of Geophysical Research*, **64**, 1959, copyright by American Geophysical Union.
Figure 2.14(a)	G. Vachaud, M. Vauclin, D. Khanji, and M. Wakil, *Water Resources Research*, **9**, 160–173, 1973, copyright by American Geophysical Union.
Figure 2.14(b)	C. Braester, *Water Resources Research*, **9**, 687–694, 1973, copyright by American Geophysical Union.
Figure 2.16	T. D. Biswas, D. R. Nielsen, and J. W. Biggar, *Water Resources Research*, **2**, 513–524, 1966, copyright by American Geophysical Union.
Table 2.1	McGraw-Hill Book Company.
Figures 3.5(a), 3.6(a), 3.6(b), 3.7	International Association of Hydrological Sciences.
Figure 3.8	D. W. Cole, *Water Resources Research*, **4**, 1127–1136, 1968, copyright by American Geophysical Union.
Figure 3.10	Natural Environment Research Council.
Figures 4.5, 4.6	Elsevier Scientific Publishing Co.
Figure 4.8	The Institute of British Geographers.
Figures 6.1, 6.7, 8.8, 6.11	R. A. Freeze, *Water Resources Research*, **7**, 347–366, 1972, copyright by American Geophysical Union.
Figures 6.12, 6.13	R. Ragan, *Water Resources Research*, **2**, 111–121, 1966, copyright by American Geophysical Union.
Figures 6.14, 6.15, 6.16	R. E. Smith and D. A. Woolhiser, *Water Resources Research*, **7**, 899–913, 1971, copyright by American Geophysical Union.

Contributors

C. V. ARDIS Jr., *Hydrologic Research and Analysis, Tennessee Valley Authority, Knoxville, Tennessee 37902, USA.*

T. C. ATKINSON, *School of Environmental Sciences, University of East Anglia, Norwich, UK.*

R. P. BETSON, *Hydrologic Research and Analysis, Tennessee Valley Authority, Knoxville, Tennessee 37902, USA.*

R. J. CHORLEY, *Department of Geography, University of Cambridge, Cambridge, UK.*

T. DUNNE, *Department of Geological Sciences, and Quaternary Research Center, University of Washington, Seattle, Wash., USA.*

W. W. EMMETT, *US Geological Survey, Denver, Colorado, USA.*

R. A. FREEZE, *Department of Geological Sciences, University of British Columbia, Vancouver, British Columbia, Canada.*

M. J. KIRKBY, *School of Geography, University of Leeds, UK.*

B. J. KNAPP, *Leighton Park School, Reading, Berkshire, UK.*

R. Z. WHIPKEY, *Agricultural Research Service, New England Watershed Research Center, South Burlington, Vermont 05401, USA.*

Contents

Series preface

Knowledge of our immediate physical environment has been gathered equally from *descriptive* and from *experimental* studies. Within descriptive studies I would include the analysis of geographical distributions, the construction of biological taxonomies, and the elucidation of the stratigraphic column. Experimental studies are mainly concerned with the study of processes at all scales. A science may develop from a purely descriptive to an experimental stage, and, in principle, the physical environment responds in a deterministic manner to external changes, and could therefore be studied as purely experimental science. In detail, however, our environment shows great variability even within a small area, so that accurate prediction often appears to be an impracticable goal. At the same time, there are, for instance, only a limited number of major climatic types, so that theories of, say, landscape formation can never be more than partially tested. With such difficulties, environmental science still seems to be a long way from a purely experimental stance, and consequently the need for collaboration between descriptive and experimental scientists will be with us for the foreseeable future.

Although it was, of course, backed by a considerable body of previous research, I believe that the present phase of quantitative geomorphology was signalled most significantly by the publication of Leopold, Wolman and Miller's *Fluvial Processes in Geomorphology* in 1964. Since that date, newer books have made this recent work accessible to schools, and while others are more specialist and advanced texts. The present series, on *Landscape Systems*, is an attempt to bring work in active research fields to a wider audience at postgraduate level, and to draw together concurrent work in the several disciplines relevant to each of the fields covered.

Each volume of the series is being written by a number of authors, each of whom is contributing within his own field of expertise, but has been asked to collaborate within an overall framework. In this way, important topics in geomorphology are covered comprehensively enough to provide a coherent advanced text without masking differences of approach between the contributors. The series covers topics in geomorphology, but is highly relevant to other disciplines. For example, the volume on *Hillslope Hydrology* is very directly applicable to workers and students in Hydrology, Forestry, Agriculture or Soil Science; and a similar breadth of relevance is intended for the second (*Soil Erosion*) and subsequent volumes.

Professor M. J. Kirkby

Preface

As a predictive science, hydrology dates from the 1930s. Surface water hydrology has long proved its worth for reservoir design and flood forecasting in particular, and this success slowed the further development of the science until the 1960s. The strength of hydrology in this period rested on the effectiveness of some rather simple concepts, especially the infiltration theory of surface runoff for estimating how much of the rainfall was effective in producing surface runoff, and the unit hydrograph model for forecasting the timing of river discharge following effective rainfall. Forecasting based on these concepts proved fully adequate for large natural catchments and small impermeable areas, but was less successful for small headwater areas.

Since 1960 two major trends have transformed hydrology. The first is the availability of increasingly-powerful computers which have allowed the development of much more complex models. The second is the multiplication of field catchment studies, with a tendency to dense instrumentation of small areas in a period of exceptional funding for all research. Both of these trends have led to a new interest in describing and modelling hydrological processes for a single soil profile or a hillslope profile. Increased computer power logically led, in one direction, towards models in which each elementary unit represents a single block of soil. Field studies showed that the implications of the earlier models are often not valid for a site, particularly for forest soils. This realization has led to a re-examination of the physical processes involved.

The collection of new observational data and the need for better information in computer models has led to a questioning of the fundamental concepts of surface water hydrology. The book is an attempt to put the results of this new work together for the first time in a research-level text. The authors who have contributed to this book are all active in the field of hillslope hydrology and bring different viewpoints to the subject which reveal the variety of opinion normal to a developing subject. Each chapter may therefore be read in isolation, and contains an introduction which sets out the author's view in the context of hillslope hydrology as a whole. Apart from this necessary duplication, the chapter topics have been designed to provide a sequential text which covers the whole of hillslope hydrology as I see it.

Chapter 1 is a broad introductory survey of hillslope hydrology, but also sets it in the context of atmospheric hydrology and of earlier hydrological work. Chapters 2 to 4 examine the physical processes of infiltration and subsurface flow together with methods for measuring them, and Chapter 5 similarly examines overland flow. Subsequent chapters begin to integrate this material and examine its implications. Chapter 6 shows how these detailed processes may be built into computer models of the hillslope flow regime, and Chapters 7 and 8 discuss the ways in which the hydrological processes may combine to produce output hydrographs from the hillslope and drainage basin, respectively. Lastly, in Chapter 9 the implications are carried a stage further, towards short-term sediment production and drainage basin formation in the long term.

Although hillslope hydrology has developed historically from specialized research, I believe that the subject has implications which are much broader than for traditional hydrology. It is only by relating large-scale hydrology to statements about water distribution which are true for a single soil profile that hydrologists can communicate with other environmental scientists to their mutual benefit. Even if our goal is not a model which incorporates all aspects of ecology, pedology and geomorphology, we must learn enough to discover what are the really important controls on and effects of the hydrological system if we are to manage it effectively: and I believe that we still have much to learn.

In editing this book, I have received a vast amount of help which I would like to acknowledge. The contribution of my co-authors is evident in the form of the chapters they have written, but I would also like to thank them for their encouragement in getting this joint project started and keeping it going to completion. My thanks also go to my wife Fiona for her direct assistance and support; to my colleagues and students at Bristol and Leeds who have read or listened to drafts of the manuscript, even if unwittingly, and have commented on it; to Margie Salisbury who kept track of the manuscript in its various stages, and to Gordon Bryant and his staff who turned many of my sketches into diagrams.

CHAPTER 1

The Hillslope
Hydrological Cycle

R. J. Chorley

Department of Geography,
University of Cambridge, UK

1.1 THE CLASSICAL MODEL

Forty years ago Robert Elmer Horton (1933), at the age of 58, first outlined in full the classical model of hillslope hydrology in terms of his infiltration theory of runoff, although he had already published on the same subject two years previously. Central to his analysis of runoff was the view that the soil surface acts as a sieve capable of separating rainfall into two basic components (Horton, 1937a):

The surface of a permeable soil acts like a diverting dam and head-gate in a stream ... with varying rain-intensity, all of the rain is absorbed for intensities not exceeding the infiltration capacity, while for excess rainfall there is a constant rate of absorption as long as the infiltration capacity is unchanged. As in the case of the dam and head-gate, there is usually some pondage which remains to be disposed of after the supply to the stream is cut off, so in the case of infiltration, surface-detention remains after rain ends. Infiltration divides rainfall into two parts, which thereafter pursue different courses through the hydrological cycle. One part goes via overland flow and stream-channels to the sea as surface runoff; the other goes initially into the soil and thence through the groundwater flow again to the stream or else is returned to the air by evaporative processes. The soil therefore acts as a separating surface and the author believes that various hydrologic problems are simplified by starting at this surface and pursuing the subsequent course of each part of the rainfall as so divided, separately (Horton, 1933).

In a stream of subsequent papers, ending only with his death, Horton (1935; 1937a; 1939a; 1939b; 1941; 1942; 1945) developed this simplistic hydrological model to such an extent that in 1946, Cook was able to refer to contemporary hydrology as the 'era of infiltration'. A particularly happy circumstance strengthened Horton's thesis in that his assumptions that the excess of rainfall intensity over infiltration capacity is the sole source of runoff quick enough to

1

produce the stream hydrograph peak and that all infiltration would pass into groundwater and was the sole source of the baseflow part of the hydrograph (Amerman and McGuinness, 1967) fitted so well Sherman's (1932) highly-influential unit hydrograph theory of basin runoff. Horton's overland flow from rainfall excess immediately became associated with Sherman's unit hydrograph, after the separation of baseflow (Horton, 1937a; Tischendorf, 1969), and the influence of this idea has persisted in sophisticated computer-based models of basin runoff (Amorocho and Hart, 1964).

Horton's ingenuity lay in the sweeping assumption that, of all slope hydrological variables, the infiltration capacity of the soil surface was the easiest to measure with accuracy, and that from it, in conjunction with rainfall-intensity data, both surface runoff and total infiltration to groundwater might be determined (Horton, 1933). Recognizing that, with prolonged rain, this infiltration capacity would decrease asymptotically due to changes in surface moisture, rain-drop packing of the surface particles, entrapment of air, closing of sun cracks, the swelling and breakdown of the soil-crumb structure and in-washing of fine particles (Horton, 1933; 1941; 1945); Horton developed the following negative-exponential decay function (1933; 1939a; 1941; 1945) which could be employed to express changes of infiltration capacity (f) throughout given drainage basins as a rainstorm continued, so that, by comparing this with changes in storm-rainfall intensity (i), the rainfall excess ($i - f$), bearing a close relation to overland flow to the stream channels (Horton; 1937a), could be calculated, after a depression storage interval (Horton, 1933; 1937a).

$$f = f_c + (f_0 - f_c)\,e^{-kt} \tag{1.1}$$

where

f = maximum rate of instantaneous in infiltration (in/hr) at which the given soil can absorb precipitation as it falls

f_c = the limiting, steady minimum infiltration rate, assumed to be more or less constant for a given soil

f_0 = the initial maximum infiltration rate at the start of the storm (at $t = 0$)

k = a positive constant of permeability for the given soil

t = time (in hours) from the beginning of the storm

Horton (1939a and elsewhere) was able to verify the form of his infiltration-capacity equation with reference to the runoff-plot exponents conducted on the Marshall and Shelby silt-loams by Musgrave (1935) and on the Putnam soil by Neal (1938).

In its simplest form, the infiltration theory of runoff predicts that prolonged rain falling on the slopes of a drainage basin having a uniform initial infiltration capacity will, if its intensity is greater than the lower limiting infiltration capacity, ultimately produce overland flow (Hortonian) more or less simultaneously over all the basin after an initial abstraction due to surface storage (Amerman and

McGuinness, 1967). This surface runoff was considered to be the sole contribution to the storm-runoff hydrograph peak and to accrue more or less uniformly from all parts of the watershed, particularly after sufficient time had elapsed for basin stream outflow to achieve a steady state with respect to the rainfall excess. At such a time, steady Horton overland flow occurs widely over all slopes, being associated with a simple increase of discharge downslope such that the depth of flow is related to the distance from the divide by a power law (Kirkby and Chorley, 1967; Haggett and Chorley, 1969; Kirkby, 1969; Carson and Kirkby, 1972) (Figure 1.1). When the

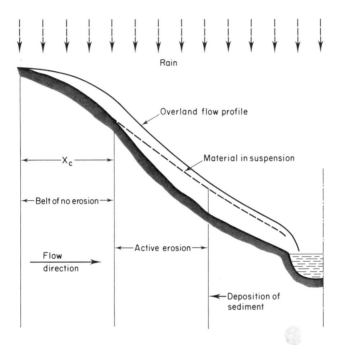

Figure 1.1   Hillslope profile depicting, overland flow processes and erosion according to the classic Horton model. X_C, Critical downslope distance (after Horton, 1945)

rain eases off and its intensity falls below the limiting infiltration capacity, the rainfall excess disappears, either by infiltration or surface runoff, first at the head of the slopes and then progressively downslope, usually within an hour or two for long slopes (Horton, 1937a). Horton (e.g. 1939b) devoted much attention to the hydraulics of the postulated thin film or sheet of laminar or mixed laminar and turbulent overland flow which he believed to occur so ubiquitously at times of rainfall excess, and which he believed to be commonly overlooked because of the effects of infiltration and of the roughness of the ground (Horton, Leach and Van Vliet, 1934), and extended his theory to include the mechanics of surface erosion as well as runoff (Horton, 1945). In the same way as Hortonian overland flow was solely responsible for the hydrograph storm peak, so it was assumed to be the sole motor of surface fluvial erosion. Under steady-state conditions, and characteristically, ignoring the possible variations due to infrequent rainfall intensities of high magnitude, Horton proposed that downslope of a critical distance from the divide (X_c) at which the depth of the sheet of overland flow is sufficient to generate a shear stress competent to entrain the surface soil particles, erosion would occur firstly in the form of rills which might subsequently coalesce to form new stream channels (Horton, 1945; Haggett and Chorley, 1969). As with the contemporary work of Sherman, it was similarly fortunate that some of the most influential work in the 'new wave' of post-war geomorphology (e.g. by Schumm, 1956a; 1956b) was carried out in small unvegetated drainage basins, with short slopes, little soil cover and low infiltration capacities. Such work reinforced the classical view of hillslope hydrology, namely, that the soil surface provides a sharp division of both hydrologic and geomorphic processes, with infiltrating water contributing only to longer-term soil- and ground-water recharge and to baseflow, and being irrelevant both as far as the storm hydrograph peak and surface erosion are concerned. Hortonian overland flow, on the other hand, was regarded as being the sole provider of the hydrograph peak in the basal stream flow and the motor of surface erosion (Kirkby and Chorley, 1967; Haggett and Chorley, 1969).

1.2   COMPONENTS OF THE HILLSLOPE HYDROLOGICAL CYCLE

Although the hydrological models of Horton and Sherman proved to be so historically complementary, they were fundamentally different in character. Whereas, in modern terms, Sherman employed sweeping assumptions to develop a 'black-box' approach to the prediction of storm runoff output from a watershed as the result of a storm input, Horton belonged to the traditional 'white-box' school of hydrology. The latter believed that the science could only be securely grounded if every component of the hydrological cycle could be investigated in terms both of its fundamental process and as to its relationships with associated components. Before seizing on the assumed fundamental rôle of infiltration as the key valve in the basin system, Horton had also researched into precipitation characteristics and movement, evaporation and transpiration, flood runoff and stream-channel hydraulics. It is therefore in a sense ironical that the research which, during the past quarter of a century, has undermined the universal application of Horton's model should have been that based on his own precepts of detailed investigation of the components of the hydrological cycle (Figure 1.2).

1.2.1   Interception

Horton (1919) produced the first definitive American report on forest interception, with particular reference to open-grown hardwood trees in New York State.

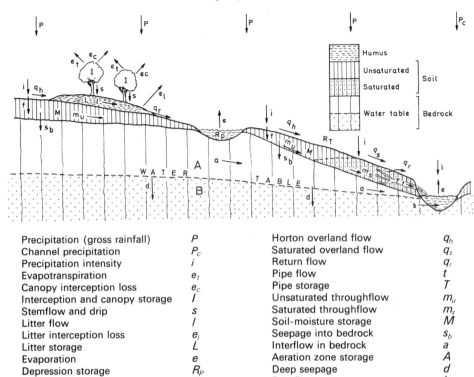

Figure 1.2   Components of the hillslope hydrological cycle

Precipitation (gross rainfall)	P	Horton overland flow	q_h
Channel precipitation	P_c	Saturated overland flow	q_s
Precipitation intensity	i	Return flow	q_r
Evapotranspiration	e_t	Pipe flow	t
Canopy interception loss	e_c	Pipe storage	T
Interception and canopy storage	I	Unsaturated throughflow	m_u
Stemflow and drip	s	Saturated throughflow	m_s
Litter flow	l	Soil-moisture storage	M
Litter interception loss	e_l	Seepage into bedrock	s_b
Litter storage	L	Interflow in bedrock	a
Evaporation	e	Aeration zone storage	A
Depression storage	R_P	Deep seepage	d
Detention storage	R_T	Baseflow	b
Infiltration	f	Groundwater storage	B

Despite his recognition of the range of interception possible, depending on vegetation species and density, Horton's work was characteristically directed towards providing rules of thumb regarding water abstraction and losses for this phase of the hydrological cycle. For a general cover of heavy-crowned, open-grown trees, Horton estimated that the first 1 mm of rainfall would be intercepted, together with some 20% of the remainder. Of the latter, some 1–5% would be expected to form stemflow and the rest would evaporate, giving a total throughfall of 75–79% in excess of the first 1 mm (Carson and Kirkby, 1972). Horton (1919) showed that total interception consisted of two parts: that satisfying the surface storage of the vegetation, and that which evaporates during rainfall (Linsley, Kohler and Paulhus, 1949). This was expressed by the equation:

$$V_i = S_i + C_p \cdot E_a \cdot t_R \qquad (1.2)$$

where

V_i = total storm vegetational interception
S_i = the storage capacity per unit of the projected area

C_p = the ratio of the vegetal surface area to its projected area
E_a = the evaporation rate per unit area of surface area
t_R = the duration of rainfall

However, for purposes of the infiltration theory of runoff, evaporation from leaf surfaces was later assumed to take place mainly after the rainstorm (Horton, 1937a) and little concern was directed to the special hydrological conditions associated with the litter-covered ground surface of the forest floor. Since Horton's death, much attention has been given to the special, but highly important, features of forest hydrology. For example, Hoover (1962) has shown how canopy interception forms a greater percentage of light showers and a smaller percentage for flood-producing storms. For storms in excess of 2 in, this interception might well be less than 0.2 in, but the litter covering the floors of conifer stands could be expected to have a field moisture capacity of up to twice this amount.

A most complete analysis of canopy and litter interception has been given by Helvey and Patric (1965) in respect of mature mixed hardwood forests of the eastern United States. Making up the gross rainfall (P), there is, firstly, a total interception loss (I) of rainfall retained in the canopy (C) and litter (L) which is evaporated without adding moisture to the mineral soil, and, secondly, the remaining infiltration into the soil. Precipitation intercepted by the canopy can, by dripping, add to that proportion of the gross rainfall which directly reaches the litter through spaces in the canopy to make up throughfall (T). Throughfall and stemflow (S) together form the sources of moisture supply for the litter, and the amount of this which enters the mineral soil has been termed the net rainfall (R). The following simple relationships have been derived:

$$R = P - I = P - (C + L) \tag{1.3}$$

$$C = P - (T + S) \tag{1.4}$$

$$L = (T + S) - R. \tag{1.5}$$

Helvey and Patric (1965) supported Horton in showing that both throughfall and stemflow show surprisingly little variation for hardwoods during the growing season, even over a wide range of canopy conditions ($T = 0 \cdot 901P - 0 \cdot 031$ and $S = 0 \cdot 041P - 0 \cdot 005$). However, it has been in terms of the influence of forest litter that recent work on forest hydrology has been most revolutionary and has led to the greatest modifications of the Horton hydrological model. Helvey and Patric (1965) showed that litter-interception losses could reach 5% of the annual precipitation and that it could be much more variable than canopy interception losses, particularly due to human intervention. More important, Tischendorf (1969) has drawn attention to the rôle of forest litter in promoting lateral, downslope flow, as distinct from interception losses or infiltration. Nutter (1968) referred to the ability of the litter layer to transmit water without changes in its moisture content and Lamson (1967) proposed that stormflow in a forested New England watershed was entirely the product of downslope flow in the litter zone.

1.2.2 Evapotranspiration

Classical hydrology treated both evaporation and evapotranspiration in terms of empirical pan and botanical observations directed towards the obtaining of average amounts of ratios of water loss from standing water or soil moisture so that subtractions from total precipitation could be made in order to arrive at estimates of precipitation effective in producing runoff. Horton (1923) himself analysed Höhnel's transpiration ratios directly observed from Austrian forests. More recent workers have been impressed both by the difficulties and inadequacies of such direct methods (Holmes, 1961; King, 1961; WMO, 1966).

Barry (1969) has described the three major types of approach to the estimation of evaporation and evapotranspiration:

1. Aerodynamic equations: involving measured temperatures, vapour pressure, etc.
2. Energy-budget equations: including terms involving solar radiation, transfer of sensible heat, transfer of latent heat, and the like, in an attempt to calculate the total amount of energy available for evaporation and evapotranspiration. A pioneer of this approach was the Russian scientist Budyko.
3. Combination methods: involving the use of both the above approaches. In this connexion the work of Penman has been of paramount importance.

Just after the death of Horton, Penman (1948; 1956) developed the formula:

$$H = E + K = R_c(1 - r) - R_B \qquad (1.6)$$

where

H = the total heat budget of the surface
E = the energy available for evaporation and evapotranspiration
K = the energy required to heat the air
R_c = the energy received from the sun
r = the surface albedo
R_B = the energy radiated by the earth

Using an aerodynamic approach, Penman obtained estimates for the transport of sensible heat and for the division of H between E and K, so as to arrive at a calculated value for the energy available for evapotranspiration (E) (More, 1967). Subsequently, Monteith (1959; 1965) improved the estimates of r in respect of vegetated surfaces. In 1963, Penman published a generalized equation for E.

$$E = \left(\frac{\Delta}{\gamma} \cdot H + E_a\right) \bigg/ \left(\frac{\Delta}{\gamma} + x\right) \qquad (1.7)$$

where

E_a = an aerodynamic expression for the drying power of the air, involving wind speed and saturation deficit
Δ = a constant, depending on temperature (being the slope of the saturation vapour-pressure curve at the mean temperature of the surface air)

γ = the constant of the wet- and dry-bulb psychrometer equation

$\dfrac{\Delta}{\gamma}$ = thus, a dimensionless factor weighting the relative effects of energy supply and ventilation on evaporation

H = the total heat budget

x = a factor depending both on the stomatal geometry of the plant cover and the length of day

Penman's calculations have been particularly impressive for estimating evapotranspiration from surfaces completely covered by short green crops under temperate conditions, where the supply of soil moisture is abundant. Under such conditions during the growing season, the vegetation can be crudely viewed as acting as an assemblage of evaporating wicks, the diurnal evapotranspiration from which is much more a function of meteorological variables than of biological ones. For example, Penman was able to show that during the summer months of 1949 some 39% of the energy received from the sun in east-central England was utilized in evapotranspiration. Such work as Penman's has been particularly valuable in calculating potential evapotranspiration on the assumption of maximum supplies of moisture, and, similarly, estimates of evaporation from bare saturated soil surfaces show approximately the same rates as for water surfaces at the same temperatures (Hoover, 1962). Difficulties arise, however, when estimates are required for conditions of sub-optimum moisture supply to soil and plant surfaces (Van Bavel, 1966); Pegg and Ward (1972), in comparing various methods of estimating evapotranspiration losses from a boulder clay catchment in Holderness, Yorkshire, have concluded that where evapotranspiration is near the potential, the Penman formula and the simple evapotranspirometer give the most accurate indirect and direct results, respectively, but that at values well below the potential, none of the methods employed are particularly satisfactory.

1.2.3 Depression storage and surface detention

The infiltration theory of runoff assumed that when the rainfall intensity exceeds the surface infiltration capacity the surface depressions are filled and then overflow to allow a thin sheet of overland flow to build up to a limiting depth at which the surface discharge at the base of the slope is in equilibrium with the rate of supply of surface runoff. Depression storage thus consists of the amount of water held in the surface depressions, none of which runs off (Horton, 1933), but which may subsequently be evaporated or infiltrated. Surface detention is that part of the rain which remains on the ground surface during the storm, gradually moving downslope by overland flow, and either runs off or is absorbed by infiltration after the rain ends (Horton, 1933; 1937 a). Surface detention is thus the storage effect due to overland flow in transit (Ven Te Chow, 1964). Cook (1946) has added the important linking idea of effective surface storage, defined as the depth of water that would have run off were it not for the existence of surface storage, this being depression storage plus the overland flow water infiltrated out in transit.

The effect of depression storage on storm runoff was considered by Horton (1937b), who made the simplifying assumptions that all depression storage must be filled before surface runoff begins and that infiltration is constant throughout the storm (Linsley, Kohler and Paulhus, 1949). It was recognized, however, that the effect of depression storage is mainly important in hillslope hydrology in that it affects the length of time from the beginning of the storm before equilibrium surface runoff can occur. Horton (1935) estimated that on moderate or gentle slopes, surface depressions can commonly hold the equivalent of 0·25–0·50 in of water, and even more for natural meadow and forest land. Bare smooth soil surfaces were later estimated to store between 0·20 and 0·10 in of water for sand and clay, respectively (Ven Te Chow, 1964). It must be remembered, however, in light of subsequent work on infiltration and the movement of soil moisture, that the above concept of the rôle of detention storage in the hillslope hydrological cycle, together with that of surface detention, depends on the simplistic assumptions underlying the infiltration theory of runoff.

Horton's work on surface detention arose primarily from his concern with the hydraulics of overland flow and he estimated that it commonly ranges from $\frac{1}{8}$ to $\frac{3}{4}$ in for flat areas and $\frac{1}{2}$ to $1\frac{1}{2}$ in for cultivated fields and natural grasslands or forests (Horton, 1935; Ven Te Chow, 1964). Even allowing for this to include depression storage, these estimates seem excessive for all but the most intense precipitation and, as will be seen, belief in the common and widespread occurrence of sheets of overland flow in soil-covered and vegetated areas is far less secure now than it was forty years ago. It is significant that the verification of Horton's general formula, involving the average depth of surface detention (δ_a):

$$q = K\delta_a^M ; \tag{1.8}$$

where

q = rate of surface runoff

K = a constant, depending on slope and surface characteristics

M = an exponent, depending on the degree of turbulance of flow

was examined under artificial conditions by Ree (1939). For runoff in a channel lined with a Bermuda grass sod, the latter obtained values of $1420/L$ for K (where L = slope length in feet) and of 5/3 for M.

1.2.4 Infiltration

Horton's infiltration theory of runoff hinged on simple assumptions regarding the controls over infiltration capacity and its temporal variation, yet of all the components of the hillslope hydrological cycle none has been subjected to the more critical scrutiny of recent years (Hillel, 1971). Not that the general decay pattern of infiltration capacity has been discredited—quite the reverse—but its detailed characteristics have been tested and questioned. Sherman (1944) showed how rates of surface infiltration are inverse functions of the volume of capillary moisture in the soil column and that surface-capillary intake decreases as the water penetrates deeper into the soil, although gravity flow in the larger channels continues to provide water at depth for lateral capillary absorption. The experimental work of Bodman and Colman (1943) on the Yolo sand and silt loams, which excluded impact effects, in-washing of colloids and compression of air below the wetting front (but not the possibility of clay swelling), confirmed that a decline of infiltration capacity still took place towards some possibly constant value, and they concluded that changes in the soil-infiltration rate during a storm are the result of the operation of physical laws governing the flow of water through the soil. This conclusion was later supported by Rubin's (1966) mathematical treatment which assumed no soil air compression, parameter hysteresis, fabric changes or areal heterogeneity. Previously, Holtan (1961) had proposed a variation of Horton's infiltration equation when supply of water at the surface is not the limiting factor, but the problems of estimating infiltration through wide ranges of surface supply remain. Smith (1972) developed a mathematical model for unsaturated soil-moisture infiltration, showing that the form of the decay curve associated with surface ponding depends not only on the initial soil-moisture content, as Horton himself recognized, but also on the rainfall rate. In general, one is still faced with the problem identified by Cook (1946) more than a quarter of a century ago, namely, the difficulty of equating observed surface runoff with measured infiltration rates (Calver, Kirkby and Weyman, 1972). It is clear that Horton's empirical infiltration equation gives poor results for short-term infiltration rates (Carson and Kirkby, 1972), which are precisely those most important in governing hillslope hydrology.

As Baver (1937) pointed out, water moves into and in the soil mass under the influence of both gravitational and capillary forces, the latter due to molecular forces between the soil particles and the water giving rise to very slow moisture movement from thicker to thinner capillary films. This concept was most effectively developed by Philip (1957–8), whose findings were verified by the mathematical model of vertical infiltration in layered soils developed by Hanks and Bowers(1962). Philip showed that, where water is supplied as fast as it can enter the soil, the upper layers are saturated such that a wetting front develops, and that the instantaneous infiltration rate (f) through time (t) can be expressed by the following equation containing two flow components (Kirkby, 1969; Carson and Kirkby; Gregory and Walling, 1973):

$$f = A + Bt^{-\frac{1}{2}} \tag{1.9}$$

The first component represents the conductivity flow under gravity occurring at a steady rate of transmission, determined by the soil character and expressed by a transmission constant (A), by unimpeded laminar flow through a continuous network of large pores (Kirkby, 1969). The second component is a diffusion term representing the slow filling-up of soil air pores from the surface downwards and from zones of low to high porosity by capillary flow, controlled by a diffusion constant (B) and subject to a declining gradient with time as water enters the soil, representing flow in very discrete steps from one pore space to the next, in a more or less random fashion (Kirkby, 1969). *A* and *B* are thus constants for a given soil, but dependent on its antecedent moisture constant. The Philip infiltration model has proved valuable in expressing the initial decay in the instantaneous infiltration rate, but has been less useful in predicting longer-term infiltration rates (Carson and Kirkby, 1972).

The hydraulics of the wetting front, conspicuously ignored by Horton, were investigated by Hansen (1955) who showed that the rate of water entry is less for dry soils than for wet ones, and that, after wetting, a transmission zone lies above a wetting zone, the latter bounded beneath by a wetting front. The transmission zone is one of essentially constant hydraulic conductivity and approximately 80% saturation, but as the wetting front is approached, the hydraulic conductivity is reduced. The movement of the wetting front is most rapid in soils of initially high moisture content but becomes more erratic as the latter is reduced, and in drier soils it advances by a jumping action. The wetting front, slowly advancing through the capillary pores, therefore became viewed as an effective restrictive layer to downward water movement and a possible promoter of lateral subsurface stormflow within the non-capillary region of the soil (Hoover, 1962; Tischendorf, 1969). Liakopoulos (1965a; 1965b) modelled the downward movement of the wetting front into sandy soil during surface infiltration and showed that drainage downwards past the initial wetting front often proceeds long after surface infiltration has ceased (Carson, 1969). Ignoring soil air resistance, he developed mathematical models corresponding to the physical problems of gravity drainage, evaporation, infiltration and capillary rise in soils, demonstrating the similarity of processes involved in the development of a wetting front downwards from infiltration and upwards from capillary rise (Carson, 1969).

The displacement of existing soil water by newly-infiltrating rainfall has long posed a problem, which has recently been attacked with both mathematical models and by experiments with radioactive tracers. Rubin and Steinhardt (1963), joined by Reiniger (1964), developed a mathematical treatment of soil-moisture movement during low-intensity rainfall by assuming the application of the Darcy- and continuity-equations, that the hydraulic conductivity and diffusivity of the soil are unique, positively- and monotonically-increasing functions of soil-moisture content, and that rainfall entering the soil can be considered as a continuous body of water. Among other results, the model predicted that the moisture contents of soil profiles during rainfall infiltration might be considerably influenced by rainfall intensity. Experiments were carried out on the Gilead sandy loam using radioactive tracers (Haskell and Hawkins, 1964; Horton and Hawkins, 1964; 1965) to try to measure the displacement of existing soil water, and it was shown that this is accomplished throughout most of the flow path by the downward displacement of water previously retained by the soil at field capacity, contrary to previous ideas of pore filling. This concept of bulk displacement of old rainwater by new rainwater was further tested by Zimmermann, Münnich and Roether (1966) who confirmed that new rainwater simply pushes old water downwards and that seeping water in a bare loamy soil slows up at a depth of 2–3 m, hours or days after heavy rainfall and is, in fact, old capillary water from the overlying soil set free by a pressure wave, the travelling velocity of which is equal to the seepage velocity of water in the wider pores. These authors went as far as to assert in 1966, that 'a single *rainfall* . . . forms a tagged layer of water that, although blurred by diffusion effects, passes downwards as a distinguishable water mass between the older rainwater below and the younger rainwater above'. It is interesting that, when applied to the lateral movement of saturated throughflow, this concept has achieved additional significance in hillslope hydrology.

The investigation of the hydraulic effects of soil layers of variable permeability has also revolutionized our view of hillslope hydrology, particularly in terms of the promotion of lateral water flow. Baver (1937) noted that infiltration rates are governed by the permeability of the least-permeable soil horizon and Swartzendruber (1960) modelled this concept mathematically and showed that the flow of ponded water through a soil profile is limited by the hydraulic resistance of the least-permeable layer. The field investigations of soil anisotropy by Reeve and Kirkham (1951) showed that, in some sites, horizontal permeability of saturated soil is greater than the vertical permeability, and, more recently, Wind (1972) has produced a hardware model simulating vertical unsaturated moisture flow in soils, with a separate vessel representing each layer. It is clear that in multi-layered soils the discontinuous decrease of hydraulic conductivity with depth may, under suitable antecedent moisture and rainfall conditions, lead to saturated conditions building up from the bases of several soil layers within which saturated lateral flow (throughflow) may occur (Kirkby, 1969; Calver, Kirkby and Weyman, 1972). This is particularly applicable to conditions in the most permeable A horizon (Betson and Marius, 1969).

It is true to say that, whereas both practical and theoretical work on infiltration processes has proceeded apace, there have continued to exist difficulties in linking infiltration rates with the excess of rainfall required to produce the overland flow presumed by Horton to be solely responsible for hydrograph peaks. Smith and Woolhiser (1971a; 1971b) produced a mathematical model for the generation of overland flow on a sloping infiltrating surface with a rainfall-ponded upper boundary. It is significant that, when this model was tested on a laboratory slope of fine sand and a natural pasture plot, artificial rainfalls as excessive as 9·9 and 7 in/hr, respectively, were used.

1.2.5 Overland flow

Although Horton placed such emphasis on overland flow (rainfall intensity–infiltration rate) as the origin of storm hydrograph peaks and as the motor of surface erosion, he was always strangely defensive regarding one's inability to commonly observe this phenomenon, particularly on vegetated and soil-covered slopes. His sequence of events when rainfall intensity exceeds the infiltration rate is listed by Cook (1946):

1. A thin water layer forms on the surface and downslope surface flow is initiated.
2. The flowing water accumulates in surface depressions.
3. When full, these depressions begin to overflow.
4. Overland flow enters micro-channels, which coalesce to form rills, which combine to form rivulets, which, in turn, discharge into small gullies, this being continued until discharge into major channels occurs.
5. Along each collecting channel, lateral inflow from the land surface takes place.

This simple model of infiltration and runoff is depicted in Figure 1.3. Cook, however, pointed to two serious deficiencies in Horton's infiltration approach to the calculation of surface runoff; firstly, that the calculation of surface runoff from rainfall intensity and infiltration rate only holds good for very small areas where time of transit can be virtually ignored and, secondly, that some surface runoff has existed for a time as subsurface flow which has been returned to the surface (Cook, 1946). Observations by Kirkham (1947) on an experimental hillslope plot in Iowa led to his view that during intense precipitation, water infiltrated downwards near the top of the hill, horizontally outwards over the middle of the slope and vertically upwards near the hillslope base due to 'artesian pressure' developed by downward seepage over the higher parts of the slope. As Kirkby (1969) has pointed out, Horton overland flow will occur instantaneously over a basin only if it is small and has areally-homogeneous soil, soil moisture, interception, depression storage and infiltration conditions. Further, although Horton overland flow is quite common where vegetation is sparse and soils thin, it is rare where there is a vegetative cover. For example, Tischendorf (1969) studied 55 storm events in the Whitehall Watershed in the Southeastern Piedmont (60 acres of vegetated surface, with regolith 30–100-ft deep) during the period January 1967 to March 1968 and observed no overland flow at all during that period, although 19 of the storms produced strong peaks in the runoff hydrograph. Similarly, Rawitz, Engman and Cline (1970), studying the water balance of a 16-hectare watershed on shallow sandstone and shale soils in east-central Pennsylvania during 10 storms, concluded that viable overland flow from a storm event there was a rare event, even though the streamflow hydrographs indicate a rapid response to rainfall and have all the characteristics usually attributed to surface runoff.

Kirkby and Chorley (1967) have suggested that where there is appreciable soil and vegetation, and especially where there is a humus or litter cover, little overland flow may be expected to occur over much of the drainage basin, except in the most extreme storms. Soil-moisture conditions, both at the start of the rain and during all but the most intense and protracted storms, are more areally and temporally variable than Horton assumed, and overland flow tends to occur where water is forced to the surface, or to remain on it by the complete saturation of the surface soil layers (Kirkby and Chorley, 1967). Many storms may be expected to produce overland flow from limited contributing areas at much lower rainfall intensities than are required to exceed the infiltration capacities over the whole basin and

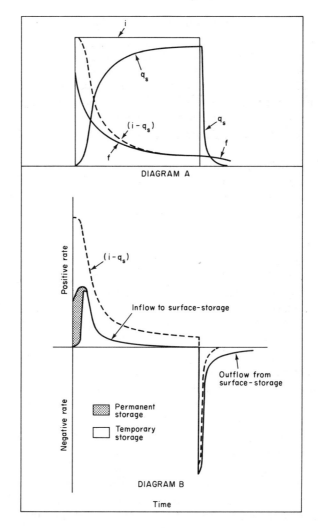

DIAGRAM A

DIAGRAM B

Figure 1.3   Variations in infiltration (f), surface runoff (q) and surface storage ($i - q$) on a slope during and after the application of rainfall of constant intensity (i), according to the Horton model (after Cook, 1946)

so to produce universal Hortonian overland flow (Haggett and Chorley, 1969). These limited areas (Figure 1.4) are:

1. Zones at the slope base, immediately marginal to stream channels where, despite the usually thicker soil, lateral soil drainage commonly produces high antecedent moisture conditions in the upper layers (Kirkby and Chorley, 1967; Jamison and Peters, 1967; Carson and Kirkby, 1972). The extent of

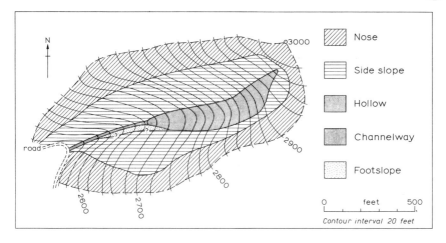

Figure 1.4 Contour map of a small valley on the west side of Crawford Mountains, Central Appalachians, showing a classification of slope areas into: Nose: runoff proportional to a function of the radius of curvature of the contours; side slope: runoff proportional to a linear function of slope length; hollow: runoff proportional to a power function of slope length; channelway: runoff proportional to a power function of channel length; foot slope: transitional between side slope and channelway (after Hack and Goodlett, 1960)

such contributing areas is initially controlled by the soil characteristics and antecedent moisture conditions, but as the storm continues the zone of saturation may extend upslope to an extent determined by the temporal pattern of storm intensity and the characteristics of the slope soil profile, including hydraulic efficiency and available soil-moisture storage (Rawitz, Engman and Cline, 1970; Calver, Kirkby and Weyman, 1972; Carson and Kirkby, 1972).

2. Concavities or topographic hollows where surface flowlines converge (Kirkby and Chorley, 1967; Dunne and Black, 1970a). Stream-head hollows are for this reason especially susceptible to surface runoff (Hack and Goodlett, 1960) (Figure 1.4).

3. Areas of thin soil cover.

This concept of limited surface runoff has resulted particularly from the past 15 years of research, with Hewlett (1961a and b) showing that lateral movement of hillslope soil moisture makes runoff most effective at the slope base and Betson (1964) that only part of a catchment may be expected to contribute to direct runoff (Calver, Kirkby and Weyman, 1972). This matter will be treated at more length later but it has led to the concept of saturation overland flow (Kirkby and Chorley, 1967) which occurs where the soil is saturated (partly by lateral flow in the soil), even though the local infiltration capacity has not been exceeded by the rainfall intensity. Such overland flow is non-Hortonian. Hewlett and Nutter (1970), recognizing that some surface runoff is commonly made up partly of water which has infiltrated and moved only a few inches or feet in the soil before seeping out downslope, have proposed to limit the term *overland flow* to rainwater that fails to infiltrate the soil surface at any point on its way from the basin to the gauging station. Thus overland flow would be largely viewed as saturation overland flow resulting from rain falling on already-saturated parts of the surface and should be logically treated as an expansion of the perennial channel system into zones of low-storage capacity (Hewlett and Nutter, 1970). Calver, Kirkby and Weyman (1972) have pointed to the spatial variability of surface runoff, together with the occurrence of throughflow and the delay between rainfall and throughflow, as an important source of non-linearity in the relationship of runoff to rainfall. It is also clear that saturation overland flow may, under certain circumstances of rainfall intensity and contributing areas, dominate the storm runoff of catchments (Dunne and Black, 1970b).

A later chapter by Emmett will describe the mechanics of overland flow, which has been estimated as moving at up to 27,000 cm/hr (Horton, 1945), or 0·25 ft/sec (10,000 ft/day) (Hewlett and Nutter, 1970). Emmett (1970) has already shown how overland flow, becoming increasingly turbulent, increases in depth downslope proportionally to some two-thirds of the power of the downslope increase in discharge, the remainder being attributable to downslope changes in velocity. This theoretical model, linked to the production of convex-straight-concave individual slope profiles, was developed as the result of field tests on sage-covered slopes in west-central Wyoming which were subjected to artificial rainfall at the significantly high intensities of up to about 8 in/hr.

1.2.6   Soil-moisture storage

The classical view of soil-moisture storage relied heavily on the concept of field-moisture capacity, the quantity of water that can be permanently retained in the soil in opposition to the downward pull of gravity (Horton, 1933). This has led to the complementary concepts of soil-retention storage and soil-detention storage, the latter consisting of that in excess of field-moisture capacity slowly draining down through large, non-capillary pores (Fletcher, 1952; Hoover, 1962). This concept, based on water content as distinct from pore water pressure, has been criticized by Liakopoulos (1965b) (*see also* Carson, 1969; Hillel, 1971). Traditional soil-moisture studies concentrated on the vertical variations within a given soil column, rather than spatial variations in terms of topographic position, but Helvey (1971) has reviewed the scanty and disparate literature on this topic prior to 1970. In 1955, Van't Woudt noted a general soil-moisture increase from the top to the base of a 50-ft, 30° slope in volcanic ash, and ascribed this to the greater depth of soil near the slope base wetted during a storm because of surface and subsurface lateral flow, and to a greater moisture retention and lower evaporation there. Stoeckeler and Curtis (1960) showed that the moisture content of the top 2 ft of soil-covered slope in part of Wisconsin is inversely related to distance above the stream channel (Helvey, 1971).

The most important work on temporal and spatial soil-moisture variations, however, has derived very recently from the observations of Helvey and his associates (1971) and Helvey, Hewlett and Douglass (1972). Soil-moisture changes were studied for $3\frac{1}{2}$ years on long hardwood-forested slopes over a wide area near Franklin, North Carolina, where relief was up to 3000 ft, rainfall between 30 and 90 in/year and soils up to 20-m thick, but averaging some 8 m. On all parts of the slopes (ridge, midslope and lower 'cove') soil moisture varied significantly with season, with the top 2 m showing an annual cycle approximating a sine wave with a maximum in mid-April and a minimum in mid-October (Helvey and Hewlett, 1962). These soil-moisture changes were naturally greater in the upper soil layers and it was shown that the changes at various depths in the profile depend mainly on lagged rainfalls. At the surface the soil moisture was found to be greatest in the lower 25% of the slope length; this was where seasonal changes were least and it was the only zone where position on the slope was found to be an important control over soil moisture.

1.2.7 Diffuse lateral soil-water movement

Although the concept of the lateral movement of water within the soil layers, so central to contemporary ideas of hillslope hydrology, was absent from Horton's classic infiltration model, its origins are almost as antique. Forty years ago, Lowdermilk (1934) referred to storm runoff as including 'shallow seepage or discharge of wet weather springs' and associated subsurface streamflow (neither true overland or groundwater flow) with the existence of certain soil profiles. Between 1936 and 1944, Hursh and his co-workers in the Southern Appalachians made important contributions to the study of what they termed *subsurface stormflow* which moved at a rate more rapid than usual groundwater flow (Hursh, 1936). Hursh and Brater (1941) suggested that in small forested catchments such stormflow along soil layers plus ground-water movement near stream channels could account for a major portion of the hydrograph peak, and that overland flow seemed of much less importance in this respect. Small plot studies revealed significant lateral flow in the top 12 in of the soil (Hursh and Hoover, 1941) and in small experimental basins it was proposed that gravitational flow of soil water could reach the stream in sufficient time to contribute to the storm hydrograph (Hursh and Fletcher, 1942). It was concluded that soil characteristics may, under certain conditions, be a more important control over runoff than basin morphometry (Hoover and Hursh, 1943) and that a dynamic form of subsurface water may result from the relatively shallow penetration of storm water into the porous upper-soil horizons and its rapid lateral flow downslope to natural outlets, independent of the position of the true water table (Hursh, 1944). Barnes (1939; 1944) also associated the hydrograph peak with what he termed *secondary base flow*; Cook (1943) was concerned with 'areas in which the capacity of the soil mantle to store infiltrated water is so small that a substantial portion of the volume of damaging flood flows is derived from subsurface runoff'; and Horner (1943) noted channel inflow taking place during the storm period from areas of high infiltration where overland flow is of little significance. Fletcher (1952) held that subsurface stormflow, moving laterally through large non-capillary pores in the soil detention storage reservoir, accounted for the major part of the total storm discharge from undisturbed forest watersheds, and a similar suggestion had been made by Roessel (1950) who pointed out that short, high-intensity rains will produce hydrograph peaks even without any surface runoff. Roessel, besides being influenced by the Swiss forest watershed observations of Burger, had also worked on tropical hydrology problems in Java and subsequent work has tended to point to lateral flow in the deep soil layers as being especially important in such cases (Burykin, 1957). Ruxton and Berry (1961) have noted the hydrological significance of the basal weathered surface of granitic debris in Hong Kong which 'usually corresponds with the lower limit of the seasonally fluctuating water-table' and that lateral mechanical eluviation can occur in the debris layers. Douglas (1969) has also noted that some soils in the humid tropics possess experimental percolation rates far in excess of probable rainfall intensities, although infiltration may be inhibited by splash effects.

It is, however, in the past 15 years that active research on 'throughflow' in soils (Kirkby and Chorley, 1967) has been intensified. In soils where there is a discontinuous decrease of hydraulic conductivity with depth, subject to suitable antecedent moisture and rainfall conditions, saturation may build up from the base of a soil layer within which saturated throughflow may occur downslope, the velocity of which is of the order of 20 cm/hr (Whipkey, 1965a), compared with average channel velocities which are about 45 cm/sec (Pilgrim, 1966). This saturated throughflow increases as the saturated layer becomes thicker and if the latter reaches the surface, saturated overland flow can result (Kirkby and Chorley, 1967; Kirkby, 1969; Calver, Kirkby and Weyman, 1972). Usually, however, throughflow occurs under unsaturated conditions, except close to flowing streams and at the base of especially permeable soil horizons (Carson and Kirkby, 1972). Discharge of saturated throughflow at the slope base increases with the lateral and vertical extent of the zone of saturation, and, as the antecedent moisture and the storm rainfall increases, the throughflow hydrograph increasingly assumes the features of the overload flow hydrograph (Calver, Kirkby and Weyman, 1972). For all but the shortest slopes, throughflow equilibrium does not have time to occur during a single storm, and for it to contribute significantly to the hydrograph peak throughflow must occur in a zone very close to stream channels (Kirkby and Chorley, 1967; Weyman, 1970), although where deep soil lies on impermeable bedrock throughflow can continue for weeks and contribute to the baseflow (Hewlett, 1961b; Carson and Kirkby, 1972).

This picture of throughflow has developed as a result of continued work in forested basins with thick-soil covers in the Southern Appalachians since about 1960 (TVA; 1960a; 1960b). One of the most notable workers in this respect has been Hewlett (1961a; 1961b) who discussed the consequences of the downslope drainage of soil moisture and showed that the lower parts of slopes could be expected to produce runoff early in the storm period while infiltration was still occurring on the higher area. He suggested that the saturated soil providing baseflow expands or shrinks in response to the interactions between recharge, soil moisture and precipitation. These observations were supported by work in Japan (Tsukamoto, 1963). In a very important paper, Hewlett with Hibbert (1963) experimentally observed throughflow in a 45-ft long trough of undisturbed soil on

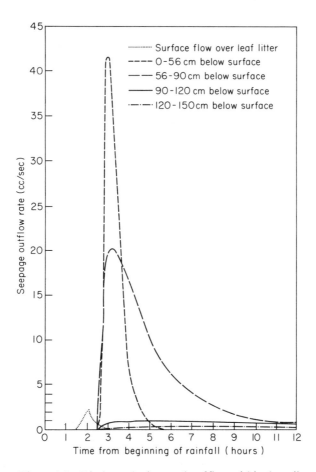

Figure 1.5   Discharge hydrographs of flow within the soil (dashed and continuous lines) resulting from a simulated storm of 5·1 cm/hour lasting two hours, occurring on a 16° slope which had previously drained for more than 4 days. The rapid, although small, surface flow of a Hortonian type (dotted) results from the initial low infiltration capacity of the dry surface soil, which rapidly increases with wetting. The lag before the throughflow begins is the time taken for rain to infiltrate vertically to the 90-cm less-permeable interface (after Whipkey, 1965a)

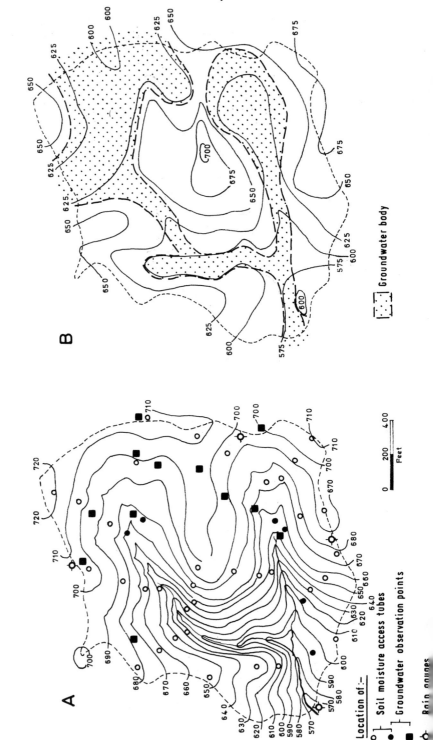

Location of :-

Soil moisture access tubes

Groundwater observation points

Rain gauges

Groundwater body

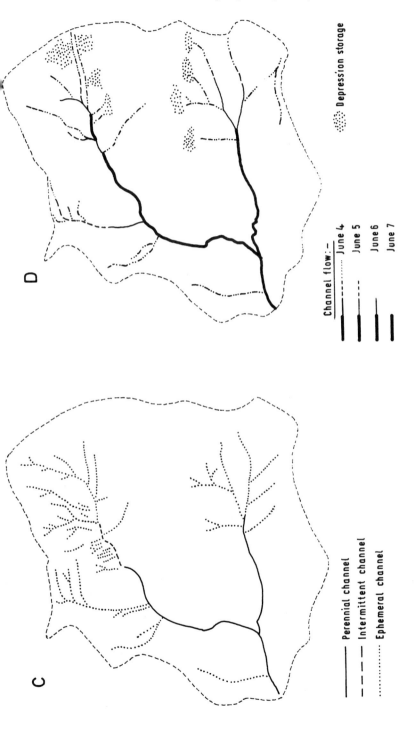

Figure 1.6 The Whitehall Watershed, Clarke County, Georgia (from Tischendorf, 1969). A, Topography (after J. L. Lambert); B, bedrock surface and groundwater body (after W. P. Neal, 1967); C, The drainage net; D, Shrinking net of water-carrying channels after 4·14 in of rain, June 1967

⌈ a 40% slope and the long-continued drainage from the unsaturated soil mass pointed to deep soils as providing a primary reservoir for sustained baseflow. Whipkey (1965a; 1965b; *see also* Whipkey and Fletcher, 1959), another important worker who is contributing to the current volume, investigated the effect of seepage of simulated rainfall of up to 5 cm/hr in a natural slope of sandy loam 17-m long. He was able to show that high-intensity storms falling on wet soil yielded the equivalent of up to 16% of the rainfall as runoff in 24 hr and that the individual soil layers produced individual seepage hydrographs with strongly-marked flood-like peaks (Figure 1.5). The same author (1967a; 1967b), referring to the work of Van Dijk (1958) and others, concluded that subsurface stormflow is a major component of total stormflow from forested catchments especially in fine-textured silt loam with impeded layers near saturation, containing channels and openings, and that long-continued flow from deep, coarse-textured soils would produce baseflow. Other experimental field studies on sandy or silt loams overlying clays showed that 1 in of simulated rainfall would drain from a 6·5-ft long plot of 6–18-in deep soil at 8% slope in 64–88 hr by subsurface flow (McDonald, 1967), and that on long slopes the amount of lateral 'recession flow' increases with the length and relief of the slope and that much of this is derived from the top inch or two of the surface soil horizon (Jamison and Peters, 1967). Similarly, Troendle (1970) and Troendle and Homeyer (1971) studied a natural soil profile isolated by ditching and noted that precipitation percolated to the lower horizons above the bedrock and then moved downslope.

During the past few years, three important catchment studies have added much to our knowledge of throughflow:

1. The Whitehall Watershed, a 60-acre forested catchment with 30–100 ft of regolith in the southern Piedmont investigated by Tischendorf (1969) who took 14,500 soil-moisture measurements at 42 moisture access tubes (Figure 1.6). Only 19 of the 55 observed storm events produced peaked direct runoff not completely attributable to channel precipitation; the maximum soil-moisture response took place in the top 4 ft of soil near stream channels in which throughflow occurred; at most, 7–8% of the basin area, mostly close to the channels, was contributing to subsurface stormflow and of this, quickflow (involving, at most, only 3–4% of the precipitation, excluding channel precipitation) all came from subsurface sources.

2. Sleepers River experimental watershed, a 0·6-acre slope, 60-ft high in Vermont where a 1–3 ft layer of sand overlies silt loam (Dunne and Black, 1970a; *see* Chapter 7). Subsurface runoff intercepted by a trench at the slope base was small and delayed for storms of recurrence intervals of 2 years, but an artificial 25-year storm (1·72 in in 2 hr) produced quick runoff which was mainly saturation overland flow plus the 'return flow' of water flowing in the upper soil horizon for a short distance in the lower concave part of the slope and then emerging at the surface. Total flow rates for the latter were estimated at 100–500 times that of wholly subsurface flow. The authors concluded that subsurface stormflow alone was too small, too late and too insensitive to

fluctuations in rainfall intensity to contribute significantly to the storm hydrograph.

3. East Twin Brook, a 0·21-sq km basin in Somerset (Weyman, 1970). Employing techniques similar to those of Whipkey (1965a), maximum throughflow peaks were obtained for the B-horizon, lagged for some hours for low rainfall intensities, and throughflow in the lower stony C-layer appeared to provide all the baseflow for the stream, even after 40 rainless days. It was concluded that soil layers must be saturated before throughflow contributions to the stream begin, that the basal saturated zone is partly supplied by unsaturated throughflow from upslope and that the instantaneous discharge from the slope depends on the hydraulic head of the saturated zone.

A general mathematical model for throughflow has been developed by adaptation of Darcy's law (Carson and Kirkby, 1972) and a specific model has been applied to flow in the layered soil of East Twin Brook (Calver, Kirkby and Weyman, 1972). A most important contribution to the field of deterministic mathematical modelling of saturated and unsaturated subsurface flow linked to streamflow has been made by Freeze (1972a; 1972b *see also* Chapter 6), whose theoretical work supports field runoff observed by Ragan (1967; 1968) and Dunne (1969; 1970) (*see also* Freeze 1969, 1971a and 1971b, together with the comments by Snyder, 1973; Knisel, 1973; Hewlett, 1974).

1.2.8 Concentrated lateral soil-water movement

A considerable indeterminacy exists within a spectrum of lateral soil-water movement between, at one extreme, the role of large non-capillary pores (Rode, 1959) and, at the other, concentrated flow in pipes and fissures (Gregory and Walling, 1973). The rôle of percolines, or seepage lines, orthogonal zones of relatively deep soil especially associated with first-order streams along which the downslope movement of moisture becomes concentrated (Bunting, 1961), allows a concentration of throughflow and suggestions have been made regarding the possible genetic relationships between percolines and soil pipes (Jones, 1976).

The hydrologic effects of root cavities and animal burrows have long been the subject of general comment and speculation (Hursh and Hoover, 1941; Reeve and Kirkham, 1951; Gaiser, 1952). Hoover (1962) noted that abundant plant and animal channels in surface soil layers increases their lateral permeability so that water movement may take place downslope before the underlying soils are completely wet. Patric, Douglass and Hewlett (1965) noted in the Coweeta Experimental Forest that the roots of most trees are concentrated in the top 6–7 ft of the thick soil profile, but little more analytical work on the hydrological rôle of biological cavities seems to have been accomplished.

The existence of interconnected pipe-like channels in soil due to eluviation of particles has also been considered to be of hydrological significance. Fletcher *et al.* (1954) noted that it may occur where an erodible layer exists above a layer of lower permeability and where there is an outlet for lateral flow. Jones (1976) has summarized the large but disparate literature on piping and has noted that it can occur in a variety of locations including flood plains marginal to stream channels and on steep valley side-slopes. A number of authors have pointed to the potential importance of pipeflow in assisting storm runoff even in unsaturated soils (Whipkey, 1969; Jones, 1971). Observations on the Plynlimon catchment have shown pipes to exist on the steeper slopes through which drainage occurs at times of heavy rainfall and initially-high moisture content (Knapp, 1970a; 1970b), and pipeflow has been shown to be a significant part of runoff from the Nant Gerig subcatchment (Gilman, 1971; Pond, 1971).

1.2.9 Deep seepage

Although Horton regarded recharge of the main groundwater body as an inevitable and complete consequence of surface infiltration, current views on the relations between infiltration and water table fluctuations are by no means as clear (Harding, 1937). Although in a uniformly-permeable rock, such as the Upper Chalk, there is a lagged but general relationship between infiltration and groundwater recharge, in cases where a marked decrease of permeability exists between the lower parts of the weathered soil mantle and the unweathered bedrock a more complex situation exists in which perched soil-water tables occur. The use of radioactive tracers has tended to emphasize the slow rate of recharge of the main groundwater body. Nixon and Lawless (1960) found in southern California that, in sand with a brush cover, after one rainless month water equal to 159% of evapotranspiration had been 'deeply translocated' to a depth of 20 ft, and that in a bare plot, 31% of the moisture supplied to the top 20 ft of soil was lost during a rainless period of 240 days. Similar work by Zimmerman, Münnich and Roether (1966) on unvegetated loamy soil suggested the layered downward diffusion of water from a given rainstorm, slowing up in a few hours or days at a depth of 2–3 m and perhaps taking years to reach the water table. In the Whitehall watershed, Tischendorf (1969) found relatively slow rates of vertical percolation in the soil (less than 3 in/day), giving response lags of up to 2 months at a 12-ft depth. He concluded that here the groundwater body was not being rapidly recharged through the soil mantle. In the East Twin Brook catchment, with bedrock of low permeability and a soil cover of some 75-cm thickness, Weyman (1970) gained the impression that the entire stream discharge originated as throughflow with no overland flow and no groundwater flow. For some environments where soil cover is deep, vegetated and much more permeable, the bedrock, Hewlett (1969) has concluded to have the following characteristics:

> Thus two fairly distinct systems supply streamflow; part of the baseflow comes at an almost constant rate from the deep groundwater aquifer, while the remainder of the baseflow and all the stormflow comes from the upper zone of the soil mantle and the open channel system. The two source systems join near the expanding and shrinking channel system, where an ephemeral rise in the groundwater table helps produce the

storm hydrograph and also sustains many days of baseflow. During and following exceptional rain storms, this ephemeral groundwater body extends rapidly upstream along intermittent channels forming a diffused, perched groundwater system that reaches maximum extent after rainfall ceases, and then retreats to the perennial channel within a few days. During winter, when reduced evapotranspiration fails to interfere with the desorption drainage of the top 6 ft of the soil mantle, baseflow is sustained for weeks by the slow drainage of unsaturated soil. During summer, early expansion of the subsurface source area under intense rainfall is shallower in depth and more rapid; consequently the retreat of the ephemeral channel and groundwater system is also more rapid, leading to the typically steep falling limb of the summer storm hydrograph.

1.3 MODELS OF HILLSIDE HYDROLOGY

Having begun with the apparently broad vistas of the Horton infiltration model of hydrology, we now find that it merely represents one end of a more or less continuous spectrum of such models. While providing a satisfactory model for the disposition of water on and within poorly-vegetated slopes having thin soil covers (the hydrologic properties of which differ little from those of the underlying unweathered rock) it is balanced by the throughflow model applicable to heavily-vegetated slopes with thick soil covers containing less-permeable layers promoting lateral flow and sharply overlying relatively-impermeable un-weathered bedrock (Kirkby and Chorley, 1967). Between these two extremes lie a variety of models in which the rapid runoff forming the hydrograph peak is variously conceived of as being composed of a varying mix of Hortonian overland flow, saturation overland flow, return flow (Dunne and Black, 1970b), unsaturated throughflow, saturated throughflow, and the mobilization en masse of 'old' soil water by the piston action of a new rainfall input (Hewlett and Hibbert, 1967) (Figure 1.7). Two lines of research have led to the development of these models during the past decade or so, firstly as we have seen, the measurement or inferences regarding the reality of throughflow, and, secondly, the recognition that it is rational to assume that the entire watershed may not contribute to storm runoff (Amorocho and Orlob, 1961; Southeastern Forest Experiment Station, 1961; Hewlett, 1974). This development of the so-called 'partial-area' view (Betson, 1964) has done much to control the use of basin area, described by Anderson (1957; *see also* Gregory and Walling, 1973) as the 'devil's own variable'.

The partial-area concept developed from the recognition that estimates of runoff, calculated from rainfall minus evaporation and infiltration, produced linear errors which could be explained by assuming that only the rainfall on a small and fairly constant part of each drainage basin is able to contribute to runoff during the hydrograph peak (Kirkby and Chorley, 1967; TVA, 1968). Betson (1964) described how the development of a non-linear mathematical model relating rainfall and runoff in small watersheds in the Southern Appalachians led to the detection of this error and how the assumption of a partial contributing area (itself

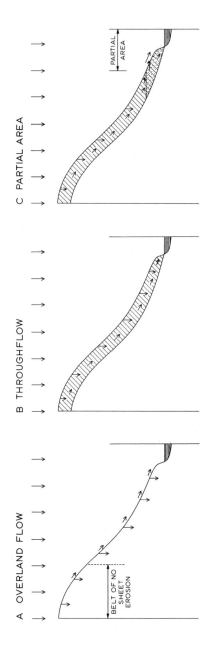

Figure 1.7   Schematic representations of the (A) overland flow, (B) throughflow and (C) partial area models of hillslope hydrology

a variable function of total storm rainfall, antecedent moisture conditions and instantaneous rainfall intensity) improved runoff predictions (TVA, 1960b, 1963a, 1963b and 1964; *see also* TVA, 1966 and 1968; Kirkby and Chorley, 1967). To these partial-area corrections, Amerman (1965) suggested adding those connected with unmeasured interflow (throughflow) and with the influence of upslope runoff on downslope runoff production. Dickinson and Whiteley (1970) later inferred analytically that rainfall/runoff non-linearities must result from runoff which varies both areally and over time, as a response to input and physiographic controls (Calver, Kirkby and Weyman, 1972). The magnitude of the partial contributing area was shown to vary significantly in the Southern Appalachians, being not more than 40% during heavy and prolonged rains (Southeastern Forest Experiment Station, 1961), 4–5% for light-to-moderate storms in forested catchments, but rising to a maximum of more than 80% where de-forestation had taken place (Betson, 1964). Subsequent work in the same vegetated region revealed that, depending on the storm rainfall and the antecedent moisture, the contributing area might be expected to vary between 5% and 20% of the total basin (TVA, 1966); and studies in Sleepers River, Vermont (Dunne, 1969; Johnson, 1969; Dunne, 1970) led to the mapping of the variable contributing area of saturated surface soil near the rivers (Dunne and Black, 1970b).

Thus, coincident with the partial-area concept, was developed its logical extension of the variability of source areas providing quick runoff, both between and within storms (Tsukamoto, 1961 and 1967; Ragan, 1967 and 1968; Nutter, 1969). These areas are envisaged as comprising bottom lands (TVA, 1965), especially those marginal to rivers (Dunne and Black, 1970b) and near watershed outlets (Betson, Marius and Joyce, 1968), and where the soil, especially the A horizon, is thin. Variations in the source area are dictated by the initial soil-moisture content, the soil-moisture storage capacity (especially of the A horizon) and the rainfall in intensity (Betson, 1964; TVA, 1965; Betson, Marius and Joyce, 1968; Betson and Marius, 1969; Dunne and Black, 1970b). When the upper soil horizon is saturated, the surface becomes a potential source of saturated overland flow and return flow, combined with saturated throughflow which yields water through the basal stream bank at a rate partly determined by the upslope extent of the saturated soil slab (Klute, Scott and Whisler, 1965), which is itself influenced by supplies of unsaturated throughflow from the slopes above. The variable source area can be viewed therefore in some hydrological respects as an expanded stream system, and Dunne and Black (1970b) have shown this effective expansion to be up to 170% of the natural channel system for Sleepers River. This notion represents an extension of empirical observations that the drainage density of flowing channels may bear a striking logarithmic relationship with discharge (Gregory and Walling, 1968 and 1973), with Carson and Sutton (1971) showing a three-fold increase of drainage density in a small forested Quebec watershed during a 104-mm rainstorm. Even though the total expanded stream network occupied less than 1% of the basin in the latter instance, it is clear that such channel area growth is associated with a disproportionately greater source area expansion, and Hanwell and Newsom (1970) showed that a five-fold increase in contributing channel length during floods in Swildon's Hole basin on the Mendip Hills went along with an increase of the contributing source area from 10% of the basin, during drier conditions, to more than 80%.

The dynamic role in hillslope and basin hydrology played by the variable partial source area under certain soil and slope conditions now forms the basis for the range of models which are coming to be recognized as lying between the extremes represented by the Horton and the pure throughflow models (Hewlett and Nutter, 1970). This dynamism may manifest itself in moisture and runoff changes on and within the surface soil of a variable source area which occur annually, seasonally, between storms and during storms (TVA, 1965; TVA, 1966; Dunne and Black, 1970b). As Tischendorf (1969) graphically put it; 'the source area is pulsating, shrinking and expanding in response to rainfall' (Figure 1.8). Although only some 2% to 10% of storm precipitation in a basin is involved in storm runoff or quickflow (Tischendorf, 1969; Rawitz, Engman and Cline, 1970), it is effectively yielded from usually a restricted source area which varies in size with the length ahd intensity of a given storm, such that the longer the rain continues the faster the runoff responds to precipitation changes (Tischendorf, 1969). The apportionment of storm runoff between the various flow sources associated with source areas on slopes remains at issue. Dunne and Black (1970b), have pointed to the ability of the

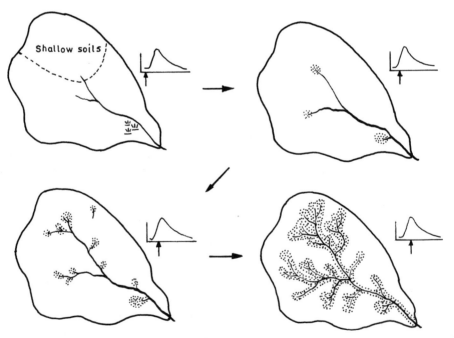

Figure 1.8   A time-lapse view of a basin showing expansion of the source area and channel system during a storm (after Hewlett and Nutter, 1970)

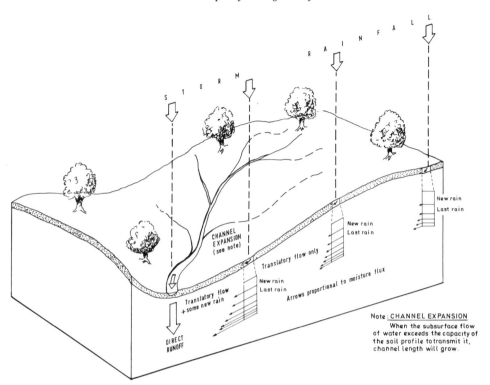

Figure 1.9   Diagram showing the source of stormflow from a forested watershed with a uniform soil cover. The interaction between the factors illustrated constitutes the variable source area concept (after Hewlett and Hibbert, 1967)

source area to generate overland flow (e.g. saturation overland flow plus return flow), whereas the remainder of the watershed acts mainly as a reservoir during storms to provide baseflow after the storm and to maintain the wet areas that will produce subsequent storm runoff. Hewlett and Hibbert (1967), while showing that laterally-blocked soil water will re-emerge as overland flow (return flow) to supplement saturation overland flow within a variable source area, postulate that, despite the relatively low velocities of throughflow, it is capable of producing storm-runoff peaks by the process of 'translatory flow'—namely, the relatively quick forcing-out of old rain at depth and near the slope base by a process of displacement, whereby distinct units of water are 'bumped' along the soil often to produce large amounts of storm runoff without appreciable overland flow (Figure 1.9). Indeed, overland flow may be regarded as an expansion of the perennial channel system into zones of low storage capacity fed from below by subsurface stormflow and by rainfall from above, such that 'the channel *reaches out* to tap the subsurface flow systems which, for whatever reason, have over-ridden their capacity to transmit water beneath the surface' (Hewlett and Nutter, 1969 and

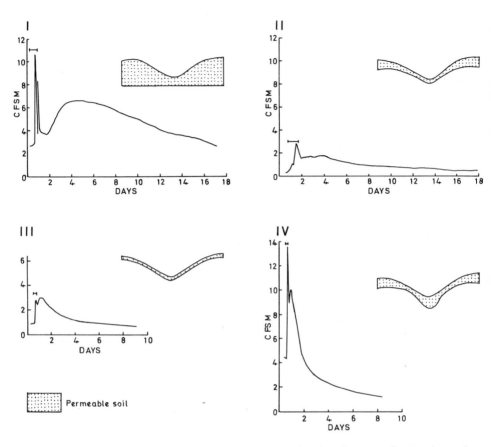

Figure 1.10   Hydrographs associated with four types of basin soil covers (after Hewlett and Nutter, 1970). The bar above each hydrograph represents the duration of the rainstorm. CFSM = cubic feet per second per square mile

Type I Deep soil. Hydrography from the Kimakia Basin, Kenya, a 130-acre catchment located on deep volcanic ash soils, associated with a rainstorm of 2·4 inches in 24 hr. Note the pronounced secondary peak some 4 days after the storm

Type II Deep soils on the divides. Hydrograph from the Wagon Wheel Gap Basin, Colorado, a 200-acre catchment, associated with a rainstorm of 3·14 inches in 24 hr. Note pronounced secondary peak

Type III Shallow soils. Hydrography from Whitehall Watershed, Georgia (see text and Figure 1.6), a 60-acre catchment, associated with a rainstorm of 1·07 inches in 8 hr. The initial peak is almost 'drowned out' by the larger secondary peak. There is a zone of restricted permeability at 3–6-ft depth

Type IV Deep soils on divides and in valley bottoms. Hydrograph from Coweeta 17 Basin, North Carolina. A high-intensity rainfall of 1·25 inches in 6 hr produces a well-marked initial peak which would have been 'drowned out' by the main rising limb if the rainfall had been of longer duration and lower intensity

1970). Given this definition of the expanded channel, the dynamic source-area concept 'visualizes the average storm hydrograph as composed of rain falling directly onto the channel plus water transmitted rapidly through wet soil adjacent to the stream. The area along the stream which contributes to the storm hydrograph shrinks and expands depending on rainfall amount and antecedent wetness of the soil. During rainless periods, streams are fed, to a large extent, by moisture migrating slowly downslope under conditions of unsaturated flow. This migration theoretically creates a gradient of increasing moisture content downslope from the ridge and provides a primed zone along the channel for quick release of water during storms' (Helvey, Hewlett and Douglass, 1972).

An idea of the range of possible models of hillslope hydrology has been presented for several differing soil conditions by Hewlett and Nutter (1970; *see also* Zalavski and Rogowski, 1969) (Figure 1.10). In these, a primary blip, produced by channel precipitation and non-Hortonian flow from the expanded source area (often drowned out by the main rising limb in all but short rainfalls of high intensity) is followed by the main hydrograph curve resulting from subsurface runoff. The small proportion of rainfall commonly involved in quickflow and the delay of most of the runoff until well after rainfall has ceased, show that the majority of stormflow is not derived from the simple overland flow (Hewlett and Nutter, 1970; Woodruff and Hewlett, 1971).

1.4 HILLSLOPE HYDROLOGY AND THE BASIN HYDROLOGICAL MODEL

In view of the advances in the understanding of hillslope hydrology during the past fifteen years or so, which have tended to sever many of the traditional links between overland flow and the unit hydrograph (Hewlett, Lull and Reinhart, 1969) and propagated the idea that infiltration is seldom limiting and that Hortonian overland flow is a special case of runoff (Hewlett and Nutter, 1970), it is strange that current models of basin hydrology are so little concerned with them. Although the rôle of soil moisture as a source of supply of baseflow or interflow is increasingly recognized (Dooge, 1968; Hall, 1968), as is that of those geomorphic factors which bear specifically on the zones of preferential soil-moisture supply (Hickok, Keppel and Rafferty, 1959; Troeh, 1964; Black, 1970), catchment runoff models such as those by Amorocho and Hart (1964) and Dawdy and O'Donnell (1965) are based on a simple Hortonian division between overland flow and baseflow controlled by infiltration capacity. The sophisticated Stanford Watershed Model (Crawford and Linsley, 1962; 1966a and 1966b), however, does include 'interflow', which is assumed to be proportional to local infiltration capacity (Crawford and Linsley, 1966a) together with an upper-zone storage, and it is interesting that two subsequent modifications of this model have been particularly concerned with aspects of hillslope hydrology. The Texas Watershed Model includes consideration of unsaturated flow of water within the soil under the influence of both gravitational and capillary forces (Moore and Claborn, 1971) and the

Kentucky Watershed Model places emphasis on indices of soil-moisture storage capacity and the rate at which infiltration occurs over the watershed (James, 1970 and 1972). Nevertheless, a runoff model which takes adequate account of post-Horton work in hillslope hydrology is still awaited.

1.5   APPENDIX

Of recent years the flood of contributions to the developing discipline of hillslope hydrology shows no sign of peaking out. Indeed the stage has been reached where it is no longer possible to provide both a complete and a coherent picture of the science. I have included in this appendix a limited number of articles which have made significant advances in the fields of infiltration, overland flow, throughflow, and dynamic partial contributing areas.

Much of recent work on infiltration has involved the construction of mathematical models. Braester (1973) provided a linear mathematical model to predict soil moisture near the surface and the rate of advance of the wetting front resulting from infiltration into a semi-infinite soil column and into a finite one with a constant water table. Parlange (1972 and 1975) has pursued similar ends employing more flexible non-linear methods. Wind and Doorne (1975) developed a computer-based mathematical model simulating non-steady unsaturated vertical flow in soils, Brustkern and Morel-Seytoux (1975) a mathematical model of one-dimensional infiltration into a homogeneous porous medium, Selim (1975) a model for steady state saturated flow through stratified hillsides of differing hydraulic conductivities, and Reeves and Miller (1975) tested a time compression approximation against hysteretic Darcy computations for estimating infiltration for erratic rainfall and showed that the former worked satisfactorily except for the greatest durations of the initial rainfall period.

It is interesting that the increasing emphasis on studies relating to throughflow and the dynamic contributing areas, reflected in Freeze's (1974) excellent analytical summary, has resulted in a decline in research specifically directed to overland flow. Overton (1974) treated the mechanics of surface hillslope runoff and Musik (1974) developed a lumped mathematical model for overland flow applied to a series of interacting reaches with unsteady uniform flow subjected to pulse inputs. In contrast, work on the field measurement, laboratory simulation and mathematical modelling of throughflow has proceeded apace. Weyman (1973 and 1974) investigated unsaturated and saturated lateral soil water flow in East Twin Brook, Somerset, where a thick soil layer is underlain by impermeable bedrock. It was confirmed that after rain unsaturated flow builds up a saturated zone in the B and C horizons within which flow is lateral according to Darcy's law such that a zone of saturation grows upslope from the slope base. A true storm hydrograph is produced here by surface, or near-surface runoff processes operating over only up to 32% of the basin area, whereas during subsequent drainage the basal saturated zone is progressively replaced by lateral unsaturated flow which supplies baseflow for up to 42 days. Throughflow on Plynlimon, Wales, was measured by Knapp

(1974), who suggested a model for its prediction. More recently, Arnett (1974 and 1976) examined lateral flow in soil on slopes in two field plots in North Yorkshire and found that some 75% of the observed variations in this flow depended on topsoil/subsoil permeability differences and that cracking and vegetative effects in the upper layers were also important, whereas slope angle, length, surface roughness and soil texture exercised little control. Anderson and Burt (1977a) used tensiometers in the field to demonstrate the control by a topographic hollow and its attendant throughflow over a storm discharge peak and the subsequent baseflow. The same authors (Anderson and Burt, 1977b) developed a laboratory model of a draining slope to determine that slope discharge is predominantly effected by saturated flow, rather than by tension gradients and hydraulic conductivities in unsaturated conditions. Two interesting mathematical models involving throughflow are those by Warrick and Lomen (1974) who examined seepage through a hillside where a curved water table is associated with high overall seepage rates or by seepage at lower elevations, and by Stephenson and Freeze (1974) who provided a mathematical model of snowmelt runoff on a layered slope.

The concept described variously as that of dynamic source areas, partial contributing areas or variable source areas has of recent years continued to suggest the need for a complex space–time model subsuming research on infiltration, overland flow and throughflow. Nutter (1973) showed how the distribution of soil water affects the contributing source area, suggested that subsurface flow produces hydrograph peaks and that unsaturated subsurface flow sustains the subsequent baseflow. Hewlett and Troendle (1975) compared the Horton runoff model with that of variable source areas and developed a complex model for the latter under storm rainfall. A very useful review of the variable source area concept was given by Dunne, Moore and Taylor (1975), whose field work led them to suggest various methods for recognizing and quantifying the seasonal and in-storm variations. Mathematical modelling has also become important in dealing with the distributional effects of channel networks and dynamic contributing areas on stream runoff (Beven and Kirkby, 1976) and in providing a dynamic contributing areas model which has been tested against measures of saturated surface area and stream discharge (Kirkby *et al.*, 1976). Finally, Beven (1977) has produced a mathematical model for transient, partially-saturated flow in a hillslope soil mantle demonstrating that the non-linear behaviour is a complex response to the interaction of topographic (especially convergence) and soil parameters, and that the initial conditions of the unsaturated zone are of major importance in governing the timing and magnitude of the hydrograph peak.

REFERENCES

(An extended list, including those from the glossary.)

Amerman, C. R., 1965, 'The use of unit-source watershed data for runoff prediction', *Water Res. Res.*, **1** (4), 499–508.

Amerman, C. R. and McGuiness, J. L., 1967, 'Plot and small watershed runoff: its relation to larger areas', *Trans. Am. Soc. Agric. Engrs.*, **10**(4), 464–466.

Amorocho, J. and Orlob, G. T., 1961, 'An evaluation of inflow–runoff relations in hydrologic studies'. *Water Resources Center Contribution 41, University of California, Berkeley*, 59 pp

Amorocho, J. and Hart, W. E., 1964, 'A critique of current methods in hydrologic systems investigation', *Trans. Am. Geophys. Union*, **45**, 307–321.

Anderson, H. W., 1957, 'Relating sediment yield to watershed variables', *Trans. Am. Geophys. Union*, **38**, 921–924.

Anderson, M. G. and Burt, T. P., 1977a, 'Automatic monitoring of soil moisture conditions in a hillslope spur and hollow', *J. Hydrol.*, **33**, 27–36.

Anderson, M. G. and Burt, T. P., 1977b, 'A laboratory model to investigate the soil moisture conditions on a drainage slope', *J. Hydrol.*, **33**, 383–390.

Arnett, R. R., 1974, 'Environmental factors affecting the speed and volume of topsoil interflow', *Institute of British Geographers, Special Publication No. 6*, pp. 7–22.

Arnett, R. R., 1976, 'Some pedological features affecting the permeability of hillside slopes in Caydale, Yorkshire', *Earth Surface Processes*, **1**, 3–16.

Barnes, B. S., 1939, 'The structure of discharge-recession curves', *Trans. Am. Geophys. Union*, 721–725.

Barnes, B. S., 1944, 'Subsurface-flow', *Trans. Am. Geophys. Union*, **5**, 746.

Barry, R. G., 1969, 'Evaporation and transpiration', in Chorley, R. J. (Ed.), *Water, Earth and Man*, Methuen, London, pp. 169–184.

Baver, L. D., 1937, 'Soil characteristics influencing the movement and balance of soil moisture', *Proc. Soil Sci. Soc. Am.*, **1**, 431–437.

Betson, R. P., 1964, 'What is watershed runoff?', *J. Geophys. Res.*, **69**, 1541–1552.

Betson, R. P., Marius, J. P. and Joyce, R. T., 1968, 'Detection of saturated interflow in soils with piezometers', *Proc. Soil Sci. Soc. Am.*, **32**(4), 602–604.

Betson, R. P. and Marius, J. B., 1969, 'Source areas of storm runoff' *Water Res. Res.*, **5**, 574–582.

Beven, K. and Kirkby, M. J., 1976, 'Towards a simple, physically-based, variable contributing area model of catchment hydrology', *Leeds University, Department of Geography, Working Paper 154*, 11 pp.

Beven, K., 1977, 'Hillslope hydrographs by the finite element method', *Earth Surface Processes*, **2**, 13–28.

Black, P. E., 1970, 'Runoff from watershed models', *Water Res. Res.*, **6**(2), 456–477.

Bodman, G. B. and Coleman, C. A., 1943, 'Moisture and energy conditions during downward entry of water into soils', *Proc. Soil Sci. Soc. Am.*, **8**, 116–122.

Braester, C., 1973, 'Moisture variation at the soil surface and the advance of the wetting front during infiltration at constant flux', *Water Res. Res.*, **9**, 687–694

Brustkern, R. L. and Morel-Seytoux, H. J., 1975, 'Description of water and air movement during infiltration', *J. Hydrol.*, **24**, 21–35.

Bunting, B. T., 1961, 'The rôle of seepage moisture in soil formation, slope development and stream initiation; *Am. J. Sci.*, **259**, 503–518.

Burykin, A. M., 1957, 'Seepage of water from soils in mountainous regions of the humid subtropics', *Pochvovedenie* (English translation), **12**, 90–97.

Butler, S. S., 1957, *Engineering Hydrology*, Prentice-Hall, New Jersey, 356 pp.

Calver, A., Kirkby, M. J. and Weyman, D. R., 1972, 'Modelling hillslope and channel flow', in Chorley, R. J. (Ed.), *Spatial Analysis in Geomorphology*, Methuen, London, pp. 197–218.

Carson, M. A., 1969, 'Soil moisture', in Chorley, R. J. (Ed.), *Water, Earth and Man*, Methuen, London, pp. 185–195.

Carson, M. A. and Sutton, E. A., 1971, 'The hydrologic response of the Eaton River Basin, Quebec, *Canad. J. Earth Sci.*, **8**, 102–115.

Carson, M. A. and Kirkby, M. J., 1972, *Hillslope Form and Process*, Cambridge University Press, 475 pp.

Cook, H. L., 1943, 'Report of committee on runoff 1942–43', *Trans. Am. Geophys. Union*, **24**, 422.

Cook, H. L., 1946, 'The infiltration approach to the calculation of surface runoff, *Trans. Am. Geophys. Union*, **27**, 726–743.

Crawford, N. H. and Linsley, R. K., 1962, 'The synthesis of continuous streamflow hydrographs on a digital computer', *Tech. Rept. 12, Dept. Civil Engr., Stanford Univ., California*.

Crawford, N. H. and Linsley, R. K., 1966a, 'Digital simulation in hydrology: Stanford Watershed Model IV, *Tech. Rep. 39, Dept. Civil Eng., Stanford Univ., California*.

Crawford, N. H. and Linsley, R. K., 1966b, 'A conceptual model of the hydrologic cycle', *12th General Assembly Internat. Assoc. Sci. Hydrology, Symposium on Surface Water, Publication 63*, pp. 573–587.

Dawdy, D. R. and O'Donnell, T., 1965, 'Mathematical models of catchment behaviour', *Proc. Am. Soc. Civil Engrs*, **91**(HY4), 123–137.

Dickinson, W. P. and Whiteley, H., 1970, 'Watershed areas contributing to runoff', *Proc. General Assembly Internat. Assoc. Sci. Hydrology, New Zealand*, pp. 12–16.

Dooge, J. C. I., 1968, 'The hydrologic cycle as a closed system', *Bull. Intern. Assoc. Sci. Hydrology*, **13**(1), 58–68.

Douglas, I., 1969, 'The efficiency of humid tropical denudation systems', *Trans. Inst. Brit. Geographers*, No. 46, 1–16.

Dunne, T., 1969, *Runoff Production in a Humid Area*, Ph.D. dissertation, The Johns Hopkins Univ., Baltimore, Maryland.

Dunne, T., 1970, 'Runoff production in a humid area', *Rept. ARS 41–160, Agric. Res. Ser. US Dept. Agric., Washington, D.C.*, 108pp.

Dunne, T. and Black, R. D., 1970a. 'An experimental investigation of runoff production in permeable soils', *Water Res. Res.*, **6**, 478–490.

Dunne, T. and Black, R. D., 1970b, 'Partial area contributions to storm runoff in a small New England watershed', *Water Res. Res.*, **6**, 1296–1311.

Dunne, T., Moore, T. R. and Taylor, C. H., 1975, 'Recognition and prediction of runoff-producing zones in humid regions', *Hydrol. Sci. Bull.*, **20**, 305–327.

Emmett, W. W., 1970, 'The hydraulics of overland flow on hillslopes', *US Geolog. Surv. Prof. Paper 662–A*, 68pp.

Fletcher, J. E., Harris, K., Peterson, H. B. and Chandler, Y. N., 1954, 'Piping', *Trans. Am. Geophys. Union*, **35**, 258–262.

Fletcher, P. W., 1952, 'The hydrologic function of forest soils in watershed management', *J. Forestry*, **50**, 359–362.

Freeze, R. A., 1969, 'The mechanism of natural groundwater recharge and discharge. I. One-dimensional, vertical, unsteady, unsaturated flow above a recharging or discharging groundwater flow system', *Water Res. Res.*, **5**(1), 153–171.

Freeze, R. A., 1971a. 'Three-dimensional, transient, saturated–unsaturated flow in a groundwater basin', *Water Res. Res.*, **7**(2), 347–366.

Freeze, R. A., 1971b, 'Influence of the unsaturated flow domain on seepage through earth dams', *Water Res. Res.*, **7**(4), 929–941.

Freeze, R. A., 1972a, 'Rôle of subsurface flow, in generating surface runoff. 1. Base flow contributions to channel flow', *Water Res. Res.* **8**(3), 609–623.

Freeze, R. A., 1972b, 'Rôle of subsurface flow in generating surface runoff. 2. Upstream sources areas', *Water Res. Res.*, **8**(5), 1272–1283.

Freeze, R. A., 1974, 'Streamflow generation', *Reviews of Geophysics and Space Physics*, **12**(4), 627–647.

Gaiser, R. N., 1952, 'Root channels and roots in forest soils', *Proc. Soil Sci. Soc. Am.*, **16**, 62–65.

Gilman, K., 1971, 'A semi-quantitative study of the flow of natural pipes in the Nant Gerig, subcatchment', *Inst. Hydrology, Subsurface Section Report 36.*

Gregory, K. J. and Walling, D. E., 1968, 'The variation of drainage density within a catchment', *Bull. Intern. Assoc. Sci. Hydrology*, **13**(2), 61–68.

Gregory, K. J. and Walling, D. E., 1973, *Drainage Basin Form and Process: A Geomorphological Approach*, Arnold, London, 456 pp.

Hack, J. T. and Goodlett, J. G., 1960, 'Geomorphology and forest ecology of a mountain region in the Central Appalachians', *US Geolog. Surv. Prof. Paper 347*, 66 pp.

Haggett, P. and Chorley, R. J., 1969, *Network Analysis in Geography*, Arnold, London, 348 pp.

Hall, F. R., 1968, 'Base flow recessions—A review', *Water Res. Res.*, **4**(5), 973–983.

Hanks, R. J. and Bowers, S. A., 1962, 'Numerical solution of the moisture flow equation for infiltration into layered soils', *Proc. Soil Sci. Soc. Am.*, **26**(6), 530–534.

Hansen, V. E., 1955, 'Infiltration and soil water movement during irrigation', *Soil Sci.*, **79**, 93–105.

Hanwell, J. D. and Newson, M. D., 1970, 'The great storms and floods of July 1968 on Mendip', *Wessex Cave Club, occasional publication* **1**(2), 72 pp.

Harding, S. T., 1937, 'Direct accretions to groundwater from rainfall', *Trans. Am. Geophys. Union*, **17**, 368–371.

Haskell, C. C. and Hawkins, R. H., 1964, 'D_2O^{24} Na method for tracing soil moisture movement in the field', *Proc. Soil Sci. Soc. Am.* **28**, 724–728.

Helvey, J. D. and Hewlett, J. D., 1962, 'The annual range of soil moisture under high rainfall in the Southern Appalachians', *J. Forestry*, **60**, 485–486.

Helvey, J. D. and Patric, J. H., 1965, 'Canopy and litter interception of rainfall by hardwoods of Eastern United States', *Water Res. Res.*, **1**(2), 193–206.

Helvey, J. D., 1971, *Predicting Soil Moisture in the Southern Appalachians*, M.Sc. dissertation, School of Forest Resources, Univ. Georgia, Athens, Georgia, 79 pp.

Helvey, J. D., Hewlett, J. D. and Douglass, J. E., 1972, 'Predicting soil moisture in the Southern Appalachians', *Proc. Soil Sci. Soc. Am.*, **36**, 954–959.

Hewlett, J. D., 1961a, 'Watershed management', *US Dept. Agric., Forest Ser., Southeastern Forest Experiment Station, Asheville, North Carolina, Report for 1961*, pp. 61–66.

Hewlett, J. D., 1961b, 'Soil moisture as a source of base flow from steep mountain watersheds', *US Dept. Agric. Forest Ser., Southeastern Forest Experiment Station, Asheville, North Carolina, Station Paper No. 132*, 11 pp.

Hewlett, J. D. and Hibbert, A. R., 1963, 'Moisture and energy conditions within a sloping soil mass during drainage', *J. Geophys. Res.*, **68**(4), 1081–1087.

Hewlett, J. D. and Hibbert, A. R., 1967, 'Factors affecting the response of small watersheds to precipitation in humid areas', in *Proceedings of the International Symposium on Forest Hydrology (1965), Pennsylvania State University*, Pergamon, pp. 275–290.

Hewlett, J. D., 1969, 'Tracing storm and base flow to variable source areas on forested headwaters', Tech. Rept. No. 2, School of Forest Resources, University of Georgia, Athens, Georgia, 21 pp.

Hewlett, J. D., Lull, H. W. and Reinhart, K. G., 1969, 'In defence of experimental watersheds', *Water Res. Res.*, **5**(1), 306–316.

Hewlett, J. D. and Nutter, W. L., 1969, *Forest Hydrology*, unpublished manual for the course 'Forest Hydrology', School of Forest Resources, University of Georgia, Athens, Georgia, 137 pp.

Hewlett, J. D. and Nutter, W. L., 1970, 'The varying source area of streamflow from upland basins', *Paper presented at Symposium on Interdisciplinary Aspects of Watershed Management, Montana State University, Bozeman*, American Society of Civil Engineers, New York, pp. 65–83.

Hewlett, J. D., 1974. Comments on letters relating to 'Rôle of subsurface flow in generating surface runoff', by Freeze, R. A., *Water Res. Res.*,

Hewlett, J. D. and Troendle, C. A., 1975, 'Non-point and diffused water sources: A variable source area problem', *Paper presented at Symposium on Watershed Management, Utah State University, Logan*, American Society of Civil Engineers, New York, pp. 21–46.

Hickok, R. B., Keppel, R. V. and Rafferty, B. R., 1959, 'Hydrograph synthesis for small arid-land watersheds', *Agric. Engr.*, **40**, 608–611; 615.

Hillel, D., 1971, *Soil and Water: Physical Principles and Processes*, Academic Press, New York and London, 288 pp.

Holmes, R. M., 1961, 'Estimation of soil moisture content using evaporation data', in *Proceedings of Hydrology Symposium No. 2. Evaporation. Department of Northern Affairs and National Resources, Ottawa*, pp. 184–196.

Holtan, H. N., 1961, 'A concept for infiltration estimates in watershed engineering', *ARS 41–51, Agric. Res. Ser., US Dept. Agric. Washington, D.C.*

Hoover, M. D. and Hursh, C. R., 1943, 'Influence of topography and soil-depth on runoff from forest land', *Trans. Am. Geophys. Union*, **24**, 693–697.

Hoover, M. D., 1962, 'Water action and water movement in the forest', in *Forest Influences, FAO Forestry and Forest Product Studies, Rome*, No. 15, pp. 31–80.

Horner, W. W., 1943, 'Rôle of the land during flood periods', *Proc. Am. Soc. Civil Engrs.*, **69**, 665–690.

Horton, J. H. and Hawkins, R. H., 1964, 'The importance of capillary pores in rainwater percolation to the ground water table', *E. I. du Pont de Nemours and Co., Savannah River Plant, DPSPU 64 30 23*, 13 pp.

Horton, J. H. and Hawkins, R. H., 1965, 'Flow path of rain from the soil surface to the water table', *Soil Sci.*, **100**(6), 377–383.

Horton, R. E., 1919, 'Rainfall interception', *Monthly Weather Review*, **47**, 603–623.

Horton, R. E., 1923, 'Transpiration of forest trees', *Monthly Weather Review*, **51**, p. 569.

Horton, R. E., 1931, 'The rôle of infiltration in the hydrologic cycle', *Trans. Am. Geophys. Union*, **12**, 189–202.

Horton, R. E., 1933, 'The rôle of infiltration in the hydrological cycle', *Trans. Am. Geophys. Union*, **14**, 446–460.

Horton, R. E., Leach, H. R. and Van Vliet, R., 1934, 'Laminar sheet-flow', *Trans. Am. Geophys. Union, 15th Annual Meeting*, 393–404.

Horton, R. E., 1935, 'Surface runoff phenomena—Part I. Analysis of the hydrograph', *Horton Hydrological Laboratory Publication 101*, Vorheesville, New York, 73 pp.

Horton, R. E., 1937a, 'Hydrologic interrelations of water and soils', *Proc. Soil Sci. Soc. Am.*, **1**, 401–429.

Horton, R. E., 1937b, 'Headwaters Control and use', in *Surface Runoff Control*, Government Printing Office, Washington, D.C.

Horton, R. E., 1939a, 'The interpretation and application of runoff-plat experiments with reference to soil erosion problems', *Proc. Soil Sci. Soc. Am.*, **3**, 340–349.

Horton, R. E., 1939b, 'Analysis of runoff-plat experiments with varying infiltration-capacity', *Trans. Am. Geophys. Union*, **20**, 693–711.

Horton, R. E., 1941, 'An approach toward a physical interpretation of infiltration capacity', *Proc. Soil Sci. Am.*, **5**, 399–417.

Horton, R. E., 1942, 'Remarks on hydrologic terminology', *Trans. Am. Geophys. Union*, **23**(1), 479–482.

Horton, R. E., 1945, 'Erosional development of streams and their drainage basins: hydrophysical approach to quantitative morphology', *Bull. Geolog. Soc. Am.*, **56**, 275–370.

Hursh, C. R., 1936, 'Storm-water and absorption', in 'Discussion on list of terms with definitions; Report of the Committee on Absorption and Transpiration', *Trans. Am. Geophys. Union*, **17**, 301–302.

Hursh, C. R. and Brater, E. F., 1941, 'Separating storm hydrographs from small drainage areas into surface and subsurface flow', *Trans. Am. Geophys. Union*, 863–870.

Hursh, C. R. and Hoover, M. D., 1941, 'Soil profile characteristics pertinent to hydrologic studies in the Southern Appalachians', *Proc. Soil Sci. Soc. Am.*, **6**, 414–422.

Hursh, C. R. and Fletcher, P. W., 1942, 'The soil profile as a natural reservoir', *Proc. Soil Sci. Soc. Am.*, **7**, 480–486.

Hursh, C. R., 1944, 'Report of the subcommittee on subsurface flow', *Trans. Am. Geophys. Union*, Part V, 743–746.

James, L. D., 1970, 'An evaluation of relationships between streamflow patterns and watershed characteristics through the use of OPSET: a self-calibrating version of the Stanford watershed model', *Res. Rep. 36, Water Res. Inst. Univ. Kentucky, Lexington*.

James, L. D., 1972, 'Hydrologic modelling, parameter estimation, and watershed characteristics', *J. Hydrology*, **17**, 283–307.

Jamison, V. C. and Peters, D. B., 1967, 'Slope length of claypan soil affects runoff' *Water Res. Res.*, **3**, 471–480.

Johnson, M. L., 1969, 'Research on Sleepers River at Danville, Vermont', *Papers presented to the Annual Meeting of the American Society of Civil Engineers and the National Meeting on Water Resources Engineering, New Orleans. Meeting Preprint 797*, 41 pp.

Jones, J. A. A., 1971, 'Soil piping and stream channel initiation', *Water Res. Res.* **7**(3), 602–610.

Jones, J. A. A., 1976, *Soil Piping and the Subsurface Initiation of Stream Channel Networks*, Ph.D. dissertation, Univ. Cambridge.

King, K. M., 1961, 'Evaporation from land surfaces', in *Proceedings of Hydrology Symposium No. 2. Evaporation. Department of Northern Affairs and National Resources, Ottawa*, pp. 55–80.

Kirkby, M. J. and Chorley, R. J. 1967, 'Throughflow, overland flow and erosion', *Bull. Intern. Assoc. Sci. Hydrology*, **12**, 5–21.

Kirkby, M. J., 1969, 'Infiltration, throughflow and overland flow', in Chorley, R. J. (Ed.), *Water, Earth and Man*, Methuen, London, Ch. 5.1, pp. 215–227.

Kirkby, M. J., et al., 1976, 'Measurement and modelling of dynamic contributing areas in very small catchments', *Leeds University, Department of Geography, Working Paper 167*, 35 pp.

Kirkham, D., 1947, 'Studies of hillside seepage in the Iowan drift area', *Proc. Soil Sci. Soc. Am.*, **12**, 73–70.

Klute, A., Scott, E. J. and Whisler, F. D., 1965, 'Steady-state water flow in a saturated inclined soil slab', *Water Res. Res.*, **1**, 287–294.

Knapp, B. J., 1970a, *Patterns of Water Movement on a Steep Upland Hillside, Plynlimon, Central Wales*, Ph.D. dissertation, Univ. Reading.

Knapp, B. J., 1970b, 'A note on throughflow and overland flow on steep mountain watersheds', *Reading Geographer*, **1**, 40–43.

Knapp, B. J., 1974, 'Hillslope throughflows observation and the problem of modelling', *Institute of British Geographers, Special Publication No. 6*, pp. 23–31.

Knisel, W. G., 1973, Comments on 'Rôle of subsurface flow in generating surface runoff. 2. Upstream source areas', by Freeze, R. A., *Water Res. Res.*, **9**(4), 1107–1110.

Lamson, R. R., 1967, *Response of Soil Moisture and Litter Flow to Natural Rainfalls in a Forested New England Watershed*, M.Sc. dissertation, Univ. Vermont, Burlington, 172 pp.

Langbein, W. B. and Iseri, K. T., 1960, 'General introduction and hydrologic definitions: *Manual of Hydrology*. Part I: General surface-water techniques', *US Geolog. Surv. Water Supply Paper 1541–A*, 29 pp.

Liakopoulos, A. C., 1965a, 'Theoretical solution of the unsteady unsaturated flow problem in soils', *Bull. Intern. Assoc. Sci. Hydrology*, **10**(1), 5–39.

Liakopoulos, A. C., 1965b, 'Retention and distribution of moisture in soils after infiltration has ceased', *Bull. Internat. Assoc. Sci. Hydrology*, **10**(2), 58–69.

Linsley, R. K., Kohler, M. A. and Paulhus, J. L. H., 1949, *Applied Hydrology* McGraw-Hill, New York, 689 pp.

Lowdermilk, W. C., 1934, 'The rôle of vegetation in erosion control and water conservation', *J. Forestry*, **32**, 529–536.

McDonald, P. M., 1967, 'Disposition of soil moisture held in temporary storage in large pores', *Soil Sci.*, **103**(2), 139–143.

Meinzer, O. E., 1923, 'Outline of groundwater hydrology, with definitions', *US Geolog. Surv. Water Supply Paper 494*, 71 pp.

Monteith, J. L., 1959, 'The reflection of short-wave radiation by vegetation', *Quart. J. Royal Meterolog. Soc.*, **85**, 386–392.

Monteith, J. L., 1965, 'Evaporation and environment', in *The State and Movement of Water in Living Organisms, 19th Symposium, Cambridge*, Society for Experimental Biology, pp. 205–234.

Moore, W. L. and Claborn, B. J., 1971, 'Numerical simulation in watershed hydrology', in Yevjevich, V. (Ed.), *Systems Approach to Hydrology*, Water Resources Publications, Fort Collins, Colorado, pp. 275–319.

More, R. J., 1967, 'Hydrological models and geography', in Chorley, R. J. and Haggett, P. (Eds.), *Models in Geography*, Methuen, London, pp. 145–185.

Musgrave, G. W., 1935, 'The infiltration capacity of soils in relation to the control of surface runoff and erosion', *J. Am. Soc. Agronomy*, **27**, 336–345.

Muzik, I., 1974, 'State variable model of overland flow', *J. Hydrol.*, **22**, 347–364.

Neal, J. H., 1938, 'The effect of the degree of slope and rainfall characteristics on runoff and soil erosion', *Missouri Agricultural Experiment Station, Res. Bull. 280*.

Neal, W. P., 1967, *The Southern Piedmont Upland of the Southeastern United States: A Geomorphic System in a Steady State of Dynamic Equilibrium*, M.Sc. dissertation, Univ. Georgia, Athens, Georgia.

Nixon, P. R. and Lawless, G. P., 1960, 'Translocation of moisture with time in unsaturated soil profiles', *J. Geophys. Res.*, **65**, 655–667.

Nutter, W. L., 1968, *Hydrologic Properties of Several Upland Forest Humus Types in the Lake States Region*, Ph.D. dissertation, The Michigan State Univ., East Lansing, 147 pp.

Nutter, W. L., 1969, 'Management implications of subsurface stormflow and the variable source concept', *Paper presented to the 1969 Annual Meeting of the Watershed Mangement Division, Society of American Foresters, October 13–16, Miami, Florida*.

Nutter, W. L., 1973, 'The role of soil water in the hydrologic behavior of upland basins', *Field Soil Water Regime*, (Soil Science Society of America), pp. 181–193.

Onstad, C. A., and Brakensiek, D. L., 1968, 'Watershed simulation by stream path analogy', *Water Res. Res.*, **4**, 965–971.

Overton, D. E., 1974, 'Mechanics of the surface runoff on hillslopes; *Proceedings of the 3rd International Seminar for Hydrology Professors*, 'Biological Effects in the Hydrological Cycle', (Purdue University, Department of Agricultural Engineering, West Lafayette), pp. 186–210.

Parlange, J.-Y., 1972, 'Theory of water movement in soils, 8, One-dimensional infiltration with constant flux at the surface', *Soil Sci.*, **114**, 1–4.

Parlange, J.-Y., 1975, 'On solving the flow equation in unsaturated soils by optimization: Horizontal optimization; *Proc. Soil Sci. Soc. Am.*, **39**, 415–418.

Patric, J. H., Douglass, J. E. and Hewlett, J. D., 1965, 'Soil water absorption by mountain and piedmont forests', *Proc. Soil Sci. Am.*, **29**(3), 303–308.

Pegg, R. K. and Ward, R. C., 1972, 'Evapotranspiration from a small clay catchment', *J. Hydrology*, **15**, 149–165.

Penman, H. L., 1948, 'Natural evaporation from open water, bare soil and grass', *Proc. Royal Soc.*, Ser. A, **193**, 120–145.

Penman, H. L., 1956, 'Evaporation—an introductory survey', *Netherlands J. Agric. Sci.*, **4**, 9–29.

Penman, H. L., 1963, *Vegetation and Hydrology*, Tech. Communication No. 53, Commonwealth Bureau of Soils, Farnham Royal, 124 pp.

Philip, J. R., 1957–8, 'The theory of infiltration', *Soil Sci.*, **83**, 345–357; 435–448; **84**, 163–177; 257–264; 329–339; **85**, 278–286; 333–337.

Pilgrim, D. H., 1966, 'Radioactive tracing of storm runoff on a small catchment', *J. Hydrology*, **4**, 289–326.

Pond, S. F., 1971, 'Qualitative investigation into the nature and distribution of flow processes in Nant Gerig', *Inst. Hydrology, Subsurface Sect. Rept. 28.*

Ragan, R. M., 1967, 'Rôle of basin physiography on the runoff from small watersheds', *Vermont Resources Research Center, Water Res. Res., Univ. Vermont, Burlington, Vermont*, Report No. 17, 25 pp.

Ragan, R. M., 1968, 'An experimental investigation of partial area contributions', *Proceedings of the General Assembly, Internat. Assoc. Sci. Hydrology, Berne*, Publication 76, pp. 241–249.

Rawitz, E., Engman, E. T. and Cline, G. D., 1970, 'Use of the mass balance method for examining the rôle of soils in controlling watershed performance', *Water Res. Res.*, **6**(4), 1115–1123.

Ree, W. O., 1939, 'Some experiments on shallow flows over a grassed slope', *Trans. Am. Geophys. Union*, **20**, 653–656.

Reeve, R. C. and Kirkham, D., 1951, 'Soil anisotropy and some field methods for measuring permeability', *Trans. Am. Geophys. Union*, **32**, 582–590.

Reeves, M. and Miller, E. E., 1975, 'Estimating infiltration for erratic rainfall', *Water Res. Res.*, **11**, 102–110.

Rode, A. A., 1959, '*Das Wassen im Boden*', Akademie Verlag, Berlin, 64 pp.

Roessel, B. W. P., 1950, 'Hydrologic problems concerning the runoff in headwater regions', *Trans. Am. Geophys. Union*, **31**, 431–442.

Rubin, J. and Steinhardt, R., 1963, 'Soil–water relations during rain infiltration. I. Theory', *Proc. Soil Sci. Soc. Am.*, **27**(3), 246–251.

Rubin, J., Steinhardt, R. and Reiniger, P., 1964, 'Soil-moisture relations during infiltration. II. Moisture content profiles during rains of low intensities', *Proceedings of the Soil Science Society of America*, **28**, 1–5.

Rubin, J., 1966, 'Theory of rainfall uptake by soils initially drier than their field capacity and its applications', *Water Res. Res.*, **2**(4), 739–749.

Ruxton, B. P. and Berry, L., 1961, 'Weathering profiles and geomorphic position on granite in two tropical retions', *Révue de Géomorphologie Dynamique*, **12**, 16–31.

Schumm, S. A., 1956a, 'Evolution of drainage systems and slopes in badlands at Perth Amboy, New Jersey', *Bull. Geolog. Soc. Am.*, **67**, 597–646.

Schumm, S. A., 1956b, 'The rôle of creep and rainwash on the retreat of badland slopes', *Am. J. Sci.*, **254**, 693–706.

Selim, H. M., 1975, 'Water flow through a multilayer stratified hillside', *Water Res. Res.*, **11**, 949–957.

Sherman, L. K., 1932, 'Streamflow from rainfall by unit-graph method', *Eng. News Record*, **108**, 501–505.

Sherman, L. K., 1944, 'Infiltration and the physics of soil moisture', *Trans. Am. Geophys. Union*, **25**, 57–65.

Smith, R. E. and Woolhiser, D. A., 1971a, 'Mathematical simulation of infiltrating watersheds', *Hydrology Paper 47, Colorado State University, Fort Collins*, 44 pp.

Smith, R. E. and Woolhiser, D. A., 1971b, 'Overland flow on an infiltrating surface', *Water Res. Res.*, **7**, 899–913.

Smith, R. E., 1972, 'The infiltration envelope: Results from a theoretical infiltrometer', *J. Hydrology*, **17**, 1–21.

Snyder, W. M., 1973, Comments on 'Rôle of subsurface flow in generating surface runoff. 1. Base flow contributions to channel flow', *Water Res. Res.*, **9**(2), 489–490.

Southeastern Forest Experiment Station, 1961, *Report for 1961*, Asheville, North Carolina, pp. 62–63.

Stephenson, G. R. and Freeze, R. A., 1974, 'Mathematical simulation of subsurface flow contributions to snowmelt runoff, Reynolds Creek Watershed, Idaho', *Water Res. Res.*, **10**, 284–294.

Stoeckeler, J. H. and Curtis, W. R., 1960, 'Soil moisture regime in south-western Wisconsin as affected by aspect and forest type', *J. Forestry*, **58**(11), 892–896.

Swartzendruber, D., 1960, 'Water flow through a soil profile as affected by the least permeable layer', *J. Geophys. Res.*, **65**, 4037–4042.

Tennessee Valley Authority, 1960a, 'Hydrology of small watersheds in relation to various covers and soil characteristics (a pictorial brochure), in cooperation with the *North Carolina State College of Agriculture and Engineering, Knoxville, Tennessee*.

Tennessee Valley Authority, 1960b, 'An analysis of the Parker Branch Watershed project 1953 through 1959 (a progress report)', in cooperation with the *North Carolina State College of Agriculture and Engineering, Knoxville, Tennessee*.

Tennessee Valley Authority, 1963a, 'A water yield model for analysis of monthly runoff data', *Res. Paper No. 2, Knoxville, Tennessee*, 78 pp.

Tennessee Valley Authority, 1963b, 'TVA Computer programs for hydrologic analyses', *Res. Paper No. 3, Knoxville, Tennessee.*

Tennessee Valley Authority, 1964, 'Bradshaw Creek–Elk River: A pilot study in area-stream factor correlation', *Office of Tributary Area Development, Res. Paper No. 4, Knoxville, Tennessee*, 64 pp.

Tennessee Valley Authority, 1965, 'Area-stream factor correlation', *Bull. Internat. Assoc. Sci. Hydrology*, **10**(2), 22–37.

Tennessee Valley Authority, 1966, '*Cooperative research project in Western North Carolina, Annual Report, Water Year 1964–65*', Knoxville, Tennessee, 31 pp.

Tennessee Valley Authority, 1968, '*The Upper Bear Creek Experimental Project 1965–67*', Knoxville, Tennessee, 49 pp.

Tischendorf, W. G., 1969, *Tracing Stormflow to Varying Source Area in Small Forested Watershed in the Southeastern Piedmont*, Ph.D. dissertation, Univ. Georgia, Athens, Georgia, 114 pp.

Troeh, F. R., 1964, 'Landform parameters correlated to soil drainage', *Proc. Soil Sci. Soc. Am.*, **28**, 808–812.

Troendle, C. A., 1970, 'Water storage, movement, and outflow from a forested slope under natural rainfall in West Virginia', *unpublished report on file at Timber and Water Laboratory, Parsons, West Virginia.*

Troendle, C. A. and Homeyer, J. W., 1971, 'Stormflow related to measured physical parameters on small forested watersheds in West Virginia (Abstract)', *Trans. Am. Geophys. Union*, **52**(4), 204.

Tsukamoto, Y., 1961, 'An experiment on subsurface flow', *J. Japanese Soc. Forestry*, **43**, 61–68.

Tsukamoto, Y., 1963, 'Storm discharge from an experimental watershed', *J. Japanese Soc. Forestry*, **45**(6), 186–190.

Tsukamoto, Y., 1967, 'Hydrologic phenomena occurring in mountain watersheds', *Ann. Rept. Tokyo Univ. Agric. Tech.*, No. 10, 11–18.

Van Bavel, C. H. M., 1966, 'Potential evaporation: The combination concept and its experimental verification', *Water Res. Res.*, **2**(3), 455–467.

Van Dijk, D. C., 1958, 'Water seepage in relation to soil layering in the Canberra District', *CSIRO, Div. Soils, Commonwealth of Australia, Report No. 5/58*, 13 pp.

Van't Woudt, B. D., 1955, 'On a hillslide moisture gradient in volcanic ash soil, New Zealand', *Trans. Am. Geophys. Union*, **36**(3), 419–424.

Ven Te Chow (Ed.), 1964, *Handbook of Applied Hydrology*, McGraw-Hill, New York.

Warrick, A. W. and Lomen, D. O., 1974, 'Seepage through a hillside: The steady water table', *Water Res. Res.*, **10**, 279–283.

Weyman, D. R., 1970, 'Throughflow on hillslopes and its relation to the stream hydrograph', *Bull. Internat. Assoc. Sci. Hydrology*, **15**(3), 25–33.

Weyman, D. R., 1973, 'Measurements of the downslope flow of water in a soil', *J. Hydrol.*, **20**, 267–288.

Weyman, D. R., 1974, 'Runoff processes, contributing area and streamflow in a small upland catchment', *Institute of British Geographers, Special Publication No. 6*, pp. 33–43.

Whipkey, R. Z. and Fletcher, P. W., 1959, 'Precipitation and runoff from three small watersheds in the Missouri Ozarks', *Univ. Missouri, Res. Publ. No. 692*, 26 pp.

Whipkey, R. Z., 1965a, 'Subsurface stormflow on forested slopes', *Bull. Internat. Assoc. Sci. Hydrology*, **10**(2), 74–85.

Whipkey, R. Z., 1965b, 'Measuring subsurface stormflow from forest soil—a plot technique', *US Forest Ser., Central State Forest Experiment Station, Res. Note, CS-29*, 6 pp.

Whipkey, R. Z., 1967a, 'Storm runoff from forested catchments by subsurface routes', *International Symposium on Floods and their Computation, UNESCO, Leningrad.*

Whipkey, R. Z., 1967b, 'Theory and mechanics of subsurface stormflow', in *Proceedings of the International Symposium on Forest Hydrology (1965), Pennsylvania State University*, Pergamon, pp. 255–259.

Whipkey, R. Z., 1969, 'Storm runoff from forested catchments by subsurface routes', *Publ. Internat. Assoc. Sci. Hydrology*, **85**(2), 773–779.

Wind, G. P., 1972, 'A hydraulic model for the simulation of non-hysteretic vertical unsaturated flow of moisture in soils', *J. Hydrology*, **15**, 227–246.

Wind, G. P. and Doorne, W. Van, 1975, 'A numerical model for the simulation of unsaturated vertical flow in soils', *J. Hydrol.*, **24**, 1–20.

Woodruff, J. F. and Hewlett, J. D., 1971, 'The hydrologic response of small basins in Georgia', *Southeastern Geographer*, **XI**(1), 1–8.

World Meteorological Organization, 1966, 'Measurement and estimation of evaporation and evapotransportation', *Tech. Notes*, No. 83.

Zalavski, D. and Rogowski, A. S., 1969, 'Hydrologic and morphologic implications of anisotropy and infiltration in soil profile development', *Proc. Soil Sci. Soc. Am.*, **33**, 594–599.

Zimmermann, U., Münnich, K. O. and Roether, W., 1966, 'Tracers determine movement of soil moisture and evapotranspiration', *Science*, **152**, 346–347.

CHAPTER 2

Infiltration and storage of soil water

B. J. Knapp

Leighton Park School, Reading,
Berkshire, UK

r	pore radius
π	3·142
h	height of a column of water that can be supported by pores of a given radius
ρ	soil-water density
τ	surface-tension force
α	angle of contact of water with soil
ϕ	total hydraulic potential
p	pressure potential
p_0	vapour pressure over free pure-water surface
z	height above free-water surface, measured +ve upward
x, y, z	Cartesian axes defined in Figure 2.7
θ	volumetric moisture content
v	moisture flux
t	time
i	infiltration rate
K	hydraulic conductivity
K_s	saturated conductivity
(f)	instantaneous infiltration rate
i_{sum}	total water infiltrated
f_c	minimum infiltration capacity at $t \to \infty$
f_0	minimum infiltration capacity at $t = 0$
c, t	constants
l	direction of flow (non-axial)
∇	differential operator
D	diffusivity
s	sorptivity
i_p	percolation rate
S_c	volumetric storage capacity of soil

q   volumetric discharge of water
T   temperature, °K
R   universal gas constant

2.1 THE SOIL AS A MATRIX FOR WATER MOVEMENT AND RETENTION

A soil may be defined as a heterogeneous collection of fragments of inorganic matter of various sizes and mineralogic composition as well as organic materials, air and water.

The mineral fraction is composed of primary minerals of igneous origin such as quartz and feldspar and also their secondary decomposition products such as the clay minerals. The primary minerals are in general larger and, together with fragments of still-cemented sedimentary rocks, form the skeleton of the soil. Between these skeletal particles are found the fine-textured materials like the clay minerals, iron oxides and organic decomposition products ('gums'). These materials act as coatings and bridges on and between the skeleton and bind it together in recognizable units of various shapes and sizes. Such macrofeatures are recognized by the term *soil structure*, each unit being called a *ped*. As water can be transmitted between peds as well as between particles, a knowledge of structure form is essential to any hydrological investigation.

Structural units vary in size and internal composition due to several factors. The particle-size distribution (texture) is one of the chief determinants. A wide spectrum of particle sizes is important for small, stable peds which develop relatively large interpedal voids and in this case; the *colloidal fraction* (clay, iron oxides, gums, etc.) has a skeleton of sand and silt particles to bind together. The relative absence of a colloidal fraction, which results in a developed structure and thus interpedal voids, becomes less important. The extreme example of this is sand, where there is no structure formation and all the water conduction is between individual particles (through *textural* voids). In other soils clay minerals act as bridges due to their high electronegative charge. Divalent metal cations then in turn act as linkages between these clay mineral particles, a process which is also true for the colloidal organic fraction. The presence of divalent metal cations in soil water is thus a prime determinant of soil structure. Clay soils will tend to aggregate (flocculate) into structural units with the addition of divalent cations (hence the use of lime containing Ca^{2+} in agriculture). In the presence of monovalent cations (i.e. Na^+, H^+) the same soil will be dispersed and the structure vanish.

The basic structural units and the relationship of interpedal (structural) and textural voids are shown in Figure 2.1. Observation indicates that a tendency to equal amounts of sand, silt and clay, plus the availability of divalent cations results in smaller and more stable peds.

The determinants of soil development are crucial in emphasizing that water conduction and storage occurs at two distinct levels. Clearly, interstructural voids

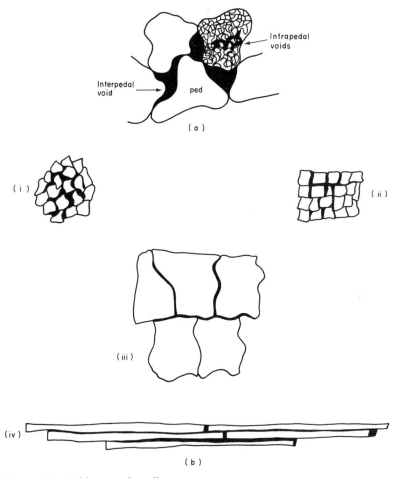

Figure 2.1 Void spaces in soils.
(a), Different scales of void spaces (shown in black).
(b), Structural types: (i), Crumb: numerous well-connected voids allowing throughflow in all directions. (ii),Blocky: many voids but regular packing keeps void sizes small. Throughflow in all directions.(iii),Prismatic: Often large peds with well-defined voids having a dominant downward orientation. Throughflow mainly in a vertical direction. (iv),Platey: Often large peds with ill-defined voids having a dominant lateral orientation.
Throughflow mainly in a horizontal direction

are likely to be larger than textural voids and thus more important for saturated throughflow. However the larger and more angular the structural units become, the less water they conduct, although their relative importance may be as great. Structural voids are mainly non-capillary and as such become unimportant over a large range of unsaturated flow conditions. More acid soils and those with a dominant 'fines' fraction develop angular peds which are often very large (sometimes greater than 15 cm) thus limiting the opportunity of interpedal void water flow. Nevertheless, interpedal flow may still be the dominant throughflow process at high moisture contents due to the small size of the textural voids. In addition to the development of textural and structural voids, tunnels are created by faunal and floral activity. Of special importance in this context are earthworms and tree roots. There are many microphotographs available showing water deposited clay linings (cutans) to such channels (e.g. Brewer, 1964) which demonstrate their effectiveness in transmitting water. A more detailed consideration of soil development under various environmental conditions is published elsewhere (Fenwick and Knapp, 1978).

Preliminary to a discussion of soil-moisture storage and infiltration must be a recognition of the heterogeneity of a soil. This can often make measurement more difficult but goes a long way to explain the nature of response of a hillside soil to rainfall. Structural development, in particular, will be shown to be a difficulty in experiments with small soil volumes. For example, a soil taken from a hillside, air dried, crushed and re-packed to the same bulk density as the original cannot be expected to have the same response characteristics as the natural soil (e.g. Hewlett and Hibbert, 1963). For this reason soils with a high-percentage coarse fraction are often chosen for laboratory experiments as they will naturally tend to have a weak structural development. Hills (1970) reports a wide range of infiltration values for an apparently uniform soil. Values quoted range from 15·5–320·0 cm/hr under old oak woodland in a clay soil and even 0·0–10·0 cm/hr for an orchard with sandy loam soil. Such variation may be due both to different packing arrangements of the same textural particles and the influence of faunal and flora activity.

A soil must be seen as a dynamic body altering with time. For example, cation availability may change with change of land-use on a hillside or by prolonged laboratory experimentation. As a result, there will be a change in the structure, transmission and storage properties of the soil. Most soils also exhibit change in water transmission due to the colloidal fraction swelling on wetting.

Soils which are uniform in their storage and transmission of water should be regarded as the exception, not the safe rule. Any property measured is the statistical average of the effect of many small responses. Within reason it is sensible for measurements to be made on sufficiently large volumes of soil to take account of this. It is the purpose of experimenters to design equipment which will establish the 'most reasonable' volume of measurements for each investigation. This will be discussed in Section 2.6.

2.2   SOIL-WATER STORAGE

Whilst being aware that hillside soils are complex it is important to establish the *basic* causes of soil responses. Thus it is necessary to propose models of an idealized soil in order to isolate and identify the multiplicity of elements involved. From this, an attempt can be made to explain real soil responses.

One such simplified model is required to show the reason for soil-moisture storage. Soil may be considered as a labyrinth of caverns of circular cross-section connected by various sizes of cylindrical tunnel. The similarity of this to a real situation can be seen by reference to Figure 2.1 and photographs in Brewer (1964). In order that water shall be stored in the soil at all, there must be a retaining force against gravity. Surface tension is the main force at air–water interfaces (menisci) whether they may be water films around particles or partially water-filled voids (often called *pores*). The action surface tension under drainage, can be visualized by considering water held in a single-pore within a complex void system (Figure 2.2)

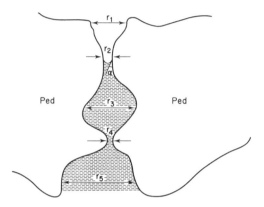

Figure 2.2   Water retention in a complex void
system. α, Angle of contact of meniscus with the
soil

connected to a free water surface (water table) at depth. Here an equilibrium state has been established such that the meniscus is at r_2. If it is assumed, for simplicity, that the pore containing water is vertically disposed and that the column of water supported is h then

$$\pi r_2^2 h \rho g = 2\pi r_2 \tau \cos \alpha \qquad (2.1)$$

where

ρ = soil-water density
g = gravitational constant
τ = surface-tension force
α = angle of contact of the meniscus with the soil

As the surface-tension force is a function of the area of the meniscus, the smaller the radius, the longer the column of water that can be supported. Therefore only voids having a controlling radius of a certain size or smaller will be water-filled. If h is increased (in this case produced by a fall of the free-water surface) then equation

2.1 will no longer be valid and the void system will empty until a suitably smaller controlling radius, say r_4, is reached so attaining equilibrium.

If a new equilibrium is established at r_4 and the free water once more rises to the position held before the drainage step took place, a reversal of drainage does not take place. In order for water to reach r_2 it must first fill the void radius r_3, but to do this requires a value of h appropriate to r_3 and this is smaller than h. It is thus evident

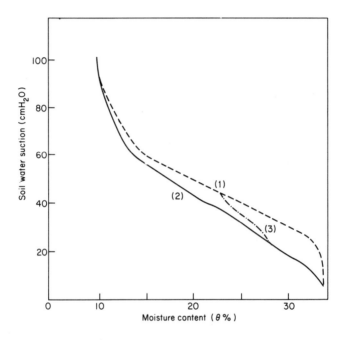

Figure 2.3   The moisture characteristic curve of a fine sand under laboratory packing conditions. The soil reveals hysteresis effects. Curve (1) is the drainage profile; Curve (2) the wetting profile. The intermediate line (3) shows a characteristic curve for partial wetting from 20–43cm water suction

that the volume of void system full of water under *drainage* (desorption) conditions is different from that under adsorption for a given soil-moisture tension (pressure potential). This property is known as *hysteresis* (Figure 2.3). For a general case, the proportion of voids filled with water is the soil-moisture content (θ), and the column of water is the pressure potential (p) (moisture tension, soil-moisture suction or capillary potential). As the void-size distribution is irregular and the surface-tension force is proportional to the cross-sectional area of the void, the relation of p to θ is non-linear and thus because as hysteresis is involved, p is not a single-valued function of θ. For the same reasons other soil properties like hydraulic conductivity (k) are also non-linear.

The property whereby some of the soil voids can retain moisture supplied to them has resulted in use of the term *field capacity*. This is defined as the stage reached in drainage from saturation, after all the larger (mainly non-capillary) pores have been drained by gravity. There is sometimes a well-marked change of slope in a graph of soil moisture against the time of drainage at the point of field capacity, and this justifies field-use of this rather nebulous concept.

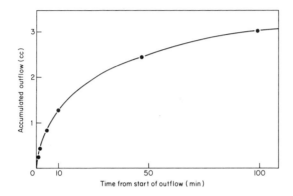

Figure 2.4   The drainage of a volume of soil. (E_a horizon of podsol; silty clay texture; sample 5-cm diameter, 1·5-cm thick)

Figure 2.4 illustrates the asymptotic nature of drainage for a silty clay Ea-horizon soil as determined in the laboratory, excluding the effects of structure. It is clear that as more water drains from the sample a larger number of voids become empty causing further water movement to take place in smaller voids along more tortuous paths. This results in a rapid reduction of outflow. Finally, at very low moisture contents, water can still move through a soil via surface films or vapour transfer although they are, by comparison, practically insignificant processes.

With no evaporation and a static free-water surface, equilibrium is eventually reached. At this time the hydraulic gradient is zero throughout and conditions are such that the pressure potential (p) at all points is equal to the height above the free-water surface. A plot of θ against p produces the *moisture characteristic curve*. An example is given in Figure 2.3

$$\phi = p + z \qquad (2.2)$$

where

ϕ = total hydraulic potential
z = height above the free-water surface ($\equiv h$) measured +ve upward
At $\phi = 0$, $-p = z$.

c

Moisture characteristic curves are common in the literature and form the basis of many calculations concerning soil-moisture storage and infiltration. For details of laboratory determinations see, for example, Black (1962). Soil pressure potential is measured in pF units where $pF = \log_{10}$ (pressure potential in centimetres of water).

2.3 SOIL-MOISTURE STORAGE MEASUREMENT

Spatial and stochastic information on the value of soil moisture is a fundamental part of any consideration of hillslope water movement. Its determination can come from a variety of sources, by direct or indirect measurements.

Direct measurement

This is often supposed to be the simplest of available techniques. It involves the removal of the volume of soil under consideration or samples to represent it. These samples are collected in special sampling tools designed to disturb the soil as little as possible. Some of the many designs available and their problems are reviewed by Fenwick (1960).

The field samples obtained using a sampling device (usually in the form of a corer) are weighed, oven-dried and re-weighed to find the moisture content. Volumetric moisture-content determinations require a knowledge of the bulk density of the soil and the volume of the core holding the sample (Black, 1962). At the end of this technique, referred to as *gravimetric determination* of soil moisture, the moisture value of a soil sample which has suffered complete destruction is obtained (Reynolds, 1970). This is clearly a technique which cannot be used in a continuing study due to its effect on the site. As a result, several indirect methods are currently in use. They all require calibration via samples obtained from gravimetric determinations but these are often performed away from the direct study site on supposedly similar soils.

One of the more popular techniques which is receiving progressively more usage is that of *neutron moderation* (Bell, 1973). High-energy (fast) neutrons penetrate into the soil body from an americium–beryllium or similar source contained within a probe which is lowered down into the soil through an access tube. The fast neutrons interact principally with hydrogen nuclei by collision, thereby losing energy. The lower-energy (slow) neutrons resulting from the collisions are therefore an indication of the soil hydrogen content. Most hydrogen is contained in soil water. The slow neutrons are detected by a device again located within the access tube. It is possible to calibrate such a piece of equipment by comparison of readings with gravimetric determinations using the same type of soil as was involved in the experiment. It will be noted that the equipment usually integrates over a volume with up to a 25-cm radius, depending on the moisture status and this varying radius can create problems near soil horizon junctions or near the surface (Bell and McCulloch, 1966). For useful application, two assumptions need to be made:

(1) That the soil horizons are of uniform depth in the area of study.
(2) That the properties of the individual horizon (e.g. density, pore-size distribution) remain constant over the whole area (Cole and Green 1966). If these assumptions can be taken as valid then a comparison is possible between moisture values from different locations other than the calibration

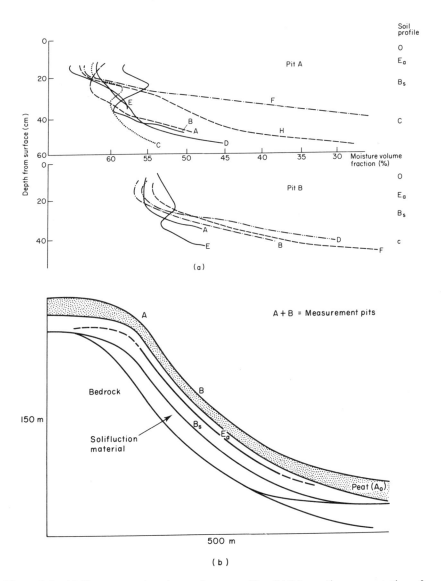

Figure 2.5 (a) Neutron moderation moisture profiles. (b) Schematic representation of soils on a slope in the Plynlimon catchments

site at given depths despite interface effects (which should be constant). Even without a calibration curve, relative differences in count-rate at a given depth in a given soil should reflect any variations in the moisture content.

Figure 2.5 shows the results of some moisture-content profiles for a hillside on Plynlimon, Central Wales. Location A contained a peaty podzol soil overlying solifluction shales whilst at location B the peaty surface horizon was absent. At both locations aluminium neutron-probe access tubes were installed in pre-cored holes down into the lower parts of the B-horizons. At location A, eight tubes were located and at location B five tubes each within a block of side 2 m.

Owing to the problems associated with surface effects the data start at a depth of 15 cm. A considerable variation in moisture content was detected at each location although the group of results may be interpreted at both locations in terms of the heterogeneity of the soil, giving different porosities and variability of the horizon depths. This provides a reminder of the problems of soil-moisture determination on hillside soils. The relatively rapid rate of downslope soil movement is likely to accentuate the heterogeneity of the soil and make interpretations from determinations at single points a somewhat hazardous procedure. Certainly this would be true of quantitative use of the data. Single-site measurements over time intervals will, however, often clearly reveal *changes* in moisture status, over time as revealed in Figure 2.5 p. 51. Reference to Figure 2.5 confirms that, despite scatter of values at any one location, major changes in absolute moisture content are still revealed. Thus, the mean moisture content at location A was consistently higher than the mean at B except at depth. The neutron-probe system would appear to be a valid instrument to measure soil-moisture differences, even in a fairly heterogeneous hillside, provided such differences lie outside the range of sampling error. It is necessary to set up several measurement tubes at each site to assess this error.

Indirect measurement

Moisture content has also been measured indirectly by means of tensiometers. A tensiometer is essentially a porous cup filled with de-aerated water and attached to a manometer. At equilibrium, the tensiometer shows the soil-moisture tension (i.e. p of equation 2.2). It is possible to obtain calibration curves for soil moisture versus tension (the soil-moisture characteristics curve) as described by Richards (1949) and others. A tensiometer is considerably less expensive than a neutron probe although it is subject to failure at pF values above 2·5–3·0 due to air leakage, freezing or interference. Pressure potentials are more likely to remain constant than moisture content as they are dependent only on the energy balance in the soil mass and not on the pore-space geometry. As with neutron moderation, calibration curves are not easy to obtain, while small changes in tension are often hazardous to interpret in terms of change of moisture content due to large standard-error values which accompany calibration curves. However they may be the only technique possible in studies, for example, near the soil surface. Their

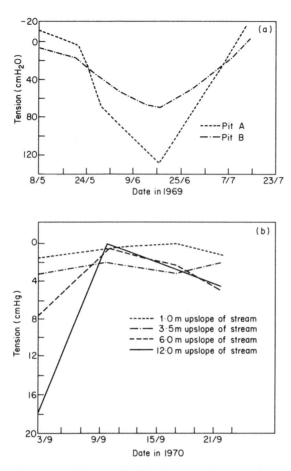

Figure 2.6 Changes of soil moisture tension over time.
(a) For A horizon of Plynlimon soils shown in Figure
2.5, at 15-cm depth. (b) For A horizon of soils near
stream bank of East Twin Brook, at 25-cm depth (by
permission of D. R. Weyman)

cheapness is a further attraction. Figure 2.6 shows tension variations for short- and
long-time periods on contrasting sites.

An alternative indirect and non-destructive technique recently adopted
concerns the measurement of vapour pressure p, of water vapour in soil air at any
point, derived from standard consideration of the behaviour of gases:

$$h = \frac{RT}{\rho g} \ln p/p_0. \qquad (2.3)$$

In this equation p/p_0 = relative humidity of the soil air. The vapour pressure of the soil air is in dynamic equilibrium with the liquid-water content of the soil and thus can be calibrated to give moisture-content values for soils.

Theoretically therefore, a psychrometer which can measure the relative humidity as a difference in temperature between wet- and dry-bulb thermometers should be a valuable instrument. The standard meterological instrument is not usable due to the bulk and consequent disturbances to the soil equilibrium. Developments of a thermocouple technique have made a more compact psychrometer possible (Richards, 1949). An initial measurement of soil air temperature is made with the thermocouple ('dry-bulb temperature'). The junction is then cooled by passing a current through it (Peltier effect) to lower the temperature below the dew point of the soil air so that condensation takes place on the junction. A measurement of the thermocouple value is then taken whilst the water droplets on the junction are still evaporating ('wet-bulb temperature').

As would be expected, the working range as shown by Richards appears to be between pF 4 and pF 8. Response times for measurement are slower towards the wet end of the range and a value of seven days to equilibrate for pF 3·5 is quoted, decreasing to 15 hr at pF 4·45. The instruments presently available are also sensitive to temperature fluctuations and are thus of more value in a laboratory under controlled conditions than in the field. As with many other techniques, calibration curves are required to permit volumetric soil-moisture determination. Consequently for the rapid response of hillslopes to rainfall under many conditions this instrument would as yet not seem of much value. Considerable practical detail on the use of some of these techniques is given in Curtis and Trudgill (1974).

Lastly, attention must be drawn to the relationship which exists between electrical resistance and moisture content. If it is assumed that the electrical conductivity of the porous material is much less than that of water, then the addition of water will fill some of the pores and a few continuous water paths will exist between the electrodes. As, with higher moisture content, more paths become continuous, the resistance measured by an alternating-current Wheatstone bridge, becomes lower. Croney et al. (1951) show this effect by placing electrodes in a Plaster of Paris block. Here the pore-size range is such that there is no change in resistance below pF 2·2 and other materials would be needed for other ranges. A considerable hysteresis effect is seen, especially at the lower end of the range (pF 2·2–4·0) giving rise to large errors if the position on the hysteresis curve is not known. In use, the Plaster of Paris block is placed in the soil so as to attain good contact between the soil pores and those in the block and this can be difficult with heavy-textured soils. The block is then left to equilibrate with the soil before readings are made with the A.C. Bridge.

The equipment chosen for field measurement clearly depends upon the purpose of the experiment, but most investigations into throughflow patterns produced by storms require rapid response and operation in the range pF 0·0–2 at least. Because of this vapour pressure and electrical resistance measurements have been less widely used than the other systems mentioned above (e.g. Knapp, 1970a).

2.4 THE DYNAMICS OF WATER MOVEMENT

The derivations described in Section 2.2 enable some of the static characteristics of the soil body to be understood. Most water in a soil system is, however, in motion as the result of imbalance of inputs, storage and outputs. Inputs can be vertically downward (infiltration and percolation) or upward (rise of free-water surface) or from upslope laterally (throughflow). Together with storage they can be formed into a continuity equation which is independent of the geometry of the system. With reference to Figure 2.7, if the change in mass flux of water due to an added

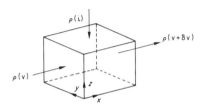

Figure 2.7   The continuity flow model

contribution of moisture from the element of soil is $\rho \delta v$ during time δt then this must be accompanied by a corresponding reduction in mass of stored water $\rho \delta \theta$. As t approaches 0 this may be expressed in a one-dimensional form as

$$\frac{-\rho \partial \theta}{\partial t} = \frac{\rho \partial v}{\partial x} \qquad (2.4)$$

where

v = moisture flux in x direction
t = time
θ = volumetric moisture content

This is for the drainage state. Added infiltration (i) for increasing soil moisture gives

$$\frac{\partial \theta}{\partial t} = -\frac{\partial v}{\partial x} + i \qquad (2.5)$$

Thus a relationship is established between inputs, storage and outputs. Attention can now be directed toward the specification and description of inputs. While inputs can be from infiltration or throughflow, discussion in this chapter will be primarily restricted to infiltration additions, throughflow being dealt with in Chapters 4 and 6.

2.5   INFILTRATION

Infiltration may be defined as the process of water entering the soil. More specifically, *infiltration capacity* is the maximum volume flux of water across the soil surface (in units of velocity). It is very intimately connected with *percolation*, the volume flux of water through the soil in a vertical sense. Because most measurements have been directed toward determining the maximum infiltration rate for erosion control, the terms 'infiltration capacity' and 'infiltration rate' have become synonymous. However, infiltration rate will here be taken to mean the volume flux across the surface and will in general be less than the maximum possible value.

The actual infiltration rate is determined by: (i), the amount of water available at the soil surface; (ii), the nature of the soil surface; (iii), the ability of the soil to conduct infiltrated water away from the surface. As explained above, the ability of a soil to admit water depends on the size, number and interconnection of voids and their potential change in size due to swelling of clay minerals on wetting. Water will only enter the soil body when a water film exists at the void entrances. If the soil has been dried out by evaporation then it must be wetted before infiltration can really commence, except into macrovoids. Soils with large structural units whose major axis lies in the horizontal plane will be particularly restrictive to infiltration. For any soil the antecedent moisture content is important as it determines the pressure potential across the soil surface. Other factors aside, a fairly dry soil will have a higher initial infiltration capacity than the same soil with a higher moisture content.

Infiltration capacity commonly decreases with time in two ways:

(1) Saturation of the soil causes a reduction in the hydraulic gradient near the surface. This can occur after long rainfall periods where horizons of low permeability underlie a permeable surface horizon. Throughflow from upslope can also cause saturation.

(2) Changes in the surface of the soil also influence the infiltration capacity, one important change being due to clay mineral swelling which reduces pore size, particularly near the surface where overburden pressure on the soil is slight. Removal of vegetation cover followed by heavy rain can also cause a surface *pan* to form as fine particles are knocked from peds by impact and are washed into the voids. Even with a vegetation cover, heavy pressure by humans, animals or machinery can break up the surface structure and close many voids. This is clearly seen, for example, near field gates and across hillsides along animal tracks.

Long-term changes in the infiltration rate can also be produced by base status changes in the soil. A new vegetation cover resulting from alteration of land-use can affect the number of free divalent cations in the soil. This in turn can affect the stability of aggregates—both directly by replacing cations like Ca^{2+} by H^+ and indirectly by changing the pH of the soil water and hence the microorganic population which produces the structure-maintaining gums. A low pH results in structure breakdown and therefore lower infiltration.

2.6 TECHNIQUES OF INFILTRATION MEASUREMENT

Three different approaches are currently in use. Parr and Bertrand (1960) in a review article classified instruments as being: (i), those in which infiltration is determined as the difference between applied water and runoff, usually employing raindrop simulators; (ii), those in which water is impounded inside a cylinder; (iii), those which attempt to determine infiltration from rainfall data.

Rainfall simulator techniques, often regarded as the most accurate available for sample plot investigation, use *sprinkler systems*, similar to those used in gardens, with the nozzles directed upward so that the drops reach the ground vertically and with as near-terminal velocity as possible. Various nozzles are employed to try to get a raindrop size close to that which occurs naturally. *Drip screens* have also been used somewhat less successfully. A wire mesh is sprayed with an overhead sprinkler system. The mesh is covered with cheese-cloth from which hang many cotton threads. The raindrops form on the ends of the threads thus giving a constant drip size. The technique is less flexible than the sprinklers and has not gained such widespread acceptance. Present American sprinkler systems wet a bounded plot of the order of 2 m × 3 m and often a buffer area is also watered. The analysis consists of three stages: (a), a calibration run with the plot covered with plastic sheet, the rainfall intensity being determined from the collected runoff; (b), a test run to wet the vegetation and soil and remove the effects of depression storage and interception, this being continued until a constant runoff is obtained from the plot; (c), an analytical run immediately after the test run, when infiltration is constant.

Cylinder infiltrometers are also widely used. They consist of cylinders of 10 cm or more in diameter which are forced a short distance into the ground to allow the water to be ponded inside. Often a cylinder of diameter up to twice that of the measuring cylinder is placed concentrically so that it too can be filled with water and produce a 'buffer' zone.

Some important problems arise from both these techniques. Essentially they provide water to only a small area while the rest of the soil remains unwetted, giving a lateral pressure potential gradient. It is the purpose of the buffer area to reduce the effects of this and it is hoped that in the central test area the flow of water is vertical, thus giving an indication of infiltration capacity. That this circumstance is rarely achieved is evidenced by the discussion in the literature for techniques of adjustment (e.g. Hills, 1970). Cylinder infiltrometers *in particular* can cause compression effects, disturb the soil, and certainly produce planes for water movement down the infiltrometer walls. The sprinkler system has the advantage of integrating over a much larger area, thus reducing the importance of edge effects, but its lack of portability and expense are severe limitations. Burgy and Luthin (1957) suggest that six cylinder infiltrometer measurements can come within 30 % of the true mean for the test site when no restricting layers in the soil are present. Slater (1957) suggests that 15 replications of cylinder infiltrometers are needed to attain the accuracy of one sprinkler infiltrometer measurement. The latter observation is, however, probably an over-estimate of the difference as in this case a

$4\frac{1}{2}$-in diameter cylinder was used without a correction factor so that edge effects must have been particularly severe.

Musgrave and Holtan (1964) give examples to show the different results obtained from the two systems (Table 2.1). These data clearly show that infiltration values at present are to be treated with a great deal of caution. In this situation it is evidently best to keep to one type of equipment in any experiment as relative values seem consistent. At the end of any analysis, a set of relative point values are known whose extrapolation to wider areas is beset with difficulty due to changes in soil depth, structure, texture, slope angle etc.

Table 2.1 Comparison of infiltration-capacity values using simulator and infiltrometer techniques (after Musgrave and Holtan, 1964)

System	Infiltration ca-pacity (in/hr)	
	Overgrazed	Grazed
Rainfall simulator	0·14	1·13
Cylinder infiltrometer	0·11	2·35

As a result of these difficulties a third system for infiltration measurement has evolved. An integrated value of infiltration for a hillside is obtained as part of a complex hydrograph separation system. Because actual rainfall values are incorporated, this method gives some idea of the effect of rainfall variability on infiltration. The techniques are described and exemplified by Musgrave and Holtan (1964).

As these techniques separate lumped components of the total hydrograph, they beg a question which is at the heart of much work on hillslope hydrology—they assume a real uniformity of infiltration and runoff production. For this reason, hydrograph-separation techniques for estimating infiltration can only be applied to flow from very small plots if the results are to be applied to the broader field of hillslope hydrology.

2.7 INFILTRATION EQUATIONS AND INFILTRATION THEORY

Clearly, data of infiltration values under varying environmental conditions are an essential part of process investigation. Equally essential is an understanding of the mechanics of infiltration. The result has been a series of attempts to model the infiltration process mathematically so that infiltration measurements can be explained more completely and extrapolated to new situations with confidence. There is no doubt that here is a situation where various complex processes interact. Solutions can thus be attempted on several levels of sophistication. At present it will be observed that there is far to go in achieving the stated goal. At the same time, much progress has been made in situations having controlled boundary conditions.

As early as 1911 Green and Ampt proposed a model for infiltration under ponded water conditions employing physically meaningful parameters expressed as

$$i = K(H + Z_f + P_f)/Z_f \qquad (2.6a)$$

grouped as

$$i = A + B/Z_f \qquad (2.6b)$$

where

i = infiltration rate
K = saturated hydraulic conductivity
H = depth of ponded water
Z_f = vertical depth of saturated zone
P_f = capillary pressure at the wetting front

Unfortunately several components of this equation are difficult to measure, especially P_f and hence the equation was applied little at the time, although interest has been revived recently due to more sophisticated measurement techniques and a better understanding of the process. For example, Swartzendruber and Huberty (1958) and Mein and Larson (1973) have been able to compute values for the parameters. A is now recognized as the wetting conductivity at residual air saturation (re-saturated hydraulic conductivity) and estimated as $0.5\,K$ for a homogeneous soil (Morel-Seytoux and Khanji, 1974). Also, as water replaces air during infiltration, there is a peak of total viscosity (Figure 2.8), causing a transition zone of the moisture-content value as distinct from an abrupt zone assumed by Green and Ampt. Modification of the equation to allow for a viscosity factor reduces relative errors of prediction, makes for a better use of the equation and recognizes that A and B are real predictable characteristics of a soil rather than largely empirical values of lumped unknowns.

Because of the difficulties mentioned above in the early stages, there was a move towards more empirical equations until the 1950s. Horton (1933) proposed a general empirical equation of infiltration

$$(f) = f_c + (f_0 - f_c)e^{-ct} \qquad (2.7)$$

where

(f) = instantaneous infiltration rate at time t
f_c = minimum infiltration capacity at $t \to \infty$
f_0 = minimum infiltration capacity at $t = 0$
c = constant for the soil,

with conditions

$$(f) \to f_c \quad \text{as } t \to \infty$$

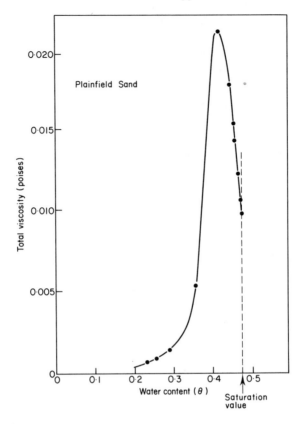

Figure 2.8   Curve of total viscosity (in poises) as a function
of water content for Plainfield Sand (after Morel-Seytoux
and Khanji, 1974)

and

$$(f) = f_0 \quad \text{at } t = 0 \tag{2.8}$$

This was developed as part of a conceptual model of overland-flow erosion and as such assumes unimpeded progress of water into the soil. It has remained the model for much experimental work due to ease of application, even though it has little physical basis. It applies also only to ponded infiltration conditions.

Subsequent detailed investigation of laboratory models by Bodman and Coleman (1944) allowed the infiltration phenomenon to be divided into three parts:

(1) A *transmission zone* which occupies the upper part of the wetted soil and once established absorbs no additional moisture only conducting water from the soil surface.

(2) A *wetting zone* below the transmission zone with a moisture gradient which increases with depth.

(3) A *wetting front* occurring as an irregular surface of very high potential gradient.

They observed too that an increased antecedent moisture content decreased the infiltration rate but increased the rate of advance of the wetting front. These were points taken up in the classic work of Philip (1957/8) concerning the physical basis of infiltration into homogeneous soils. This pioneer work has subsequently been the subject of many attempts at infiltration modelling, especially using computer techniques for numerical solutions.

A theoretical consideration of infiltration begins, in common with most soil-water flow theory, with the general equation of flow combining Darcy's law

$$v = -K\frac{\partial \phi}{\partial l} \tag{2.9}$$

where

l is the direction of flow, with the continuity Equations 2.2 and 2.4 to obtain:

$$\frac{\partial \theta}{\partial t} = \nabla \cdot (K\nabla p) + \frac{\partial K}{\partial z} \tag{2.10}$$

where

z is measured vertically downward from the surface.

Difficulty of solution results from the non-linear relation of θ with K and p. The non-linear partial differential equations that develop can only be solved by rigorous application of boundary conditions. Amongst them is the requirement of a homogeneous soil whose surface is in a horizontal plane.

Philip's solution for infiltration into a semi-infinite homogeneous one-dimensional soil with uniform initial moisture content was a power series in $t^{\frac{1}{2}}$. Bhuiyan *et al.* (1971) present results which are characteristic of this approach. They used the following boundary conditions:

(1) Water at atmospheric pressure continuously ponded on the surface in a layer of negligible thickness.

(2) A uniform initial pressure potential throughout, corresponding to a uniform moisture content.

(3) An impermeable layer at the bottom of the soil mass. Figure 2.9 illustrates the results of simulation for a soil with known K,θ: ϕ,θ relationships.

Hysteresis does not form a part of the discussion as infiltration only is concerned. Using the transformation

$$D = K(\mathrm{d}p/\mathrm{d}\theta) \tag{2.11}$$

allows (2.10) to be re-written as

$$\frac{\partial \theta}{\partial t} = \nabla(D\nabla\theta) + \frac{\partial K}{\partial \theta}\cdot\frac{\partial \theta}{\partial z} \tag{2.12}$$

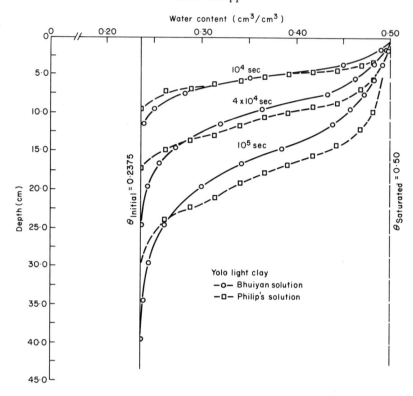

Figure 2.9   Infiltration profiles calculated for Yolo Light Clay with uniform
pressure potential (after Bhuiyan *et al.*, 1971)

D is known as the *moisture diffusivity* (dimensions $L^2 T^{-1}$) and shows soil-moisture movement to be related to many other diffusion phenomena like heat and gases. Also it reveals that the speed of advance of water depends on the moisture gradient as well as on the effect of gravity.

It is clear the nature of infiltration from rainfall of intensity such that i is greater than f_c, is dependent on the ratio i/f_c (called the relative rainfall intensity). Rubin (1966) has predicted the depth of the wetting front and time of overland flow for Rehovet sand. The penetration depth of the wetting front and time before overland flow commences were both shown to be functions of i and f_c. With incipient ponding conditions and with $i/f_c = 1.50$ and 5.25, the times before overland flows were calculated to be 581.7 and 21.7 sec respectively, with corresponding cumulative water uptakes of 11.6 and 1.52 cm^3/cm^2. As expected, there is a non-linear relation between i/f_c and response depth and time.

Data calculated similarly to the above examples enables a better understanding of hillslope soil responses. They are a basic requirement for indicating features of soil-water movement that are not readily measured, such as (i), the reaction of moisture profiles to infiltration at small times; (ii), the nature of the wetting front; and (iii), the effect of antecedent moisture conditions. A theoretical treatment using infiltration theory has, for example, shown that the Horton model does not adequately describe the infiltration process especially at small time values. Philip (1957/8) proposed a new algebraic equation to replace the Horton model, derived from his theoretical work

$$i = \tfrac{1}{2}st^{-\frac{1}{2}} + A \tag{2.13}$$

$$i_{sum} = st^{\frac{1}{2}} + At \tag{2.14}$$

where

i_{sum} = cumulative
A = a constant, in general close to the K value at the surface for $t > 0$
s = a sorptivity value obtained from the rate of penetration of the wetting front

with conditions

$$t = 0 \; i \to \infty ; t \to \infty \; i \to A \tag{2.15}$$

In practice there is a finite value to the initial infiltration capacity. Despite this limitation the equation fits experimental data well, especially at small time values. The Horton equation 2.7 is seen to under-estimate (f) at $t = 0$ and approaches f_c at $t = \infty$ in a more abrupt way (Figure 2.10).

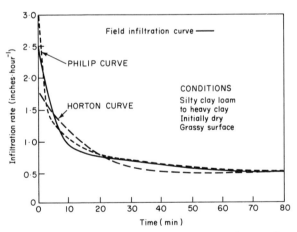

Figure 2.10 Comparison of observed and best-fit theoretical curves of infiltration capacity (by permission of K. K. Watson)

The basic assumption made above is that of a semi-infinite soil where storage effects are unimportant and have no importance on the penetration of the wetting front. Clearly, this is not going to be the case in many actual circumstances.

The effect of a storage component can be seen by considering a soil with a percolation rate i_p (assumed constant). For infiltration (f) (where (f) is less than the infiltration capacity) the soil will increase in moisture content at a rate

$$(f - i_p) \tag{2.16}$$

If the total storage possible under these conditions is S_c, then the time taken to reach saturation is

$$t_c = S_c/(f - i_p) \tag{2.17}$$

When a total of

$$S_t = f \cdot S_c/(f - i_p) \tag{2.18}$$

has infiltrated. Figure 2.11 shows this sort of curve which is similar to many infiltration tests for values less than the infiltration capacity and is consistent with the data of Rubin (1966) reported above.

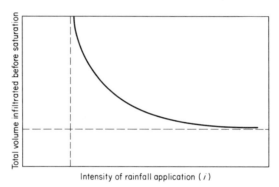

Figure 2.11   The effect of a soil storage limit on the total volume infiltrated before saturation at different intensities (i) of application

The Philip's model has no provision for the limited storage often found. Assuming constant diffusivity, a storage control on Philip's equation 2.13 would be represented by Figure 2.12. This would give a sharper transition to a constant infiltration-rate value and decrease the amount by which the Philip equation would underestimate the approach to asymptotic conditions (Figure 2.10).

Infiltration is a time-dependent process resulting in continual changes of moisture profiles. Further insights into soil-moisture behaviour patterns can be obtained from steady-state conditions (i.e. time independent). Only simple cases can be thus treated although some are adaptable to hillside conditions.

Consider the steady-state situation of a uniform soil with constant non-ponding infiltration, and with a stable free-water surface at a finite depth. Darcy's law can be used in the form

$$V = -K\left(\frac{dp}{dz} + 1\right) \tag{2.19}$$

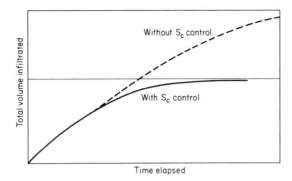

Figure 2.12   The effect of a soil storage limit on the
overall form of the infiltration curve

Substituting from (2.11) gives

$$V = -K - D\,d\theta/dz \qquad (2.20)$$

If z is taken as vertically upward, then $V = -i$ (constant) and (2.20) can be
written

$$z = \int_{\theta_{sat}}^{\theta} \frac{D}{q - K}\,d\theta \qquad (2.21)$$

and a moisture profile obtained directly on substituting, as D and K are functions of
θ. This solution attributed, to Richards (1931), shows that as z increases, θ

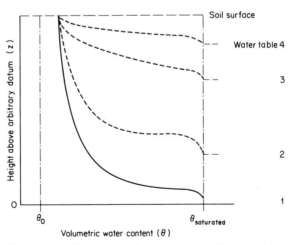

Figure 2.13   Changes in the moisture profile associated
with a rising water table

decreases, since K is always greater than q. Also, as K tends to q, z tends to infinity, resulting in a profile as in Figure 2.13 with the moisture tending asymptotically to the value for which K is equal. Childs (1967), in an excellent review article, presents the type of solution applicable for a stratified soil. Reference to Figure 2.14a (after Vachaud *et al.*) for $t = 0$ shows this situation esperimentally. The discontinuity at horizon interfaces arises from the different moisture contents obtained for the

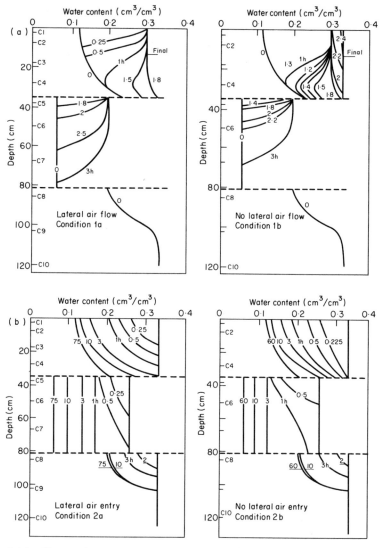

Figure 2.14   Changes in water content profiles in a layered soil: 1. during constant flux infiltration; 2. during gravity drainage from saturation to a water table at $z = 136$ cm (after Vachaud *et al.*, 1973)

different horizons at a given pressure potential. With the increasing height of the water table the moisture profile of Figure 2.13 alters, as shown by the dashed lines and if $i = K_{sat}$ the whole profile becomes saturated. The profiles of Figure 2.13 could be interpreted as the vertical soil-moisture profile due to increasing throughflow contributions while i remains constant. Experimentally, Whipkey (1965) for example clearly shows how saturated throughflow can occur above horizon boundaries and how overland flow is primarily dependent on conditions pertaining in the upper horizons of a soil.

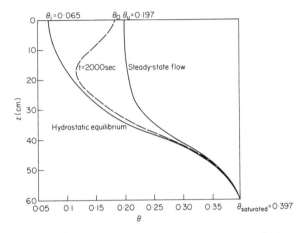

Figure 2.15 Theoretical moisture profiles during infiltration (after Braester, 1973)

If i increases with a static water table, the moisture profile becomes steeper until it is vertical at $i = k$ and the whole profile becomes saturated at the same time (Figure 2.15).

On a hillside, a 'wedge' of saturation can be envisaged, as described by Weyman (1970) in a way which is consistent with the development of Figure 2.14. In a stratified soil the development of a water table in an upper horizon causes ponding infiltration conditions to develop in a less-permeable lower horizon. As a result, a wetting front moves down through the lower horizon. On a hillside, saturated throughflow can then also occur in the impeding lower horizon above its wetting front, although this flow is of a lesser magnitude than the throughflow in the more permeable horizon above.

The theoretical profiles and characteristics for a variety of infiltration conditions have been established, partly due to the application to drainage problems in engineering. Relatively little attention has been focused on the re-distribution of water following the cessation of rainfall. Biswas *et al.* (1966) provide an experimental indication of the non-evaporative re-distribution of moisture in a soil

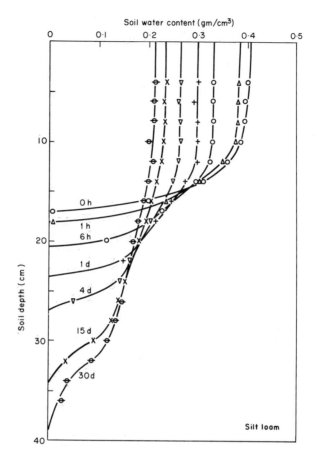

Figure 2.16   Redistribution of moisture in a vertical soil
column of Columbia silt loam after cessation of infiltration
(after Biswas *et al.*, 1966)

of 'infinite depth'. Figure 2.16 shows some of the results obtained. In a non-evaporative experiment the requirement is for conservation of the mass of water. The gradient of the moisture profile remains constant in the upper portion of the soil (the transmission zone at cessation of rainfall) whilst the wetting front becomes less pronounced. This is shown for the stratified case in Figure 2.14b.

2.8   THE REDISTRIBUTION OF INFILTRATED WATER ON HILLSLOPES

The previous discussion reveals the large gap that exists between infiltration theory and its application. Attention has been focused primarily on the evaluation of basic theory and its application to the solution of simple problems, especially for engineering practice. Relatively little has been published on the infiltration and re-distribution of soil moisture on sloping sites. It is for this reason that the work of Hewlett and Hibbert (1963) has received so much attention. In Figure 2.17 an interpretation of the re-distribution of soil moisture is given for their uniform slab of soil. Moisture values are integrated for the thickness of slab and hence do not reflect gradients normal to the plane of the slope. The graphs represent the experimental solutions of equation 2.10. They show that whilst the upper part of the slope rapidly de-saturates and asymptotically approaches its equilibrium value, the lower part soon develops a remarkably stable moisture-content value approximating saturation.

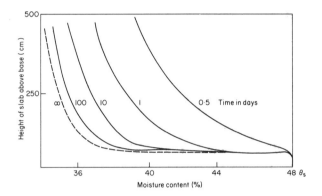

Figure 2.17   Moisture profiles for a soil slab on drainage from saturation (based on data given in Hewlett and Hibbert, 1963)

A uniform slab such as this has many of the characteristics of a vertical soil column and the nature of the response is very similar to Figures 2.14 and 2.15. When ignoring the moisture gradient normal to the slope, it would seem that the theory of drainage can be applied without too much difficulty. Techniques of dimensional similitude can be applied to scale-up the models to give an idea of true hillside response times (Stallman, 1964).

To summarize work on uniform soil slab models, theoretical considerations and laboratory models reveal the following:

(1) After drainage the moisture content of the slope is highest at the base and progressively lower towards the top. Moisture deficits are thus lowest at the slope base where further infiltration will cause saturation (and hence overland flow) most quickly. This will be supported by throughflow from upslope. The slope will drain towards values determined by the moisture characteristic curve of the soil (Figure 2.17).

(2) The thinner the soil the smaller the i sum required to satisfy the moisture deficit $(\theta_s - \theta_i)$ and cause saturation. The thinnest soils, however, tend to be at the top of a hillside where slopes are convex and conducive to rapid drainage (shedding sites) so that, in practice, Equation 2.17 is satisfied first at the slope base.

(3) Stratified soils drain differently from those with uniform soil properties (Figure 2.14). An impeding layer can cause a restriction in water distribution resulting in the development of temporary water tables within the soil. These perched water tables may rise to give more rapid saturated conditions at the surface and initiate overland flow. As with (2) the closer such an impeding layer is to the surface, the sooner surface saturation is reached. With such an impeding layer, Equation 2.17 is only appropriate for the depth of soil above it.

(4) Infiltration occurs at all points on a hillside where the moisture content is less than the saturation value. This gives a characteristic moisture profile with a well-defined wetting front, although stratified soils show marked breaks in this profile, each section being characteristic of the relevant soil properties. Such profiles cannot be expected to persist because they are soon modified by throughflow.

(5) On a real hillside these patterns of response will be affected by variability in soil depth, soil hydraulic properties and slope. From Equations 2.5 and 2.19 it may be seen that the direction of movement of water is determined by pressure (suction) and gravity gradients. Hillsides are normally convexo–concave in profile. Because the top of the upper convex section and the base of the lower concave slopes are the least steep parts of the hillside, re-distribution will be slowest at these points and moisture contents will attain high values before other sloped regions. This can be reinforced (a), in the slope base concavity which may be permanently at or near saturation (assuming the stream is approximately at water table level) and receives the throughflow from the whole slope; (b), in local three-dimensional concavities produced by uneven soil movement on the slope or by the headward development of streams (Kirkby and Chorley, 1967); and (c), by the finer texture of soils at the base of the slope consequent on translocation of fines downslope in association with throughflow. The tendency towards an increase in fines on the lower slope reduces their permeability and promotes ponding of water and consequently overland flow at higher levels on the hillside.

REFERENCES

Bell, J., 1973, 'Neutron probe practice', *Inst. Hydrology Rept.*, No. 19.

Bell, J. and McCulloch, J. S. G., 1966, 'Soil moisture estimation by the neutron scattering method in Britain', *J. Hydrology*, **4**(3), 254–66.

Bhuiyan, S. I., Hiler, E. A., van Bareland, C. H. M. and Aston, A. R., 1971, 'Dynamic simulation of vertical infiltration into unsaturated soils', *Water Res. Res.* **7**(6), 1597–1606.

Biswas, T. D., Nielsen, D. R. and Bigger, J. W., 1966, 'Redistribution of soil water after infiltration', *Water Res. Res.* **2**(3), 513–24.

Black, C. A. (Ed.), 1962, *Methods of soil analysis. Part 2*, American Society Agronomy, Madison, 770 pp.

Bodman, G. B. and Coleman, E. A., 1944, 'Moisture and energy conditions during downward entry of water into soils', *Proc. Soil Sci. Soc. Am.*, **8**, 116–22.

Braester, C., 1973, 'Moisture variation at the soil surface and the advance of the wetting front during infiltration at constant flux', *Water Res. Res.* **9**(3), 687–94.

Brewer, R., 1964, *Fabric and Mineral Analysis of Soils*, John Wiley, 470 pp.

Burgy, R. H. and Luthin, J. N., 1957, 'Discussion of paper by Burgy and Luthin called *A test of the single and double ring types of infiltrometers*', (Trans. Am. Geophys. Union, 1956), **37**, *Trans. Am. Geophys. Union*, **38**, 189–192.

Childs, E. C., 1967, 'Soil moisture theory', *Adv. Hydrosci.*, **4**, 73–117.

Childs, E. C., and Collis-George, N., 1950, 'The permeability of porous materials.' *Proc. Royal Soc. Ser. A*, **201**, 392–405.

Christiansen, J. E., 1944, 'Effect of entrapped air upon the permeability of soils', *Soil Sci.*, **58**, 355–65.

Cole, J. A. and Green, M. J., 1966, 'Measuring soil moisture in the Brenig catchment. Problems of using neutrons scattering equipment in soil with peaty layers', in *Symposium on Water in the Unsaturated Zone*, Wageningen.

Croney, D., Coleman, J. D. and Curver, E. W. H., 1951, 'The electrical resistance method of measuring soil moisture', *Brit. J. Appl. Phys.*, **2**(4), 85–91.

Curtis, L. F. and Trudgill, S., 1974, 'The measurement of soil moisture', British Geomorphological Research Group, *Tech. Bull. no. 13*, 70 pp.

Fenwick, I. M., 1960, *Some Soil Permeability Investigations on the Thames Floodplain*, unpublished M.Sc. thesis, Univ. Reading.

Fenwick, I. M. and Knapp, B. J., 1978, *Introduction to Pedology*, Duckworth, London.

Freeze, R. A., 1972, 'The rôle of subsurface flow in generating surface runoff from upstream source areas', *Water Res. Res.*, **8**(5), 1271–1283.

Green, W. H. and Ampt, G. A., 1911, 'Studies on soil physics. 1. The flow of air and water through soils', *J. Agric. Soils*, **4**, 1–24.

Hewlett, J. D. and Hibbert, A. R., 1963, 'Moisture and energy conditions within a sloping soil mass during drainage', *J. Geophys. Res.*, **68**, 1081–1087.

Hills, R. C., 1970, 'The determination of the infiltration capacity of field soils using the cylinder infiltrometer', *BGRG Tech. Bull. No. 3*, 24 pp.

Hills, R. C., 1971, 'The influence of land management and soil characteristics on infiltration and the occurrence of overland flow', *J. Hydrology*, **13**, 163–181.

Horton, R. E., 1933, 'The rôle of infiltration in the hydrological cycle', *Trans. Am. Geophys. Union*, **14**, 446–60.

Kirkby, M. J. and Chorley, R. J., 1967, 'Throughflow, overland flow and erosion', *Internat. Assoc. Sci. Hydrol. Bull.*, **12**(3), 5–21.

Klute, A. A., 1952, 'A numerical method for solving the flow equation for water in unsaturated materials', *Soil Sci.*, **73**, 105–116.

Knapp, B. J., 1970a, *Patterns of Water Movement on a Steep Upland Hillside, Plynlimon, Central Wales*, unpublished Ph.D. thesis, Univ. Reading, 213 pp.

Knapp, B. J., 1970b, 'A note on throughflow and overland flow in steep mountain catchments', *Reading Geogr.*, **1**, 40–43.

Knapp, B. J., 1973, 'A system for the field measurement of soil water movement', *BGRG Tech. Bull.*, No. 9, 26 pp.

Mein, R. G. and Larson, C. L., 1973, 'Modelling infiltration during a steady rain', *Water Res. Res.*, **9**(2), 384–94.

Morel-Seytoux, H. J. and Khanji, J., 1974, 'Derivation of an equation of infiltration', *Water Res. Res.*, **10**(4), 795–800.

Musgrave, G. W. and Holtan, H. N., 1964, *Handbook of Applied Hydrology*, Chow, V., (Ed.), Ch. 12, McGraw-Hill, New York.

Parr, J. F. and Bertrand, A. R., 1960, 'Water infiltration into soils', *Adv. Agronomy*, **12**, 311–63.

Philip, I. R., 1957/8, 'The theory of infiltration', *Soil Sci.*, **83**, 345–57 and 435–58.

Philip, J. R., 1969, 'Theory of infiltration', *Adv. Hydrosci.*, **5**, 215–296.

Reynolds, S., 1970, 'The gravimetric method of soil moisture determination. Parts 1, 2 and 3', *J. Hydrology*, **11**, 258–330.

Richards, L. A., 1931, 'Capillary conduction of liquids through porous mediums', *Physics*, **1**, 318–333.

Richards, L. A., 1949, 'Methods of measuring soil moisture tension', *Soil Sci.*, **68**, 95–112.

Rubin, J., 1966, 'Theory of rainfall uptake by soils initially drier than their field capacity and its applications', *Water Res. Res.*, **2**(4), 739–49.

Slater, C. S., 1957, 'Cylinder infiltrometers for determining rates of irrigation', *Proc. Soil Sci. Soc. Am.*, **21**, 457–60.

Stallman, R. W., 1964, 'Multiphase fluids in porous media—a review of theories pertinent to hydrologic studies', *Prof. Paper US Geolog. Surv*, 411E, 51 pp.

Swartzendruber, D. and Huberty, M. R., 1958, 'Use of infiltration equation parameters to evaluate infiltration differences in the field', *Trans. Am. Geophys. Union*, **39**(1), 84–93.

Vachaud, G., Vauclin, M., Khanji, D. and Wakil, M., 1973, 'The effects of air pressure on water flow in an unsaturated stratified vertical column of sand', *Water Res. Res.*, **9**(1), 160–173.

Watson, K. K., 1959, 'A note on the field use of a theoretically derived infiltration equation', *J. Geophys. Res.* **64**(10).

Weyman, D. R., 1970, 'Throughflow on hillslopes and its relation to the stream hydrograph', *Bull. Int. Assoc. Scientific Hydrology*, **15**(3), 25–33.

Whipkey, R. Z., 1965, 'Subsurface flow from forested slopes', *Bull. Int. Assoc. Scientific Hydrology*, pp. 74–85.

CHAPTER 3

Techniques for measuring subsurface flow on hillslopes

T. C. Atkinson,

School of Environmental Sciences, University of East Anglia, Norwich, UK

3.1 INTRODUCTION

In order to measure the rate or magnitude of any phenomenon it is necessary first to understand what it is one is trying to measure. At least it is essential to have a good working hypothesis of how the phenomenon to be measured operates. Therefore, this introduction is devoted to a brief outline of the types of subsurface flow on hillslopes, as a preparation to describing how the rate and magnitude of each type may be measured.

Except in deserts and high mountains, most hillslopes are mantled by a layer of soil. Rain falling on the soil may either infiltrate through the surface or run off as overland flow. The latter does not concern us here. The water which infiltrates into the soil may then follow a number of paths (Figure 3.1). If the rainfall is light it may simply remain within the soil and be evaporated later through the leaves of plants. Under heavier rainfall, there is a much greater flux of water across the soil surface, soil moisture content increases and water percolates vertically through the profile and laterally downslope. This movement of soil moisture may be through the *matrix* of the soil, in the inter-granular pores and smaller structural voids, or it may be through larger voids. The most common of these larger voids are *pipes* which are open passageways in the soil, of any size from one or two centimetres up to several metres in diameter. They are commonly circular in cross-section and the smaller ones are often reminiscent of animal burrows. In semi-arid areas and in thick upland peats in humid areas, the pipes may be large enough for a man to explore (Fletcher *et al.*, 1954; Darlrymple *et al.*, 1968; Ingram, 1967).

We may therefore distinguish between *matrix flow* and *pipe flow* when discussing water movement on hillslopes. The question arises of the size at which a void such as a crack between peds should be considered as a pipe rather than a matrix void.

73

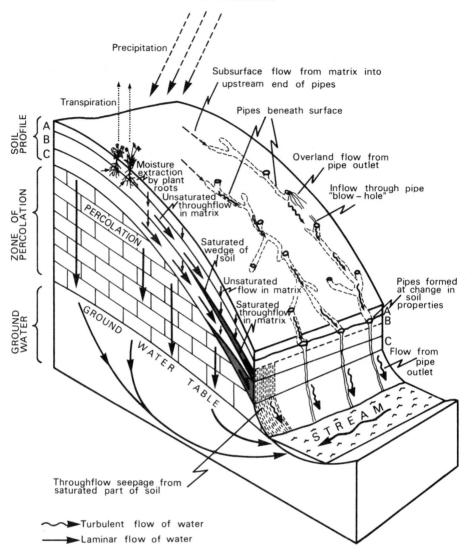

Figure 3.1   Flow routes followed by subsurface runoff on hillslopes

Probably the best definition is to consider all flow which is of turbulent hydraulic regime as pipe flow and all laminar flow as matrix flow. Different techniques will be needed to measure pipe flow, which is concentrated in discrete channels, and matrix flow which is dispersed.

3.1.1   Matrix flow

Matrix flow may be divided into *downslope* and *vertical* components and may occur under both *saturated* and *unsaturated* conditions. In all cases, movement of the soil moisture occurs in response to a gradient of *hydraulic potential* (Figure 3.2) and, if it is laminar, conforms to Darcy's law. At every point in the soil over the whole of hillslope profile, the soil moisture possesses a certain amount of energy which varies from one point to another. For moisture within the matrix, which moves very slowly, this energy is almost entirely potential energy. It is made up of

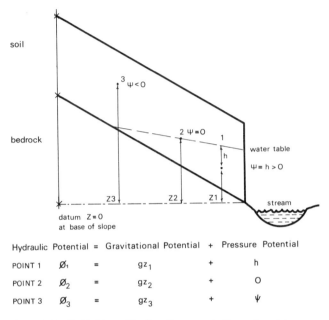

Figure 3.2   Definition of hydraulic potential for soil moisture on a hillslope

two components, a *gravitational potential* due to the height of the point above some fixed datum (usually taken as the base of the slope) and a *pressure potential*, ψ, which is the difference between the pore-water pressure at a particular point and atmospheric pressure. A third component is the *osmotic potential* which depends upon variations in the chemical composition of the soil solution, but this will be neglected in this account. The pressure potential may be expressed in terms of the head of water required to produce the observed pore-water pressure and measured in centimetres of water. The *total hydraulic potential* per unit weight of water, ϕ, is the sum of gravitational and pressure potentials in centimetres

$$\phi = z + \psi \tag{3.1}$$

where

z = the height of the point under consideration above the base of the slope.

The pressure potential is measured with reference to atmospheric pressure. If the pore-water pressure is greater than atmospheric pressure, the potential is positive. In this case, the soil is saturated and the point under consideration lies beneath the water table, the positive pressure being caused by the weight of the column of water above it. At the water table the pressure potential is zero since the pore-water pressure there is atmospheric. Above the water table the soil is unsaturated and surface-tension forces between the soil moisture and the walls of the pores create a suction, so that the pressure within the soil moisture is less than atmospheric. The difference between the pore-water pressure and atmospheric pressure is often called the *tension* or *soil* or *negative pressure potential*. Here we shall adopt the convention that if the pore-water pressure is greater than atmospheric pressure, the pressure potential is positive, if less than atmospheric, negative.

The magnitude of the surface-tension forces within a particular pore is inversely proportional to its radius. Thus, the pressure potential at a particular locality in the soil is determined by the size of the largest pores that are filled with water at that locality. At large negative values of pressure potential, only the smallest pores are filled with water, whereas when pressure potential is close to zero all but the largest pores are filled. Clearly, for a soil with a given distribution of pore sizes there should be a definable relationship between pressure potential and moisture content. This relationship may be determined for soil samples in the laboratory, using a pressure-plate apparatus to expel water from the sample until it reaches the required negative value of pressure potential (Richards, 1947; Croney and Coleman, 1954). However, it is found that the relationship is different when the soil sample is progressively wetted from the case in which it is progressively dried. Moreover, the variability, or *hysteresis*, of the relationship depends upon the previous history of wetting and drying (Poulovassilis, 1962). This makes the relationship of limited application for precise field measurements of soil-moisture movements.

Within the matrix, moisture flows from points of high to points of low total hydraulic potential. The flux from one point to another is proportional to the difference in potential between them, or to the *hydraulic potential gradient*.

$$\bar{F} = k \text{ grad } \phi \tag{3.2}$$

in which $\bar{F}$ is the flux of moisture per unit area and k is a coefficient called the *hydraulic conductivity* of the soil. The value of hydraulic conductivity is not constant but varies with the moisture content. It is at a maximum when the soil is saturated and declines to zero in dry soil. The relationship is not linear because the permeability is affected not only by the proportion of the total pore space which is filled with water, but also by the sizes of the water-filled pores. When the moisture content is low only the smallest pores are filled and these transmit water much less efficiently in relation to their total cross-sectional area than do larger pores. Moreover, the relationship between moisture content and hydraulic conductivity frequently shows hysteresis, depending partly upon the previous history of wetting and drying of the soil (Poulovassilis, 1969; Pavlakis and Burden, 1972). For this reason it is not always practical to determine the hydraulic conductivity of the soil in the laboratory and then to apply the results in measuring moisture movement in the field. Also, the values of hydraulic conductivity determined for small samples in the laboratory may not be representative of larger masses of soil in the field, where fissures and cracks between peds have a significant effect.

In the determination of the downslope component of flow, the hydraulic gradient parallel to the slope is substituted for 'grad' in Equation 3.2, whereas for the measurement of vertical flow it is the vertical hydraulic gradient which is important.

For further details of the physics of moisture movement in the soil matrix the reader is referred to the standard accounts by Hillel (1971), Baver (1956) and especially Childs (1969).

3.1.2 Pipe flow

By definition, pipe flow has a turbulent regime and occurs in channels and pipes which are larger than the capillary-size pores of the matrix. Thus, the appropriate equations for describing the flow are those normally employed for open channels or pipes. These are all variations of the Darcy–Weisbach Equation

$$q^2 = \frac{8Rg}{f} \cdot \frac{d\phi}{dl} \tag{3.3}$$

where

q = discharge per unit cross-sectional area

R = hydraulic radius

f = a friction factor or flow resistance

$\dfrac{d\phi}{dl}$ = the hydraulic gradient in the case of a water-filled pipe, or the slope of the water surface in the case of uniform flow in a partially-filled pipe (Alberson and Symons, 1964).

However, there are almost no observations on the velocity of flow in natural pipes and none at all on the values of f, so that an application of this type of equation to the problem of measuring pipe flow in the field is unlikely to be fruitful. Instead, a direct method of measurement is required.

3.1.3 Types of Field Measurement

Methods of measuring moisture movement on slopes fall into three categories.

(1) *Methods involving interception of the flow*, in which all or part of the flow is intercepted and channelled into a measuring device to determine its discharge. These are applicable to both matrix and pipe flow.

(2) *Methods involving the addition of tracers*, in which radioactive compounds or fluorescent dyes are added to the flow and detected at a point some distance downslope. In this way the velocity can be determined and, under ideal

circumstances, the discharge. Tracer methods are applicable to both matrix and pipe flow.

(3) *Indirect methods*, in which measurements are made of moisture content and hydraulic potential over the whole of a slope profile or experimental plot. The results are used to calculate the moisture flux in the matrix of the soil by applying Darcy's law and the Equation of Continuity (see p. 101). This method is suited only to matrix flow.

In the remainder of this chapter, examples of each of these groups of methods will be discussed. Different techniques will be briefly described and, where possible, an example will be given of the type of results to be expected from each.

3.2 METHODS INVOLVING INTERCEPTION OF THE FLOW

3.2.1 Downslope flow

Throughflow gutters and troughs

A number of studies have been published comparing the storm hydrograph formed on hillslope plots with that formed in adjacent channels, the intention being to explain the channel hydrograph in terms of processes occurring on the hillslope. In particular, the discovery by Hewlett (1961), Hewlett and Hibbert (1963) and Whipkey (1965) that in permeable soils substantial volumes of water may flow downslope during and after storms has stimulated interest in measuring the velocity and discharge of throughflow in other hillslope situations. Most of these studies have involved exposing a vertical face of soil and collecting the water which flows out of it. This can be done by digging a pit or trench across the hillslope or by exposing a vertical section at the stream bank. However, it should be noted that any technique which is based upon collecting water seeping from a free face will collect only saturated throughflow. This is because water at the free face must be at atmospheric pressure in order to leave the pore space of the soil and flow away. Thus, the soil at the face must be saturated. Inevitably, if the soil at the face itself is saturated, a wedge of saturated soil will extend upslope, perhaps into soil which would not normally be saturated had an artificial free face not been constructed.

The creation of an artificial free face may induce saturated conditions, especially on the lower parts of slopes where unsaturated downslope fluxes may be appreciable. Consider the sketch in Figure 3.3a which shows a greatly simplified section of a slope profile before a pit is dug. Let there be a downslope flux of moisture throughout the whole thickness of the soil under unsaturated conditions. At the vertical plane A, where the upslope wall of the pit is to be located, the potential varies from ϕ_A at the surface to ϕ_B at the base of the soil. Pressure potential is uniform throughout the soil.

$$\phi_B = z + \psi, \quad \psi < 0 \tag{3.4}$$

$$\phi_A = (z + h) + \psi. \tag{3.5}$$

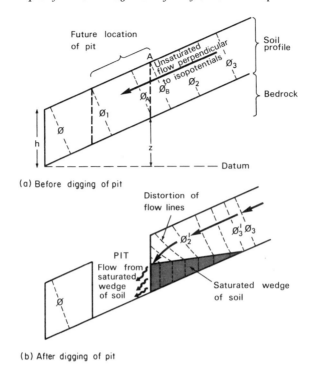

Figure 3.3 Effects of a pit in distorting unsaturated
downslope flow

Now, in Figure 3.3b which shows the situation after the pit has been dug, water
has been unable to leave the free face until sufficient flow from upslope has
accumulated to saturate the soil, so that $\psi \geqslant 0$. Because the soil at the free face is
saturated, its hydraulic conductivity is greater than before, so that it is unnecessary
for water to flow out over the whole thickness of the soil. Accordingly, saturated
conditions extend up the face only so far as it necessary to accommodate the flux
from upslope. This saturated part of the face must be supported by a wedge of
saturated soil extending upslope. In several studies it has been found that the extent
and thickness of the wedge is dependent upon the flux from upslope and changes
as the flux changes (Hewlett and Hibbert, 1963; Weyman, 1970, and 1974). This is
important for the following reason: The conditions sketched in Figure 3.3b
represent a new steady state, accommodating the influence of the pit, in which the
downslope flux is the same as in Figure 3.3a, although the net of hydraulic potential
is distorted by the wedge of saturated soil. However, in unsteady conditions, the
situation shown in Figure 3.3b requires a constant re-distribution of moisture over
the slope and soil profiles, by extending or contracting the wedge of saturation. The
pattern of this re-distribution is quite different from that which would occur in

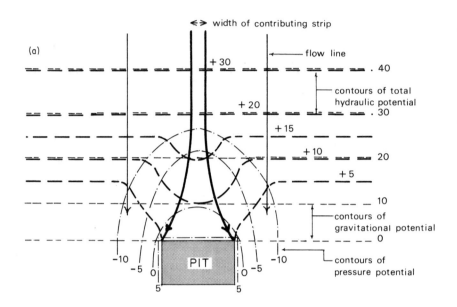

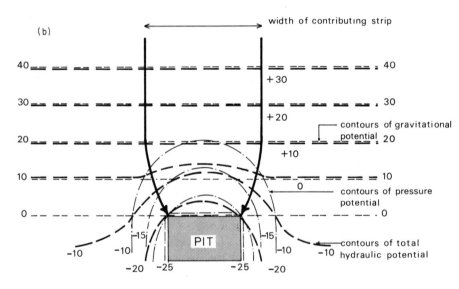

Figure 3.4 Variations in area contributing throughflow to a pit due to distortions of the unsaturated flow net. (a) Effect of a saturated wedge, upstream of a pit, on unsaturated natural throughflow: the contributing strip is narrower than the pit face. (b) Effect of a pit upon saturated conditions on slope: greater drying of the soil around the pit leads to lower pressure potential; the contributing strip is wider than the pit face

Figure 3.3a, where a change in flux from upslope may be achieved by changing the moisture content equally over the whole soil profile. Thus, the shape of the hydrograph produced during a change from one steady flux to another will be different for Figures 3.3a and 3.3b *because of the presence of the pit.* The direct measurement of throughflow by means of collection at a free face will give distorted hydrographs except in situations where there is a naturally occurring wedge of saturation. One such situation occurs at the stream bank. The magnitude of these distortions is not known, and will presumably vary with circumstances, but their possibility should be recognized when designing experiments involving throughflow pits in the middle of a slope profile.

A second undesired effect of pits is that they may not only distort the hydrograph that they are designed to measure, but also distort the net of hydraulic potential on the slope so that they receive drainage from areas which are not directly upslope from them. Knapp (1973) has described how the hydraulic potential net is altered so that downslope drainage on either side of a pit is directed inwards towards the free face (Figure 3.4b) when the natural throughflow on the slope is unsaturated, and outwards around the pit when a wedge of saturated soil is formed (Figure 3.4a). Once again, this effect will vary with the discharge and mode of the flow and may lead to distortions of the hydrograph recorded in the pit. It can be corrected by using tensiometers to record the hydraulic potential at a network of points upslope of the pit, and using these readings to construct equipotential lines and flow-lines de-limiting the contributing area. These are shown in Figure 3.4. The contributing area will vary with the recorded discharge into the pit, and some hysteresis may be expected to occur in the relationship between them.

Direct measurements of throughflow seepage from the soil have been reported by Van't Woudt (1954), Tsukamota (1961), Hewlett and Hibbert (1963), Whipkey (1965), Weyman (1970 and 1974), Dunne and Black (1970) and Knapp (1973 and 1974). In most cases, the apparatus employed consisted of a series of gutters which were dug into a vertical or stepped face of soil in the manner shown in Figure 3.5. To ensure good hydraulic contact between the guttering and the soil, the gutters were packed with soil or pea-gravel and protected and sealed at the front by wooden shuttering. Polythene sheeting was dug into the face at an appropriate height for each gutter, to ensure that each drew its flow from only a single layer of the soil. It is often supposed that throughflow builds up just above an impeding layer in the soil (*see* e.g., Kirkby and Chorley, 1967; Kirkby, 1968; Calver, Kirkby and Weyman, 1972). For this reason, throughflow gutters should be located so as to drain the soil at horizons where there is a marked change in bulk density, as well as the boundaries of the pedogenic horizons.

Most of the authors listed above used gutters draining a free face which was only one or two metres wide, with an upslope plot between 17 m and 300 m long. However, Dunne and Black (1970) describe how they monitored flow from a whole hillside, using a trench dug across the base of the slope. The throughflow was collected in tile drains in which it was monitored by small slot weirs. To ensure

D

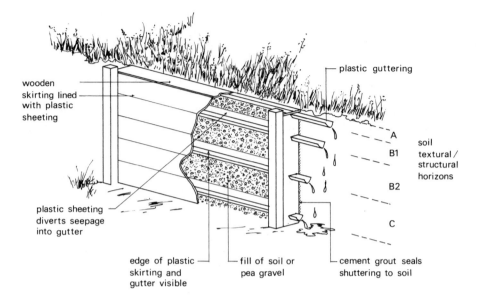

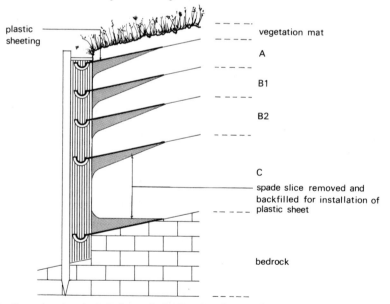

(b) Section showing installation of plastic sheeting and gutters

Figure 3.5   Construction of throughflow troughs (Figure 3.5a redrawn after Whipkey,
1965)

hydraulic continuity the trench was back-filled with soil, and to eliminate end effects the monitored section, which was 52·5 m long, was sandwiched between two 7·5 m sections of ungauged trench. The total area of hillside drained was 0·24 ha. By gauging the drainage from such a large area, Dunne and Black avoided the problem of fluctuating contributing areas, as the total area drained is much greater than the probable fluctuations.

Of all the studies cited, only those by Weyman (1970 and 1974) monitored flow emerging from a natural stream bank. The presence of his throughflow gutters did nothing to alter the effect of a natural free-soil face. As a result, distortions of the hydrograph and contributing area were probably at a minimum in Weyman's experiments. It is worth stating as a general principle that throughflow gutters should ideally be placed on *natural* soil faces at stream banks or the base of slopes, and should be representative of general conditions at the stream bank. If it is desired to monitor throughflow at points above the slope base, pits should be dug, but it is essential to make simultaneous measurements of the hydraulic potential net upslope of each pit, so that fluctuations in contributing area can be calculated and corrected (Knapp, 1973).

Discharge rates from the gutters may be monitored manually using a stopwatch and measuring cylinder, or automatically. Automatic devices vary from calibrated, slotted weirs and water-level chart-recorders to tipping-bucket gauges. Knapp (1973) gives detailed instructions for the construction of a system of tipping buckets. Each tip of a bucket activates a reed switch which in turn causes a record to be made of the tip on a multi-channel event logger. In this way, several gutters can be monitored simultaneously.

The magnitude and flashiness of flow to be recorded varies considerably. Figure 3.5 shows results recorded from the work of Whipkey (1965) who employed stage recorders to monitor the level of accumulated throughflow in a collecting drum, while Figure 3.6 displays the work of Weyman (1970) who used the manual-stopwatch and measuring-cylinder technique.

Multiple stream gauging and dilution gauging

Throughflow troughs and pits normally measure the flow from a very small area of slope. Even those used by Dunne and Black (1970) collected throughflow from an area of only 0·24 ha. When it is desired to obtain a representative average figure for subsurface flow per unit width of slope over the whole of a catchment, gaugings must be made of flow from much larger areas. One way to do this is to gauge a steam at several points along its course in reaches where it receives no tributaries. Any downstream increase in discharge must then be due to contributions from hillslopes on either side of the gauged reach (neglecting the effect of precipitation directly into the channel).

Normally, information on the flow from hillslopes will be required for various discharges and at different stages in the hydrograph, so that continuous gauging is needed. Standard methods should be employed, using thin-plate weirs or flumes in

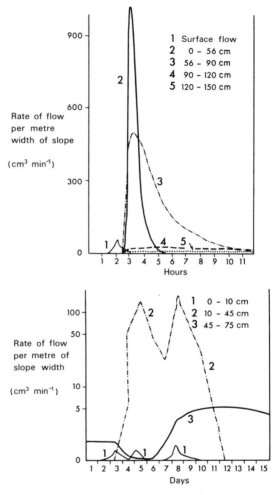

Figure 3.6 Throughflow from a Brown Earth soil,
East Twin Brook, Somerset, England (redrawn after
Weyman, 1970)

conjunction with water-level recorders (British Standards Institution, 1964). An example of the technique is provided by Weyman (1970) who used two V-notch weirs, 400-m apart, to estimate the contribution of throughflow from valley sides to the discharge of the East Twin Brook, Somerset. By subtracting the flow at the upper weir from that at the lower, he was able to estimate the overall response of throughflow to storm rainfall (Figure 3.7). Allowance must be made for the finite time taken by water to flow through the reach. This can be measured simply by colouring the stream with fluorescein dye, or by dissolving common salt in the stream at the upper weir and detecting its arrival at the lower weir by measuring the

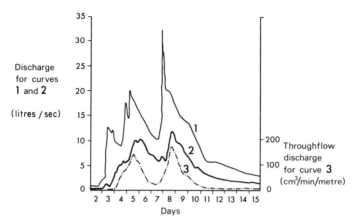

Figure 3.7　Throughflow and streamflow in East Twin Brook, Somerset, England (redrawn after Weyman, 1970)
Curve 1 Discharge from 0.11 km^2 first-order basin.
Curve 2 Discharge from hillslopes in downstream 400 m of basin (obtained by difference of hydrographs).
Curve 3 Throughflow discharge from plot within lowest 400 m

conductivity of the water. A number of suitable portable conductivity meters are available commercially. The time of travel through the reach will normally vary with discharge, and allowance or calibration should be made for this.

Where continuous stream gauging is not feasible or if only instantaneous data are required, an alternative method of estimating flow from the banks into a reach is to measure the progressive downstream dilution of a tracer. This method is only suited for steep, rough, turbulent, upland streams in which the tracer will mix evenly over the stream cross-section in a very short distance. If the mixing length is not very much shorter than the reach over which measurements are made, serious errors may result. The dilution method of gauging has been described by Cobb and Bailey (1965) and Hosegood, Sanderson and Bridle (1969). In outline, the principle and procedure are as follows. The tracer is injected into the stream in a solution of known concentration, c, at a point some distance upstream of the head of the reach under investigation, so that mixing is complete by the time the first sample point is reached. The injection should be continuous at a steady known discharge, q, which can be achieved using a calibrated Marriotte bottle. Concentrations of tracer at any point downstream will rise initially from zero and then reach a plateau value which will depend upon the degree of dilution which the tracer has received. If this plateau concentration is C_i and the stream discharge is Q_i, then by conservation of the tracer.

$$C_i = cq$$

and therefore

$$Q_i = cq/C_i \tag{3.6}$$

Thus, the concentration of tracer at any point i can be used to calculate the discharge at that point. Clearly, by sampling at a number of points downstream, any increase in discharge due to contributions from the stream banks can be detected. Moreover, any variation in C_i through time must be due to variation in Q_i provided that c and q are constant. Thus, short hydrographs may be recorded continuously by injecting tracer for several hours and taking frequent samples. Using this method, Institute of Hydrology personnel were able to detect a 150 % increase in streamflow over a few hundred metres in a small upland catchment in Wales (Greenland, 1971). Kilpatrick (1968) considers that the tracer method of gauging is at least as accurate as the use of calibrated weirs.

Various tracers are suitable for dilution gauging. In the USA fluorescent dyes are widely used, though at concentrations so low that they remain invisible. Procedures for analysing water samples for the presence of these dyes are described by Wilson (1968). In Britain and Europe, sodium dichromate is used as a tracer (Hosegood et al., 1969). An alternative is common salt, which may be detected and measured by titration for the chloride ion or by means of its effect upon conductivity. Full instructions for gauging by the latter method are given by Aastad and Søgnen (1954) and Østrem (1964).

3.2.2   Direct measurement of vertical moisture movement in the soil

In some situations, such as those in which there is very little throughflow on a slope, and precipitation either runs off on the surface or percolates to groundwater, it may be desired to measure directly the amount of vertical percolation in the soil. If the percolation is thought to occur mainly when the soil is saturated, this can be achieved quite simply by intercepting and diverting it, using impermeable plates. Parizek and Lane (1970) describe how they studied deep percolation by digging a trench 5·5-m deep. Galvanized metal pans (30 × 38 cm) were driven into the walls of the trench at 30-cm intervals of depth, with the deepest at 5·1-m below the surface. Each pan has a rim which prevented water from leaving it except via a tube which projected from the wall of the trench. The percolate from each was collected in a bottle and measured at intervals. Like throughflow troughs, this method will only collect water under saturated or nearly saturated conditions.

A more flexible but less-simple approach is to employ a tension lysimeter (Cole, 1958 and 1968; Figure 3.8). This instrument consists of a disc of porous ceramic saturated with water and installed horizontally within the soil beneath an undisturbed profile. The disc is connected to a vacuum reservoir while holds it at a negative value of pressure potential (tension). Since it is permeable, water which percolates vertically to the surface of the disc may be sucked throught it and into the collecting reservoir, as shown in Figure 3.8. The lower surface of the disc is sealed so that water may only enter it from the soil above. To measure normal percolation, the pressure potential of the soil at 'field capacity' is defined (usually as -300 cm but sometimes as little as -50 cm) and the plate is maintained at that potential. Whenever the moisture content of the soil above the plate rises above 'field

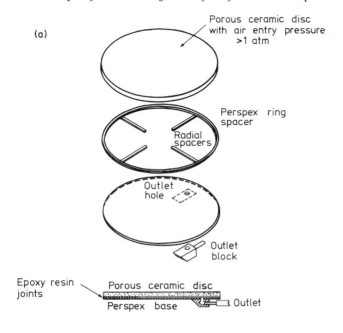

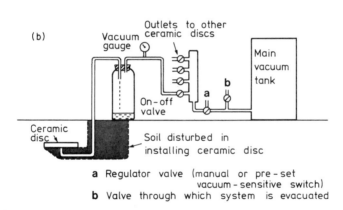

Figure 3.8 (a), Construction of the tension lysimeter (redrawn after Cole, 1968). (b), Installation of the tension lysimeter

capacity', percolation occurs and the percolate is collected and its volume measured at intervals. Cole (1968) describes a system based upon these principles which was used for continuously recording percolation amounts and water quality. An improvement might be suggested whereby the plate is maintained at the same pressure potential as the adjacent soil at the same depth. In this case, all downward-moving soil moisture would be abstracted by the plate and not just 'gravitational water'.

3.2.3   Methods of measuring pipe flow

Flow in natural pipes has been described as an important process by Dalrymple *et al.* (1968), Fletcher *et al.* (1954), Jones (1970) and others. However, the literature contains very few reports on how to measure downslope flow in pipes and to separate its contribution to stream discharge from other sources. In spite of this, an extensive body of experience in mapping pipes and measuring their flow has been accumulated by workers at the Institute of Hydrology, England (Pond, 1971; Gilman, 1971) who have monitored pipe flow as one of the major hydrological processes occurring in the small mountain catchment of Nant Gerig, a headwater tributary of the River Wye. Their investigations fall into two stages, both of which would be essential to any future work on pipes in other areas. They are the initial mapping of pipe systems and the monitoring of flow from individual pipes.

Pond (1971) reports that in the grass-covered Nant Gerig catchment, pipes located less than 30 cm below the surface could be detected in dry weather by slight

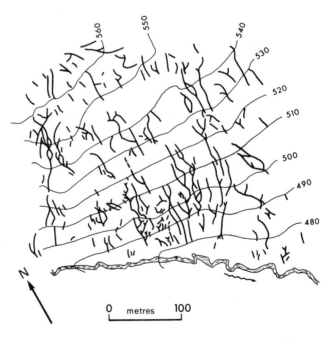

Figure 3.9   Plan of pipe networks on a slope in the River Wye catchment, Plynlimon, Wales (unpublished, Institute of Hydrology). Contours are in metres above mean sea level

depressions in the hillslope surface and also by the presence of vegetation, such as bilberry (*Vaccinium myrtillus*) which prefers good drainage conditions within the depression. In addition, pipes frequently break through to the surface and water will spout from many such breaks during storms. After a storm, but before pipe flow has ceased, a careful observer can follow the line of a pipe by listening to the water gurgling beneath the surface!

Pipes mapped on the basis of these features showed a branching, anastomotic pattern, with the main lines of the network running up and down the slope (Figure 3.9). However, it may be difficult to establish whether the network is continuous or not, and if not what degree of discontinuity occurs. To determine this, dye tracers should be employed (*see* Section 3.3). Note should also be taken of the fact that the pipe network in Figure 3.9 showed little regard for minor corrugations of the hillslope surface, so that the boundaries of sub-catchments of a hillslope defined on the basis of surface topography may be crossed and re-crossed by pipes beneath the surface. Of course, this makes it extremely difficult to define the area actually contributing flow to a short reach of stream at the foot of such a slope.

The anastomosing pipe networks which Pond (1971) has described from the Nant Gerig catchment have been observed by the writer to be typical of many areas of upland Britain, especially where the soils are podsolic. Pipes are commonly found above impeding layers in the soil, such as B_{fe} or B_t horizons, and immediately above the Ea horizon in the peaty surface layers of many upland podsols. Jones (1970) shows that pipes are often formed within, or just below, horizons of greater aggregate stability than the rest of the profile. Burrowing by animals may be partially responsible for the initial formation of these pipes, an origin which would explain the numerous connections and 'blow-holes' to the surface. The writer has seen fresh mole burrows discharging water to the surface during heavy rain in the Nant Gerig catchment, while the underlying mole-run functioned as a pipe.

The hydrographs from upland pipes are often flashy, and special equipment is needed to record them accurately. The best example of such equipment is that developed for use in the Nant Gerig catchment and described by Gilman (1971) and Pond (1971 Appendix). It is illustrated in Figure 3.10 and consists of a flexible hose which is inserted into a pipe outlet and sealed with clay, so that all of the pipe flow passes into the hose. For low flows the water is channelled into a cistern (Figure 3.10b) the outlet from which is via a U-bend pipe let into the bottom. The outflow level is well above the base of the cistern and is capped by a perspex air bell. As the tank fills up, water rises in this bell, pushing air out through the U-tube. When the water level in the bell reaches the lip of the U-tube and spills over, a siphoning action is set up and the tank is rapidly emptied. An air bleed in the side of the bell ensures that the same volume of water is siphoned on each occasion. If the volume of water siphoned on each cycle is known, then the discharge over a period of, say, fifteen minutes can be found by counting the number of cycles. The calibration is not quite linear, as water continues to flow into the cistern while it is siphoning. In fact, at discharges greater than about 0.35 litre $\cdot$ sec^{-1} the Institute of Hydrology instruments simply siphon continuously and are thus of no further use.

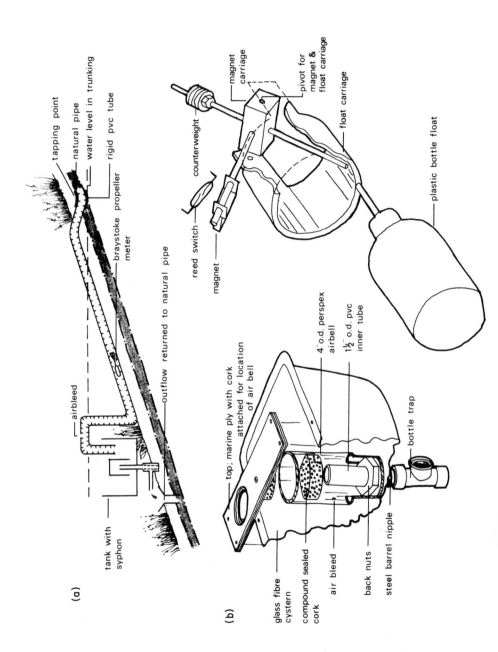

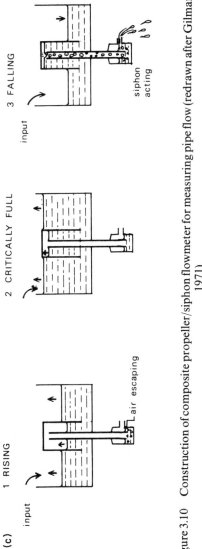

Figure 3.10   Construction of composite propeller/siphon flowmeter for measuring pipe flow (redrawn after Gilman, 1971)

Recordings on the siphon may be made by means of a float mounted on a cylindrical carriage, which is in turn carried on a horizontal pivot (Figure 3.10c). As the float rises, its carriage bears against an arm which is itself independently pivoted about the same axis as the float, and which carries a counter-weighted magnet. When the rising water in the tank reaches a certain level the counterweight tips the magnet rapidly past a reed switch. The switch closes briefly and then opens again as the magnet passes, transmitting a brief on–off pulse of electric current to a counter. When the siphon empties, the float drops, and the rear end of its carriage bears against the counterweight arm of the magnet carriage, pushing the magnet slowly past the reed switch, which thus transmits a much longer pulse of current to the counter. Finally, the float drags the magnet carriage past its balance point and it drops back into the primed position. Thus each siphoning cycle is represented by two pulses of current, one of about 0·5 sec and one of 4 or 5 secs. They may be recorded on a simple event logger or on a GPO counter, using a simple transistor circuit to prevent arcing across the reed switch.

At discharges of more than 0.35 litre $\cdot$ sec^{-1} the siphon tank ceases to operate. For higher discharges Gilman (in Pond, 1971) reports the use of a propeller mounted in a rigid perspex section jointed into the flexible hose. The propeller used by Gilman was a Braystoke Flowmeter, which is equipped with an integral reed switch activated by a magnet mounted on the spindle. Each revolution of the propeller produces a single pulse of current which can be recorded on a counter or an event logger. Calibration is almost linear, and in a 7·5-cm diameter tube the meter will register discharges of between 0.3 and 10 litres $\cdot$ sec^{-1} quite satisfactorily.

As far as is known to the author, almost the only measurements of pipe flow in upland catchments are those made by the Institute of Hydrology in Wales. Gilman (1971) reports extremely flashy hydrographs ranging from zero to 10 litres $\cdot$ sec^{-1}.

In contrast to the flashy regime of pipes formed in upland podsols and peats, pipes occurring elsewhere may have much less variable flows. In the East Twin Brook, Somerset (Weyman, 1970) there are a number of seeps fed by pipe flow which discharge directly into a small stream from slopes of 12° to 20°. These seepages are almost all associated with minor landslips and show a slow variation in discharge, with up to 36-hr lag from peak rainfall to peak discharge. Their flow regime is similar to that of throughflow collected in troughs in the same catchment and like throughflow seeping from the base of the soil, they may persist for weeks or months after runoff-producing rain has ceased (Finlayson and Waylen, 1973). The discharge of these pipes can be measured quite successfully simply be channelling their flow into a gutter and measuring it with a stopwatch and measuring cylinder. Alternatively, a small V-notch weir can be used in conjunction with a float which pivots a magnet past a series of reed switches. As the float rises or falls, different reed switches are closed by the proximity of the magnet, and each brings into circuit a different resistor. A potentiometric recorder is used to measure the resistance of the circuit and thus to indicate which reed switch is closed at a particular time and the level of water in the V-notch (Knapp, 1973).

3.3 METHODS OF TRACING THE FLOW

The hydrology of subsurface flow on a hillslope is essentially a 'black-box' problem. It is relatively easy to measure inputs to the box (rainfall) and outputs of streamflow or throughflow at the base of the slope. But when it comes to directly measuring the actual paths taken by water within the soil, one is often forced to use instruments which disturb the very things one is trying to measure, and disturb them in a continuously fluctuating manner so that a whole series of secondary observations are necessary in order to establish correction factors for the initial measurements. This situation can sometimes be avoided by using tracers to label the inputs to the system, and recording their presence and concentration at the outflows. The time taken by the tracer to pass through the system gives a measure of flow velocity, and changes in its concentration and distribution may be used as evidence in inferring the existence of certain types of pathway through the system. Tracers have been widely used to solve problems of groundwater flow in karstic limestone areas (Atkinson *et al.*, 1973) but have so far had only a limited application in hillslope hydrology.

There are two main types of tracer which may be used on hillslopes: radioactive compounds and fluorescent dyes. A review of the use of both in groundwater problems is given by Knutsson (1968).

Radioactive tracers have been very little used in hillslope studies although Pilgrim (1966) reports their use in determining the times of concentration from points in rills on the periphery of a small drainage basin. The radioactive isotope may be incorporated in a soluble salt, or in some cases the radioactive isotope of hydrogen, tritium, may be used. The latter, in the form of ordinary water containing some tritium hydroxide, is greatly superior as a tracer because it behaves exactly like water and is not subject to adsorption or precipitation on clays or organic matter in the way in which other radioactive salts and chemical dyes are. Artificially-introduced slugs of tritium have not been used in hillslope studies although Smith (1972, and 1974) has reported their use in pollution problems. One reason for this is that they would obscure 'natural' variations of tritium in rainfall. The amount of tritium in rainfall has greatly increased since the first hydrogen-bomb tests in 1954, like the amounts of other radioactive isotopes. It fluctuates seasonally and with each new outburst of nuclear folly. Thus, for example, Biggin (1971) was able to use measurements of natural tritium made during a rainstorm to establish that most of the water forming the actual flood hydrograph of the Nant Gerig stream in Wales had been displaced from storage in upland peat bogs, a result which substantially altered hydrologists' views of the runoff process in that catchment. Such measurements might not have been possible had the area been previously contaminated by artificial doses of tritium. Finally, it should be noted that tritium is a beta-emitter and is consequently difficult to monitor in the field. In most cases, samples must be taken and analysed in a specialist laboratory.

Fluorescent dyes are substances which strongly absorb light of certain wavelengths and then re-emit the absorbed energy as light of somewhat longer wavelength. Each dye has its own characteristic spectrum of absorption and fluorescence. At large dilutions the fluorescence which occurs at a given intensity of illumination is linearly proportional to the concentration of the dye. Consequently, the concentration may be accurately estimated by measuring the fluorescence intensity in a device known as a fluorometer. A sample of the dye solution is irradiated with light which has been filtered so that only the wavelengths which the dye strongly absorbs are present. If dye is present in the sample, it fluoresces and some of the fluorescent light is passed through a second filter which cuts out the primary light. The filtered fluorescent light falls onto a photo-electric surface and produces an electric current. This current is amplified, measured potentiometrically and recorded. The fluorometer is calibrated to relate instrument readings to concentration for each particular dye.

A detailed account of fluorometric procedures is given by Wilson (1968) while brief accounts are supplied in articles by Brown and Ford (1971) and Atkinson et al. (1973). A filter fluorometer which is both sensitive and suitable for field-use is manufactured at the time of writing by G. K. Turner Associates of Palo Alto, California.

Fluorescent dyes are all strongly adsorbed on clays and soil organic matter and may also be de-colourized by the same substances. Experiments by Scott, Norman and Fields (1969), Reynolds (1966), Corey (1968) and the writer have shown that two dyes are generally less severely adsorbed and de-colourized than most. These are the yellow–green pyranine (manufactured by Bayer Co. Ltd.) and rhodamine WT (Du Pont Co. Ltd.) which fluoresces in the red–orange range of the spectrum. Optical filters for use with both dyes in the Turner fluorometers are recommended by Wilson (1968) and Atkinson et al. (1973).

Because of their strong adsorption on some soils, fluorescent dyes are best used for estimating flowpaths and water velocity only, as far as hillslopes are concerned. Emmett (1970) describes how he used fluorescein as a visible marker to determine the velocity of overland flow on trial plots. He also established that even in sheet runoff there existed threads of water in which the velocity was greater than elsewhere. These threads formed an anastomosing or braided pattern up and down the length of the hillslope.

The writer has employed pyranine to measure the velocity of throughflow on a 20° slope adjoining a first-order stream in the East Twin Brook catchment, Somerset (Weyman, 1970 and 1974). Before this experiment, it had already been established by Weyman that saturated throughflow occurred over the lowest 10–20 m of the slope profile, the precise width of the contributing zone varying with discharge. Most discharge came after storms from the base of the B horizon of a 75-cm thick brown earth soil. Baseflow was sustained by flow from the B/C horizon, which was denser and finer textured than the B horizon. In the tracer experiment, 1500 gm of pyranine in 5 litres of water was poured into a pit 30 cm in diameter, which was located 6·5 m upslope of a set of throughflow troughs, with its base on the boundary between B and B/C horizons. The pit was then re-filled and the turf replaced. The throughflow troughs were monitored for the presence of dye by taking water samples. Results are shown in Figure 3.11 from which it is clear that the dye emerged eight days later, only 28 hr or so after heavy rain. As expected from previous observations on the slope, flow was faster in the B than in the B/C horizon, and a greater concentration of dye emerged from the B than from the B/C horizon indicating that a greater proportion of the storm throughflow passed through the upper horizon. Average velocity in the B horizon was 0.8 m·day^{-1} and 0.5 m·day^{-1} in the B/C horizon. However, if one assumes that very little movement of tracer took place before the heavy rain, velocities of 0.23 m·hr^{-1} (B) and 0.05 m·hr^{-1} are indicated.

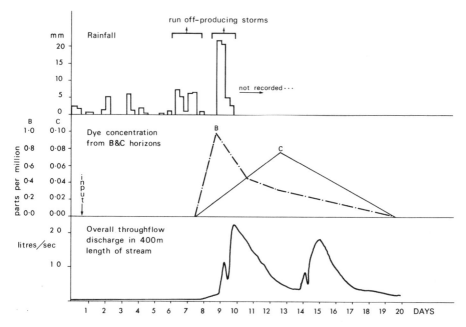

Figure 3.11 Results of tracing throughflow with pyranine dye, East Twin Brook, Somerset, England

The East Twin hillslope is thus an example of a system in which a single-input point feeds two outflows, with different velocities along the various routes within the system. In principle, one could determine the proportion of labelled water following each route by monitoring both the concentration of dye (c_i) and the discharge (q_i) at each outflow. The proportion (%) of tagged water following each of n routes is then,

$$P_i = \frac{\displaystyle\int_0^\infty c_i q_i \cdot \mathrm{d}t}{\displaystyle\sum_1^n \int_0^\infty c_i q_i \cdot \mathrm{d}t} \cdot 100 \tag{3.7}$$

However, the use of this equation involves assuming that either no tracer at all is lost by adsorption, or that equal proportions of the tracer following each route are adsorbed. We know that the first assumption is unjustified and indeed it can be checked by comparing the quantity of tracer recovered $\left(\sum_{1}^{n} \int_{0}^{\infty} c_i q_i \cdot \mathrm{d}t \right)$ with the quantity injected. The second assumption has no general justification, and its validity will depend upon the type and variety of adsorbing materials present in each soil horizon.

The use of tracers described so far involves injection at the upstream end of the system, after which the tracer is not seen again until it emerges from the downstream end. For matrix flow in the soil, it is possible to recover samples of soil moisture from both saturated and unsaturated soils by using ceramic samplers. These samples can be analysed for tracer and used to study the movements of labelled water within the 'black-box' system. The samplers consist of a plastic or aluminium tube with a hemispherical ceramic cup cemented into one end (Figure 3.12A). The other end is stoppered, with two outlet tubes: one extends to the bottom of the ceramic cup while the other is short and is merely used to evacuate the apparatus with a hand pump. The sampler is installed so as to fit snugly into an auger hole in the soil, made with a coring auger or guide tube like that described by Eeles (1969). The cup is first saturated with water and then installed in good hydraulic contact with the soil at the required depth. The top of the auger hole should then be packed with soil to prevent rainwater from trickling down to the cup. Once installed, the sampler is evacuated with a hand-pump and the tap closed. Provided that the suction is great enough to overcome the negative pressure potential of the soil moisture, water will be drawn from the soil through the ceramic. When it is desired to recover the sample, the vacuum is released and air is pumped into the sampler, forcing the sample out through the long tube into the collecting flask.

Batteries of moisture samples can be installed at various depths along a soil–slope profile, and samples taken at intervals from each and analysed for the presence of tracer. In Figure 3.12B the battery at A can be used to monitor vertical movement of the dye while those at B can be used to establish its downslope routes and velocities. Care should be taken to see that the samplers do not form points of very anomalous pressure potential which distort the net of hydraulic potentials on the slope. If a tensiometer is installed close to each sampler, the pressure potential of the soil moisture can be measured at the depth of the sampler's ceramic cup, and the vacuum adjusted so that it is only slightly greater than the ambient soil suction. Quite small samples (6–20 cc) are sufficient for fluorometric analysis, but the suction required to extract them within a given time will vary with the soil permeability.

If velocity measurements are made with tracers at the same time as measurements of soil moisture content, the two sets of data can be combined to estimate moisture fluxes through the different horizons of a soil-slope profile. Thus, in Figure 12c, dye is injected at point I and its vertical velocity established (V_v) using

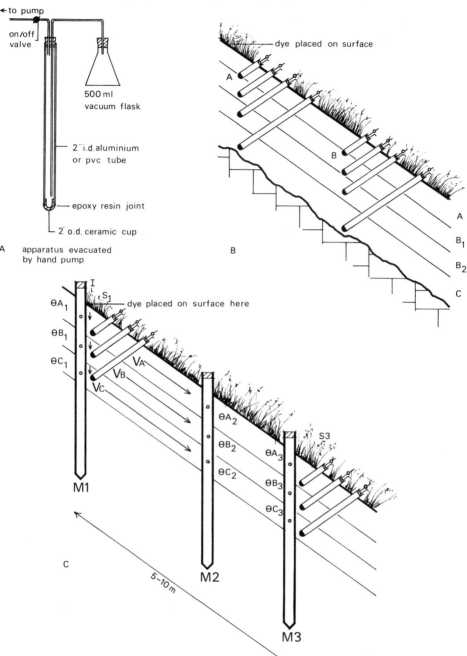

Figure 3.12 A, Construction of soil moisture sampler; B, use of banks of samplers for throughflow tracing; C, use of samplers in combination with neutron probe moisture measurements for calculation of fluxes

the moisture samplers at S1. As close as possible to S1 is a neutron moisture probe access tube M1. If measurements of moisture content (θ) are made in the A, B and C horizons at M1 then the vertical flux (F) of moisture from the A to B horizon is given by

$$F_{(A \to B)} = V_{v(A \to B)} \cdot \frac{(\theta_{A1} + \theta_{B1})}{2} \tag{3.8}$$

Similarly, the flux from the B to C horizon is

$$F_{(B \to C)} = V_{v(B \to C)} \cdot \frac{(\theta_{B1} + \theta_{C1})}{2} \tag{3.9}$$

The downslope flux in each horizon can be determined from the downslope velocity of tracers between samplers at S1 and samplers in the same horizons at S3. For the arrangement shown in Figure 12c the flux in the A horizon is given by

$$F_{A(S1 \to S3)} = V_A \cdot \frac{(\theta_{A1} + \theta_{A2} + \theta_{A3})}{3} \tag{3.10}$$

Similar equations would give fluxes in the B and C horizons. In making measurements of flux in this way, one should bear in mind that the measurements are of *average* velocity of the *tracer*, which may well be slower than the actual velocity of the water because of the adsorption.

Tracers have been used very little in the study of pipe flow although their potential as a tool for measuring velocities and establishing linkages in anastomosing pipe networks is considerable. Almost nothing is known of velocities occurring within natural pipes. Trial observations by the writer showed that values of 10 to 20 cm·sec^{-1} were not uncommon in ephemerally-flowing pipes in the Nant Gerig catchment. These velocities occurred only during the peak period of the storm hydrograph, and velocities during the recession period were much smaller. These results were obtained by locating a number of 'blow-hole' connections between the surface and pipes up to 30 cm below, and pouring up to 100 cc of concentrated pyranine solution into the hole furthest upslope (10 % pyranine in 5 % Sodium hydroxide solution). A rapid visual inspection of nearby holes revealed the direction in which the tagged water was flowing. Holes at which it appeared visually were marked by numbered pegs and the time of appearance noted with a stopwatch. After the storm, the pegs could be surveyed and distances and velocities established.

A more elaborate alternative to this technique is to install a fluorometer in the field, so as to continuously monitor the fluorescence of the outflow from a small stream basin, preferably at a point where discharge is also being continuously gauged. As a preliminary to observations on pipes, dye tracers can then be used to determine velocities and times of travel in the stream channel and how they vary with discharge. Then, during subsequent storms, dye slugs can be injected into pipes on the slopes and the times of travel of the slugs to the basin outflow measured. The velocity of flow while on the slope can be obtained by subtracting stream-channel travel time from overall travel time and dividing by the appropriate distance of the injection point from the channel. Repetition under different conditions of discharge and for different input points would allow one to collect a variety of velocity data which could form the basis of a hydraulic model of pipe runoff on the slopes concerned.

Tracers can also be used to establish topology of pipe networks. As was pointed out in Section 3.1, pipe networks consist of interconnected, anastomosing channels. If one relies solely upon 'blow holes' and topographic expressions in locating the paths of pipes, serious errors can result. Tracers can be used to define the important routes through an anastomosing network. Experiments should begin with the location and marking of as many 'blow holes' as can be found on the slope of interest. Into each 'blow hole' should be inserted a small nylon bag about 2 or 3 cm in diameter, containing about one teaspoonful of activated, granular charcoal. (These bags may be conveniently manufactured from ladies' nylon stockings.) Each bag carries a plastic tag identifying the 'blow hole' in which it was placed, and is located on the floor of the pipe where it will be wetted if flow occurs. Crystalline or powdered dye is then placed in the 'blow hole' which has been selected as an input and which should normally be near the top of the slope. The investigator then retires and waits for rain. During the next runoff-producing storm, any flow in the pipe network will dissolve the dye and disperse it through the main routes of the system. If the dye is carried in solution past any of the activated-charcoal detector bags, some will be adsorbed on the charcoal and after the storm the bags can be collected and taken to the laboratory. The charcoal should be dried and covered with a 5 % solution of potassium hydroxide in 1-butanol (Smart and Brown, 1973) which desorbs the dye and brings it back into solution. The presence of large concentrations of dye will be indicated by the solution's colour, but sub-visible quantities can be detected by fluorometric analysis. Care should be taken not to confuse fluorescence caused by the dye (especially green dyes such as pyranine or fluorescein) with natural fluorescence due to dissolved humic or fulvic acids and other organic substances in the soil. If green dyes are being used it may often be wise to rely upon visible concentrations only, or to perform a 'dummy-run' experiment in advance so as to establish levels of background fluorescence.

In the meantime, a plane-table map of the 'blow holes' which were monitored can be prepared and the positive results plotted onto it, thus de-limiting the paths actually taken by the main flows through the network. Repeated experiments with different input points will allow the important pipe routes to be de-limited over an entire slope. Such information could be most useful if it were gathered before installing equipment to monitor pipe discharges. Figure 3.13 shows a pipe-flow network established in this way for two input points on a slope feeding a tributary of the River Usk, Wales (Wilson and Smart, 1973).

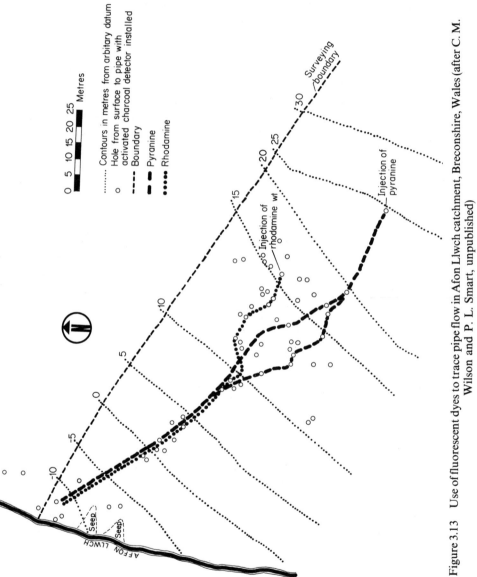

Figure 3.13   Use of fluorescent dyes to trace pipe flow in Afon Llwch catchment, Breconshire, Wales (after C. M. Wilson and P. L. Smart, unpublished)

3.4 MEASUREMENTS OF MOISTURE FLUX WITHIN THE SOIL MATRIX

3.4.1 Principles of measurement

In Section 3.1 it was pointed out that the moisture flux at any point in the soil matrix depends upon the hydraulic conductivity of the soil and the ambient hydraulic potential gradient (equation 3.2)

$$\bar{F} = k \operatorname{grad} \phi$$

It is clear that if the field of hydraulic potential could be defined over the whole of a soil-slope profile, the flux could be predicted at any point, provided that we also knew the corresponding field of hydraulic conductivity. However, hydraulic conductivity is not constant but varies with soil-moisture content. Because of the considerable hysteresis that may exist between the two the estimation of values of hydraulic conductivity from moisture content in the field is probably not practicable. Instead, we may attempt to define the field of flux over the soil-slope profile by considering the equation of continuity for each point

$$\frac{d\theta}{dt} = k_x \cdot \frac{\partial^2 \phi}{\partial x^2} + k_y \cdot \frac{\partial^2 \phi}{\partial y^2} + k_z \cdot \frac{\partial^2 \phi}{\partial z^2} \tag{3.11}$$

where

$\dfrac{d\theta}{dt}$ = the rate of change of moisture content at the point under consideration

k_x, k_y, k_z = the hydraulic conductivities in directions x, y and z, respectively.

The terms in the right-hand side of (3.11) each show the differences between fluxes towards the point under consideration and fluxes away from it in these three mutually-perpendicular directions. For the two-dimensional case of a soil-slope profile, Equation 3.11 reduces to a vertical (z) and a horizontal (x) direction

$$\frac{d\theta}{dt} = k_x \cdot \frac{\partial^2 \phi}{\partial x^2} + k_z \cdot \frac{\partial^2 \phi}{\partial z^2} \tag{3.12}$$

If Equation 3.2 is resolved into the same two components, we have

$$\bar{F}_x = k_x \cdot \frac{\partial \phi}{\partial x}; \qquad \bar{F}_z = k_z \cdot \frac{\partial \phi}{\partial z} \tag{3.13}$$

and by substitution,

$$\frac{d\theta}{dt} = \frac{\partial \bar{F}_x}{\partial x} + \frac{\partial \bar{F}_z}{\partial z} \tag{3.14}$$

If Equation 3.14 is solved for all points on the soil-slope profile, the fluxes $\bar{F}_x$ and $\bar{F}_z$ can be calculated for any point. Unfortunately, neither Equations 3.12 nor 3.14 can be solved rigorously to give point values for these fluxes on natural slopes. They may, however, be treated by means of a finite difference technique which will lend itself readily to translation into field terms. Imagine that the point under consideration is expanded into a finite quadrilateral area as shown in Figure 3.14. If

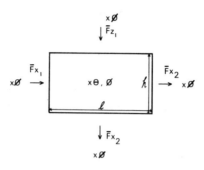

Figure 3.14 Definition sketch of moisture fluxes through a small element of a soil-slope profile

the area of the quadrilateral is A, its total moisture content at any time t_1 is $A\theta_{t_1}$. If at some later time the total moisture content has changed to $A\theta_{t_2}$ then,

$$A(\theta_{t_2} - \theta_{t_1}) = (t_2 - t_1)[(\bar{F}_{z_2} - \bar{F}_{z_1})l + (\bar{F}_{x_2} - \bar{F}_{x_1})h]$$

or

$$\frac{A(\theta_{t_2} - \theta_{t_1})}{(t_2 - t_1)} = (\bar{F}_{z_2} - \bar{F}_{z_1})l + (\bar{F}_{x_2} - \bar{F}_{x_1})h \qquad (3.15)$$

where

l = length of rectangle in Figure 3.14

h = height of the rectangle in Figure 3.14.

This equation is analogous to (3.14). The problem now is to find the values of the fluxes across the boundaries of the rectangle, assuming that we know the moisture content at times t_1 and t_2. This can only be done if we assume some specified relationship between them, or if we know a specified relationship between three of them and the actual value of one. We can define a relationship between them by using the values of hydraulic gradient across each boundary, since

$$\bar{F}_{z_1} = k_{z_1} \cdot \frac{\partial \phi}{\partial z}; \qquad \bar{F}_{x_1} = k_{x_1} \cdot \frac{\partial \phi}{\partial x}$$

and

$$\bar{F}_{z_2} = k_{z_2} \cdot \frac{\partial \phi}{\partial z}; \qquad \bar{F}_{x_2} = k_{x_2} \cdot \frac{\partial \phi}{\partial x} \qquad (3.16)$$

The potential gradients can be established by measuring total hydraulic potential at the centre of the area A and at four points around it as shown in Figure 3.14. If it is then assumed that any three of the permeabilities k are equal (or are related by fixed, known ratios), and one flux $\bar{F}$ is specified, the values of the other three fluxes can be found.

If we now take an array of elements making up a soil-slope profile, as shown in Figure 3.15, we can specify the fluxes shown as $\bar{F}$ as being the prevailing rate of

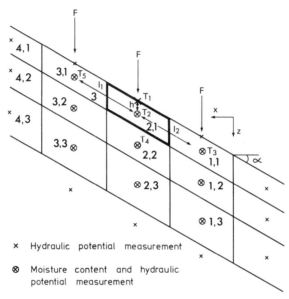

Figure 3.15   Measurements of moisture content, hydraulic potential, surface flux and topography on a soil-slope profile

precipitation or evaporation, providing that in the case of evaporation we make the additional assumption that plant roots do not penetrate deeper than the top row of elements. Measurements must be made of moisture content and hydraulic potential at all the points marked $\otimes$ and of hydraulic potential at all of the points marked 'X' on Figure 3.15. For any surface element, say that labelled '2, 1' the total flux $\bar{F}_{z_1}$ from above is specified. The hydraulic gradient across each boundary of the element is known and so is the change in moisture content of the element over time. If we assume that the hydraulic conductivity in each element is isotropic and homogeneous, $k_{x_1} = k_{x_2} = k_{z_1}$, then the ratio

$$\bar{F}_{z_1} : \bar{F}_{x_1} : \bar{F}_{x_2} = \frac{\partial \phi}{\partial_{z_1}} : \frac{\partial \phi}{\partial_{x_1}} : \frac{\partial \phi}{\partial_{x_2}} \tag{3.17}$$

Evidence has been reported by Evans (1962) and Basak (1972) to suggest that soil is frequently anisotropic in this respect, because of the tendency for platy particles to align themselves parallel to the slope of the ground. Similarly, if the soil has a prismatic or platy structure, anisotropic behaviour will result from the preferred alignment of the peds and the fractures between them. However, some such simplifying assumption is required if Equation 3.15 is to be solved at all, and in the absence of actual measurements of hydraulic conductivity it is probably preferable to assume isotropic behaviour than to specify the degree of anisotropy. It is also true that the moisture content of the soil will change from the upslope to the downslope end of the element, leading to differences between k_{x_1} and k_{x_2}. These must be ignored, a step which is justified by making the element short enough for downslope changes to be small.

Having made the assumption of isotropy, the flux $\bar{F}_{z_1}$ in element '2, 1' (Figure 3.15) can be taken to be equal to the precipitation in a given period (say, 30 mins) during a runoff-producing storm in which overland flow does not occur. If tensiometers are used to measure the pressure potential at points $T_1, T_2, \ldots T_5$, and the slope has been previously surveyed, the total hydraulic potential can be calculated for each point for times t_1 and t_2. Then, knowing $\bar{F}_{z_1}$, the values of $\bar{F}_{x_1}$ and $\bar{F}_{x_2}$ in element '2, 1' can be calculated from,

$$\bar{F}_{x_1} = \bar{F}_{z_1} \cdot \frac{(\phi_{T_5} - \phi_{T_2})h}{(\phi_{T_1} - \phi_{T_2})l_1}$$

$$\bar{F}_{x_2} = \bar{F}_{z_1} \cdot \frac{(\phi_{T_2} - \phi_{T_3})h}{(\phi_{T_1} - \phi_{T_2})l_2} \tag{3.18}$$

The flux $\bar{F}_{z_2}$ from the lower surface of the element can be calculated by difference, using

$$\bar{F}_{z_2} = \Delta S - \bar{F}_{x_2} + \bar{F}_{x_1} + \bar{F}_{z_1} \tag{3.19}$$

It is, of course, necessary to know the value of ΔS, the change in moisture content of the element between times t_1 and t_2. This is achieved by measuring the moisture content at point T_2 in the centre of the element, using a neutron soil-moisture probe.

This procedure can be repeated for all of the uppermost row of elements in Figure 3.15. Of course, if each element is taken separately then the $\bar{F}_{x_2}$ of one element, say '2, 1' should be equal to the $\bar{F}_{x_1}$ of the next element downslope, '1, 1.' If this is not the case, it will be necessary to average the two values and repeat the analysis for each element until the values converge. Alternatively, the dissonant values may be averaged and any resulting imbalance in the element's moisture budget assigned to the downward flux $\bar{F}_{z_2}$, which will then take its final value after the first iteration.

Once a satisfactory solution has been achieved for the uppermost row of elements the next row downwards can be considered. In this row the known flux is that from the row above. Thus, for element '2, 2' $\bar{F}_{z_1}$ is equal to $\bar{F}_{z_2}$ of element '2, 1'. Repeating the procedure outlined above allows Equation 3.19 to be solved for all fluxes in the second row of elements. The analysis can then pass on to the third row, and so on.

Thus, by measuring moisture content and hydraulic potential at an array of points over the soil-slope profile, it is possible to calculate the distribution of fluxes over the entire array, providing that the surface flux is known. Once these fluxes have been calculated they can be combined with the moisture content of any single element to give the average vertical or downslope velocity of water, using

$$\bar{v} = \frac{\bar{F}}{\theta} \tag{3.20}$$

where

$\bar{v}$ = the actual water velocity
$\bar{F}$ = the flux per unit area of boundary
θ = moisture volume fraction of the soil.

Similarly, substitution of the flux per unit area into Darcy's law gives the value of the hydraulic conductivity. For example,

$$k_{x_1} = \bar{F}_{x_1} \cdot \frac{\partial x_1}{\partial \phi} \tag{3.21}$$

This relationship gives a field estimate of the hydraulic conductivity at a specific soil moisture content, and the results of repeated calculations for different time intervals can be used to plot the relationship between these two parameters within each element and its variation over the slope. However, the assumptions of the present method should be borne in mind when considering such results particularly as it is not known how sensitive determined values of hydraulic conductivity are to the various approximations made in the method.

Finally, and most important, the downslope fluxes from the lowest column of elements on the slope represent the throughflow hydrograph from the slope base. They can be checked against field measurements made with throughflow troughs. If agreement between predicted and measured values is satisfactory, then it may be assumed that the calculated distribution of fluxes over the slope profile and through time are an accurate model of the throughflow process. In that case, the argument outlined above provides a satisfactory rationale for measuring the movement of moisture on the slope without disturbing the pattern one is trying to measure.

3.4.2 Field Methods

Firstly, an accurate survey must be made of the slope profile on which the measurements are to be made. Since the experiment would normally be carried out on a narrow plot, a single level and staff traverse up and down the slope should be adequate. Errors should be distributed around the traverse in the normal way. Secondly, the soil profile along the surveyed line should be established by augering immediately to one or both sides of the trial plot. Auger holes should be spaced at close intervals up and down the slope so as to give a complete picture of soil horizons over the whole profile. Thirdly, throughflow troughs should be installed at the base of the slope as described in Section 3.2. Before installing them, the bulk density of each horizon in the soil profile should be measured and troughs located at any sharp changes in bulk density as well as at other horizon boundaries. The troughs should have a width which is accurately known, one metre being a suitable minimum value. Note that if the contours on the slope are not parallel, the trial plot may become wider or narrower upslope. If this effect is marked (i.e. the plot is on a pronounced ridge or hollow) it may be necessary to take account of the changing width of the plot in the analysis of results.

Before any instruments can be installed on the experimental plot, the size and location of the slope elements to be used in the analysis of results must be decided upon. The writer found that a spacing between instruments of 2·5 m measured along the ground surface was satisfactory for a plot 30-m long. The upper and lower boundaries of the elements should coincide with the positions of throughflow gutters at the slope base. Higher up the slope, the boundaries should coincide with the same soil horizons as the gutters.

Once the slope has been surveyed, and the boundaries of elements decided upon, tensiometers and access tubes for a soil-moisture probe should be installed. Since repeated measurements are required, soil-moisture content is best measured using the *neutron scattering technique*. In this method a probe containing a source of fast neutrons is lowered into a permanent vertical access tube in the soil. The tube is sleeved with aluminium and sealed at its base to prevent flooding. The fast neutrons are irradiated into the soil surrounding the tube, where they are slowed to thermal speeds by collision with hydrogen atoms, which in mineral soils occur mainly in the soil moisture. Many of the thermal neutrons are scattered back towards the probe, forming a cloud of neutrons around the access tube. The cloud density is proportional to the intensity of emission of fast neutrons and also to the water content of the soil. Thus, the instrument measures the cloud density by means of a counter which is sensitive to thermal neutrons but not to fast neutrons. The counter is mounted alongside the fast-neutron source in the probe and the reading it gives may be calibrated in terms of the soil-moisture content.

Soil-moisture probes based upon the neutron-scattering technique are made to a number of designs in different countries. In Britain, the Wallingford Probe designed at the Institute of Hydrology is the most widely used. Instructions for its use and calibration are given by Holdsworth (1971) while Bell (1974) provides a comprehensive review of the principles, technique and application of neutron scattering methods. The techniques used to install access tubes are described by Eeles (1969). It should be noted that the Wallingford Probe should ideally be calibrated for each soil type in which it is to be used, although in practice, little accuracy will be lost by using standard calibration curves for different types of mineral soils. Also, the probe cannot easily be used at depths of less than 15 cm because at smaller depths some fast neutrons escape through the soil surface, producing an erroneously-low reading of moisture content.

The neutron-scattering technique estimates the moisture content over a finite sphere of soil rather than at a point. Normally this sphere of measurement is not greater than 30 cm in diameter and decreases in size in wetter soils. Thus, a single measurement at the centre of an element of the soil-slope profile should give a representative value for the whole element. Accordingly, vertical aluminium access tubes should be installed at the mid-point of each element, measured along the ground surface. The tubes should be long enough to penetrate the base of the soil or deeper. Details of their construction and installation are given by Eeles (1969) and in the maker's instructions with the Wallingford Probe. Each tube is inserted into a hole made by means of a hammered core auger. To prevent them from filling with rainwater they are stoppered when not in use.

Readings of soil-moisture content are taken for each element by lowering the probe to such a depth that it is opposite the centre of the element concerned. In the case of surface elements, however, this is not possible and the probe must be modified by using it in combination with a *surface extension ring*. This is a tray of undisturbed surface soil and vegetation, 15-cm deep and 45-cm in diameter. It is normally kept in a shallow pit a few feet away from the access tube and with the soil surface in the tray flush with the ground surface. The base of the tray is perforated so that the soil in it is in hydraulic contact with that beneath. It is punctured by a sleeved hole at its centre, which fits around an access tube projecting from the ground. The whole tray can be lifted up and lowered over the access tube, thus doubling the thickness of the surface soil. The probe can then be placed at the level of the true ground surface and a measurement made which gives the average content of moisture in the soil in the tray and the true surface soil. An alternative is to take repeated samples of soil from the surface layer, to be analysed gravimetrically for their water content. However, such sampling is destructive, especially as quite large numbers of samples must be taken on each occasion in order to establish a representative value (Reynolds, 1970).

The hydraulic potential should also be measured at the centre point of each element. The gravitational component of hydraulic potential depends upon the height of the element's centre above the base of the slope and is established from the survey. The pressure potential is best measured using a *tensiometer*. This instrument consists of a porous ceramic cup, saturated with water, which is installed in good hydraulic contact with the soil moisture at the required depth. The cup is connected to a reservoir of water which is initially at atmospheric pressure. If the soil moisture is at less than atmospheric pressure (i.e. pressure potential is negative) water tends to flow from the reservoir through the cup and into the soil. If it is restrained from doing so, the magnitude of the restraining force required is a measure of the pressure potential. Figure 3.16a shows a design for a simple tensiometer which can be constructed quite cheaply. It consists of a PVC or

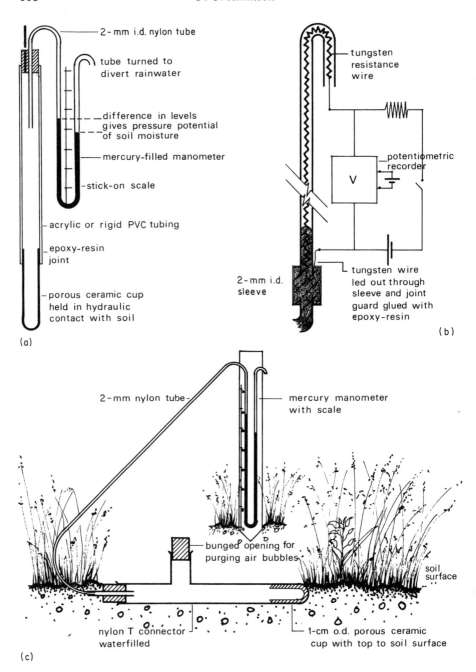

Figure 3.16 Construction of tensiometers. (a) Standard pattern for measurements at depths greater than 10 cm; (b)detail to show principle of recording manometer; (c)small tensiometer for near-surface measurements

acrylic plastic tube to the bottom of which is cemented the porous ceramic cup. The pores of the cup are sufficiently small to prevent air from bubbling through the saturated ceramic even when the pressure difference on opposite sides is one bar. The top of the tube is stoppered with a small subsidiary stopper which can be removed to release any air which collects within the instrument and to refill it with fresh water. From within the barrel of the instrument, a 2-mm bore nylon tube is led out through the stopper and looped down the outside to form a mercury manometer on which pressure differences of up to 70 cm Hg between the inside and outside of the tensiometer can be read. Measurements of pressure potential have been made up to 2 m below the surface using instruments of this type. The tensiometer must be calibrated before use by standing the instrument vertically with the ceramic cup exactly half submerged in distilled water. The difference in mercury levels is then read and recorded. Any greater difference when the instrument is installed in the ground indicates a negative pressure potential, while a smaller difference indicates positive pore-water pressures at the depth of the cup.

Such a tensiometer can easily be converted into a recording instrument by passing a length of high-resistance wire down one limb of the manometer (Figure 3.16b). The resistance of the whole limb to an electric current can then be measured using a potentiometric recorder. The greater the resistance, the lower the level of mercury in that limb.

Designs for similar tensiometers are given by Knapp (1973), Richards (1965) and Webster (1968). Very few authors, however, have considered the physical details of tensiometer design. At least in theory, account should be taken of the elasticity and thermal expansion coefficient of both the tensiometer materials and the water and mercury inside it. Moreover, the tensiometer reading may be altered by an osmotic potential between the soil moisture and the water within the cup, if these contain different concentrations and compositions of dissolved solutes. While these factors are often ignored, perhaps with justification in view of the imprecision of other hydrological measurements, account should certainly be taken of the time taken by the tensiometer to respond to a change in the soil's pressure potential. Klute and Gardner (1962) show that the response time is determined by the hydraulic conductivity of the ceramic cup, the area of the cup relative to the volume of water flowing through, and the change in the volume of water in the instrument per unit change in pressure potential. These are all internal factors of the tensiometer's design, however, and in the field the response time is often determined by the hydraulic conductivity of the soil in supplying water to, or removing it from, the ceramic cup.

Tensiometers based upon the manometer can only measure suctions of up to one bar. Higher suctions may be measured by using a sealed reservoir in which is placed a pressure transducer. When the reservoir is in hydraulic contact with the soil moisture the pressure within it will be the same as that in the soil. Since the reservoir is sealed, water cannot leave it even though the suction may be greater than one bar. The reservoir pressure is recorded electrically by means of the transducer, the only limitation being imposed by transducer design and the air-entry value of the ceramic. The latter may be as high as 15 bar. Tensiometers of this general design are currently under development at the Institute of Hydrology, Wallingford.

Returning to the experimental soil-slope profile, a calibrated tensiometer should be installed with its cup at the centre of each element. Also, tensiometers should be placed at the lower surface of the deepest elements and as close as possible to the soil surface, as in Figure 3.16c. They should also be installed upslope of the highest column of elements and as close as possible to the throughflow troughs. (The surface tensiometers must necessarily be very compact and a suitable design is shown in Figure 3.16c.) When this has been done and the access tubes installed, the plot is ready for measurements to begin.

The actual process of measurement is fairly simple although time-consuming. Repeated readings should be taken of all tensiometers, soil-moisture contents and throughflow troughs. Readings should be taken as nearly simultaneously as possible and the intervals between them should depend upon the rapidity of events being measured. Thus readings taken once or twice daily, or even less often, are sufficient for monitoring slow drainage during dry periods, while the interval may be an hour or less during a runoff-producing storm. Clearly, such a labour-intensive system could not be used for routine monitoring. Rather, it is intended to be used for short periods of intensive observation with the aim of discovering the precise mechanisms of flow on an individual hillside.

The instrument network described is adequate for determining moisture movements within the soil. The flux across the surface should be measured using ground-level rain gauges (Rodda, 1967) with a continuous recording gauge to give precipitation intensities and times. Several ground-level gauges may be needed on a long slope. Upward fluxes due to evapotranspiration are best measured using an automatic climate station (Strangeways and McCulloch, 1965) and assumed to be constant over the whole slope. However, since evapotranspiration is difficult to measure accurately over short periods, it is probably preferable to conduct experiments in winter. Evapotranspiration errors can then be ignored, or evapotranspiration itself assumed to be zero during short storm periods, at least in Britain.

3.4.3   Treatment of results

Data collected in the way outlined above is extremely suitable for computer processing. Indeed, hand calculation of the results from any sizeable experiment would be extremely laborious. For computer processing it is most convenient to arrange the data in matrices or arrays and a method of doing this will now be outlined. Many of the arrays employed will have three dimensions, a, b and c. The a dimension corresponds to the vertical on the actual soil-slope profile, whereas the b dimension corresponds to the direction up and down the slope. Thus the element of the array of moisture content at $a = 3$, $b = 1$ refers to the moisture content of the third soil-slope profile element down from the surface in the first column up from the base of the slope (*see* Figure 3.17). On the other hand, the same element of the

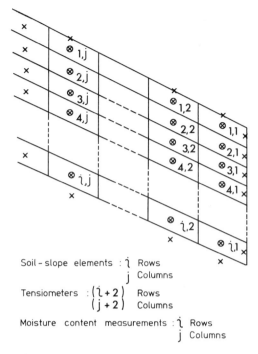

Soil - slope elements : i Rows
                        j Columns

Tensiometers : $(i+2)$ Rows
               $(j+2)$ Columns

Moisture content measurements : i Rows
                                j Columns

Figure 3.17   Notation system for identifying
soil-slope profile elements

array of tensiometer readings would refer to the third tensiometer down from the surface in the column of tensiometers next to the throughflow troughs, because there are more rows and columns of tensiometers on the slope than there are moisture-measurement points.

The dimension c of the arrays refers to time. Since much of the calculation of fluxes rests upon changes occurring in the intervals between readings a value of, say, $c = 3$ will often refer to the period between the third and fourth sets of readings. In some arrays, however, especially those of the primary measured quantities of moisture content and hydraulic potential, $c = 3$ would refer to the third reading of the quantity concerned.

It is clear from the description of the principles of calculating the moisture flux in Section 3.4.1 that all readings of moisture content and pore-water pressure should be taken simultaneously over the whole soil-slope profile. Of course, this is seldom possible in the field, and data will be collected at different points at slightly different times. Thus, before being cast into matrices or arrays, the readings of tension and soil moisture content from each soil-slope element should be plotted against time and values interpolated for certain standard times. All further calculations should be based upon these interpolated values. Linear interpolation may be adequate if values change only slowly with time, but parabolic interpolation may be more accurate during swiftly changing events. Parabolic interpolation is the technique of using three adjacent data points to define the equation of a parabola relating the variable under consideration to time. The interpolated value is then estimated from the equation at the time desired.

Initially it is possible to define the following arrays which form the input for the calculation of fluxes. Note that the soil-slope profile is assumed to contain i rows and j columns of elements (Figure 3.17). A total of n simultaneous readings are made. The notation $[X_{ijn}]$ indicates an an array of i rows and j columns with n ranks:

$[A_{ij}]$ areas of elements
$[h_{(i+1)(j+2)}]$ vertical distances between adjacent tensiometers
$[l_{(i+2)(j+1)}]$ distances between the adjacent tensiometers sub-parallel to the slope
$[\alpha_{(i+1)j}]$ angles of element boundaries to horizontal
$[\theta_{ijn}]$ soil-moisture contents at each measurement point
$[H_{i(j+1)}]$ lengths of vertical element boundaries
$[L_{(i+1)j}]$ lengths of element boundaries sub-parallel to slope
$[\Delta t_{(n-1)}]$ intervals between readings
$[z_{(i+2)(j+2)}]$ gravitational potential at tensiometer cups
$[\psi_{(i+2)(j+2)}]$ pressure potential from tensiometer readings.

From these arrays, it is possible to calculate the values of further arrays, shown below, which are in turn used to calculate the array of fluxes over the whole soil-slope profile. In what follows a standard notation is adopted, thus:

$$[X_{ijn}] = [Y_{abc} + Z_{def}] \quad \begin{array}{ll} a = 1, i & d = 2, i+1 \\ b = 1, j & e = 2, j+1 \\ c = 1, n & f = 2, n+1 \end{array}$$

This means that the array of X's has i rows, j columns, repeated in ranks n times. Each element is the sum of two elements in the arrays of Y's and Z's shown on the right hand side of the equation. The location of the relevant element in the Y array is defined by the coordinates abc (i.e. the ath row, bth column and cth rank of the array) while the relevant element in the Z array is defined by coordinates def. The range through which each coordinate varies is shown on the right of the bracket. Thus, a varies from 1 to i, b from 1 to j, and so on. The order in which they appear indicates that a varies most slowly, c varies fastest, while b varies at an intermediate speed. Thus in this case the first element in the first row is selected for the first rank, then for the second rank and so on to the nth rank in the Y array and the $(n + 1)$'th rank in the Z array. Then the second element of the first row is selected for each rank in turn, and so on until the j'th element of the first row is reached. Then the first element in the second row is selected for each rank in turn, followed by the second element of the second row, and so on. The positioning of the co-ordinates def level with *abc* indicates that *d* is varied simultaneously with *a*, *e* simultaneously with *b*, and *f* simultaneously with *c*.

The following arrays may now be calculated in the order shown below:

Hydraulic potential at tensiometer cups

$$[\phi_{(i+2)(j+2)n}] = [\psi_{abc} + z_{ab}] \qquad \begin{array}{l} c = 1, n \\ a = 1, i + 2 \\ b = 1, j + 2 \end{array}$$

Change in moisture content for each element between readings

$$[\Delta\theta_{ij(n-1)}] = [\theta_{abd} - \theta_{abc}] \qquad \begin{array}{ll} c = 1, n - 1 & d = 2, n \\ a = 1, i \\ b = 1, j \end{array}$$

Change in storage of moisture in each element between readings

$$[\Delta S_{ij(n-1)}] = [\Delta\theta_{abc} \times A_{ab}] \qquad \begin{array}{l} c = 1, n - 1 \\ a = 1, i \\ b = 1, j \end{array}$$

Differences between hydraulic potentials at adjacent tensiometers in a direction sub-parallel to the slope

$$[\Delta\phi l_{(i+2)(j+1)n}] = [\phi_{adc} - \phi_{abc}] \qquad \begin{array}{ll} c = , n \\ a = 1, i + 2 \\ b = 1, j + 1 & d = 2, j + 2 \end{array}$$

Differences between hydraulic potentials at adjacent tensiometers in a vertical direction

$$[\Delta\phi h_{(i+1)(j+2)n}] = [\phi_{dbc} - \phi_{abc}] \qquad \begin{array}{ll} c = 1, n \\ a = 1, i + 1 & d = 2, i + 2 \\ b = 1, j + 2 \end{array}$$

Hydraulic gradients between pairs of tensiometers sub-parallel to the slope

$$[d\phi/dl_{(i+2)(j+1)n}] = \frac{[\Delta\phi l_{abc}]}{[l_{ab}]} \qquad \begin{array}{l} c = 1, n \\ a = 1, i + 2 \\ b = 1, j + 1 \end{array}$$

Hydraulic gradients between pairs of tensiometers in a vertical direction

$$[d\phi/dh_{(i+1)(j+2)n}] = \frac{[\Delta\phi h_{abc}]}{[h_{ab}]} \qquad \begin{array}{l} c = 1, n \\ a = 1, i + 1 \\ b = 1, j + 2 \end{array}$$

Average values of hydraulic gradient in each time interval

$$[\text{mean } d\phi/dl_{(i+2)(j+1)(n-1)}] = [d\phi/dl_{abd} - d\phi/dl_{abc}] \qquad \begin{array}{ll} c = 1, n - 1, & d = 2, n \\ a = 1, i + 2 \\ b = 1, j + 1 \end{array}$$

E

$$[\text{mean } d\phi/dh_{(i+1)(j+2)(n-1)}] = [d\phi/dh_{abd} - d\phi/dh_{abc}] \qquad \begin{matrix} c = 1, n-1 & d = 2, n \\ a = 1, i+1 \\ b = 1, j+2 \end{matrix}$$

Ratio of hydraulic gradient upslope of an element to hydraulic gradient from above

$$[R\phi_{us_{ij(n-1)}}] = \frac{[\text{mean } d\phi/dl_{abc}]}{[\text{mean } d\phi/dh_{abc}]} \qquad \begin{matrix} c = 1, n-1 \\ a = 2, i+1 & d = 1, i \\ b = 2, j+1 \end{matrix}$$

Ratio of hydraulic gradient downslope of an element to hydraulic gradient from above,

$$[R\phi_{ds_{ij(n-)}}] = \frac{[\text{mean } d\phi/dl_{abc}]}{[\text{mean } d\phi/dh_{dec}]} \qquad \begin{matrix} c = 1, n-1 \\ a = 2, i+1, & d = 1, i \\ b = 1, j & e = 2, j+1 \end{matrix}$$

At this stage we may calculate the fluxes for the topmost row of soil-slope elements. To do this it is necessary to define a further array, $[\bar{F}_{\text{surf}_{j(n-1)}}]$, which is the total volume of flux across the ground surface into each element, during each time interval. In addition, the value of the coordinate a is held equal to one so that only the topmost row of each array is considered, corresponding to the surface elements. For $a = 1$, the flux from above each element is given by

$$[\bar{F}_{u_{aj(n-1)}}] = [\bar{F}_{\text{surf}_{bc}}] \qquad \begin{matrix} c = 1, n-1 \\ b = 1, j \\ a = 1 \end{matrix}$$

The flux from upslope of each element

$$[\bar{F}_{us_{aj(n-1)}}]$$

$$= \frac{[\bar{F}_{u_{abc}} \cdot R\phi_{us_{abc}} \cdot H_{adc} \cdot (\cos \alpha_{abc} + \cos \alpha_{adc})]}{2[L_{abc} \cdot \cos \alpha_{abc}]} \qquad \begin{matrix} c = 1, n-1 \\ b = 1, j \\ a = 1 \end{matrix} \quad d = 2, j+1$$

The flux downslope from each element

$$[\bar{F}_{ds_{aj(n-1)}}]$$

$$= \frac{[\bar{F}_{u_{abc}} \cdot R\phi_{ds_{abc}} \cdot H_{abc} \cdot (\cos \alpha_{abc} + \cos \alpha_{adc})]}{2[L_{abc} \cdot \cos \alpha_{abc}]} \qquad \begin{matrix} c = 1, n-1 \\ b = 1, j \\ a = 1 \end{matrix} \quad d = 2, j+1$$

It is now necessary to check that the flux into an element from upslope is equal to the flux out from the adjacent element. If,

$$[\bar{F}_{us_{abc}} \neq \bar{F}_{ds_{adc}}] \qquad \begin{matrix} c = 1, n-1 \\ b = 1, j-1 & d = 2, j \\ a = 1 \end{matrix}$$

then

$$\bar{F}_{us_{abc}} := \frac{[\bar{F}_{us_{abc}} + \bar{F}_{ds_{adc}}]}{2} \qquad \begin{array}{l} c = 1, n-1 \\ b = 1, j-1 \\ a = 1 \end{array} \quad d = 2,j$$

and

$$[\bar{F}_{ds_{adc}}] := [\bar{F}_{us_{abc}}].$$

We must now check that water can leave from the free surface at the downslope boundary of the lowest element in the row (in which $b = 1$). This is only possible if the pressure potential at the boundary is equal to or greater than zero. If,

$$\psi_{a1c} < 0$$

then

$$\bar{F}_{ds_{a1c}} = 0.$$

We can now calculate the downward flux from each element,

$$[\bar{F}_{d_{aj(n-1)}}] = [\bar{F}_{u_{abc}} + F_{us_{abc}} + \bar{F}_{ds_{abc}} - \Delta S_{abc}] \qquad \begin{array}{l} c = 1, n-1 \\ b = 1, j \\ a = 1 \end{array}$$

This procedure defines all four fluxes for each element in the surface row of the soil-slope profile ($a = 1$). It is now necessary to repeat the procedure for the second row by setting $a = 2$. Now,

$$[\bar{F}_{u_{aj(n-1)}}] = [\bar{F}_{d_{1bc}}] \qquad \begin{array}{l} c = 1, n-1 \\ b = 1, j \\ a = 2 \end{array}$$

and values of the second rows in the arrays $[\bar{F}_{us_{ij(n-1)}}], [\bar{F}_{ds_{ij(n-1)}}]$ and $[\bar{F}_{d_{ij(n-1)}}]$ are calculated by following the previous procedure. The third row follows in turn, and so on.

The arrays of fluxes $\bar{F}_{us}, \bar{F}_{u}, \bar{F}_{ds}$ and $\bar{F}_{d}$ can be combined in diagrams of the soil-slope profile drawn for particular values of c, the time-dimension of the arrays. These diagrams give 'snap-shot' pictures of the fluxes over the whole soil-slope profile. The accuracy of the results can be checked at this stage by comparing the values of elements $[\bar{F}_{ds_{a1c}}]$, which represent successive volumes discharged from each horizon at the base of the slope, with actual volumes measured using throughflow troughs. The correspondence between estimated and actual values can be expressed as a correlation coefficient.

The procedure for calculating fluxes which has just been described might aptly be termed the 'brute force' approach to computation. If followed to the letter it would certainly be most inefficient and wasteful of computer space and time. In practice, a computer program incorporating these calculations could be greatly shortened by calculating fluxes for only one time increment at a time and by calculating and storing values for only parts of intermediate arrays such as

$[d\phi/dl_{(i+2)(j+1)n}]$, and replacing them after use with fresh values. The techniques of doing this belong to the field of computer programming and are beyond the scope of this chapter.

It may often be desired to know the flux per unit area rather than the gross flux of moisture from one soil-slope element to another. They can easily be calculated by,

$$[\bar{F}'_{u_{ij(n-1)}}] = \frac{[\bar{F}_{u_{abc}}]}{[L_{abc} \cdot \cos \alpha_{abc}]} \qquad \begin{array}{l} c = 1, n-1 \\ a = 1, i \\ b = 1, j \end{array}$$

$$[\bar{F}'_{d_{ij(n-1)}}] = \frac{[\bar{F}_{d_{abc}}]}{[L_{d_{abc}} \cdot \cos \alpha_{dbc}]} \qquad \begin{array}{ll} c = 1, n-1 & d = 2, i+1 \\ a = 1, i \\ b = 1, j \end{array}$$

$$[\bar{F}'_{us_{ij(n-1)}}] = \frac{[\bar{F}_{us_{abc}}] \cdot 2}{[H_{abc} \cdot (\cos \alpha_{adc} + \cos \alpha_{abc})]} \qquad \begin{array}{ll} c = 1, n-1 & d = 2, j+1 \\ a = 1, i \\ b = 1, j \end{array}$$

$$[\bar{F}'_{ds_{ij(n-1)}}] = \frac{[\bar{F}_{ds_{abc}}] \cdot 2}{[H_{abc} \cdot (\cos \alpha_{abc} + \cos \alpha_{adc})]} \qquad \begin{array}{ll} c = 1, n-1 & d = 2, j+1 \\ a = 1, i \\ b = 1, j \end{array}$$

The rate of flux per unit area is given by,

$$[\bar{F}_{u^*_{ij(n-1)}}] = \frac{[\bar{F}'_{u_{abc}}]}{[\Delta t_c]} \qquad \begin{array}{l} c = 1, n-1 \\ a = 1, i \\ b = 1, j \end{array}$$

with similar formulations for $[\bar{F}_{us^*}]$, $[\bar{F}_{d^*}]$ and $[\bar{F}_{ds^*}]$.

Once flux rates per unit area have been calculated, they can be obtained with hydraulic gradients to give estimates of hydraulic conductivity in each element. Thus,

$$[k_{ij(n-1)}] = \frac{[\bar{F}_{u^*_{abc}}]}{[\text{mean } d\phi/dh_{abc}]} \qquad \begin{array}{l} c = 1, n-1 \\ a = 1, i \\ b = 1, j \end{array}$$

or

$$[k_{ij(n-1)}] = \frac{[\bar{F}_{us^*_{abc}}]}{[\text{mean } d\phi/dl_{dec}]} \qquad \begin{array}{ll} c = 1, n-1 & d = 2, i+1 \\ a = 1, i & e = 2, j+1 \\ b = 1, j \end{array}$$

Once the array $[k_{ij(n-1)}]$ has been calculated, the values of hydraulic conductivity can be compared with the values of $\bar{\theta}$ or mean soil moisture in an element, defined as the array,

$$[\theta_{ij(n-1)}] = \frac{[\theta_{abc} + \theta_{abd}]}{2} \qquad \begin{array}{ll} c = 1, n-1 & d = 2, n \\ a = 1, i \\ b = 1, j \end{array}$$

The results of this comparison should indicate the general trend of the relationship between soil moisture content and hydraulic conductivity.

The velocity of throughflow may be estimated by multiplying the flux rate per unit area by the moisture volume fraction; for example,

$$[\bar{v}_{ds_{ij(n-1)}}] = \frac{[\bar{F}_{ds^*_{abc}} \cdot \bar{\theta}_{abc}]}{100} \qquad \begin{array}{l} c = 1, n-1 \\ a = 1, i \\ b = 1, j \end{array}$$

3.5 CONCLUSIONS

The methods of measuring subsurface flow which have been described in this chapter vary from the extremely simple to the formidably complex. Clearly, some are more suitable for certain applications than others. If it is desired merely to measure the overall contribution of hillslope flow to stream channels, a system of gauging the downstream increase in stream discharge would be best. If the object is to make an empirical study of subsurface flows on a particular slope, then a network of throughflow troughs and pits would be suitable, supplemented if necessary by devices for measuring pipeflow. In deciding where and how to install such a network, tracer studies may be useful. Finally, for detailed analysis of throughflow processes and groundwater re-charge, a network of neutron-probe access tubes and tensiometers should be installed and the values read from them used to calculate moisture fluxes in the soil matrix. The results can be compared with measured discharges from throughflow troughs at the base of the same slope, or perhaps with velocities of flow determined by means of simultaneous tracer experiments. Clearly, such a complex and laborious experimental technique is only warranted in investigations of the fundamental mechanisms of hydrology on hillslopes.

REFERENCES

Aastad, J. and Søgnen, R., 1954, 'Discharge measurements by means of a salt solution', *Int. Assoc. Sci. Hydrol., Gen. Assoc. Rome 1954,* **3**, 289–292

Albertson, M. L. and Symons, D. B., 1964, 'Fluid mechanics', in Chow, V. T. (Ed.), *Handbook of Applied Hydrology* Ch. 7, McGraw-Hill, New York.

Atkinson, T. C., Smith, D. I., Lavis, J. J. and Whitaker, R. J., 1973, 'Experiments in tracing underground waters in limestones', *J. Hydrology,* **19**, 323–349.

Basak, P., 1972, 'Soil structure and its effects on hydraulic conductivity', *Soil Sci.,* **114**, 417–422.

Baver, L. D., 1956, *Soil Physics,* 3rd Edn., John Wiley, 489 pp.

Bell, J. P., 1973, 'Neutron probe practice', Rept. No. **19**, Inst. Hydrology, Wallingford, England.

Biggin, D. S., 1971, 'The use of natural tritium in hydrograph analysis', *Subsurface Hydrology, Rep.* No. **26**, Inst. Hydrology, Wallingford, England.

British Standard Institution, 1964, 'Methods of measurement of liquid flow in open channels', *British Standard BS 3680.*

Brown, M. C. and Ford, D. C., 1971, 'Quantitative tracer methods for investigation of karst hydrologic systems', *Trans. Cave Res. Gp. GB,* **13**, 37–51.

Calver, A., Kirkby, M. J. and Weyman, D. R., 1972, 'Modelling hillslope and channel flows', in (Ed.) Chorley, R. J., *Spatial Analysis in Geomorphology,* Methuen, London.

Childs, E. C., 1969, *An Introduction to the Physical Basis of Soil Water Phenomena*, John Wiley, London.

Cobb, E. D. and Bailey, J. F., 1965, 'Measurement of discharge by dye-dilution methods', *US Geolog. Surv., Surface Water Techniques Book 1, Hydraulic Measurement and Computation*, Ch. 14.

Cole, D. W., 1958, 'The alundum tension lysimeter', *Soil Sci.*, **85**, 293–296.

Cole, D. W., 1968, 'A system for measuring conductivity, acidity, and rate of water flow in a forest soil', *Water Res. Res.*, **4**, 1127–1136.

Corey, J. C., 1968, 'Evaluation of dyes for tracing water movement in acid soils', *Soil Sci.*, **106**, 182–187.

Croney, D. and Coleman, J. D., 1954, 'Soil structure in relation to soil suction (pF)', *J. Soil Sci.*, **5**, 75–84.

Dalrymple, J. B., Blong, R. J. and Conacher, A. J., 1968, 'An hypothetical nine-unit landsurface model', *Zeitschrift für Geomorphologie*, **12**, 60–76.

Dunne, T. and Black, R. D., 1970, 'An experimental investigation of runoff production in permeable soils', *Water Res. Res.*, **6**, 478–490.

Eeles, C. W. O., 1969, 'Installation of access tubes and calibration of neutron moisture probes', Inst. of Hydrology, Rept. No. 7, Wallingford, England.

Emmett, W. W., 1970, 'The hydraulics of overland flow on hillslopes', *US Geolog. Surv. Prof. Paper* **662-A**.

Evans, H. E., 1962, 'A note on the average coefficient of permeability for a stratified soil mass', *Geotechnique*, **12**, 145–146.

Finlayson, B. and Waylen, M., 1973, Personal communication.

Fletcher, J. E. and Carroll, P. H., 1948, 'Some properties of soils that are subject to piping in southern Arizona', *Proc. Soil Sci. Soc.*, **13**, 545–547.

Fletcher, J. E., Harris, K., Peterson, H. B. and Chandler, V. N., 1954, 'Piping', *Trans. Am. Geophys. Union*, **35**, 258–263.

Gilman, K., 1971, 'A semi-quantitative study of the flow of natural pipes in the Nant Gerig sub-catchment', *Subsurface Hydrology*, Rep. No. **36**, Inst. Hydrology, Wallingford, England.

Greenland, P. C., 1971, 'Dilution gauging in the Nant Gerig', *Subsurface Hydrology*, Rep. No. **27**, Inst. of Hydrology, Wallingford, England.

Hewlett, J. D., 1961, 'Soil moisture as a source of base flow from steep mountain watersheds', *Southeastern Forest Experiment Station, Asheville, N. Carolina, US Dept. Agric.–Forest Ser.*, Station Paper No. **132**.

Hewlett, J. D. and Hibbert, A. R., 1963, 'Moisture and energy conditions within a sloping soil mass during drainage', *J. Geophys. Res.*, **68**, 1081–1087.

Hewlett, J. D. and Hibbert, A. R., 1965, 'Factors affecting the response of small watersheds to precipitation in humid areas', *Internat. Symp on Forest Hydrology*, Pennsylvania State University.

Hillel, D., 1971, *Soil and Water: Physical Principles and Processes*, Academic Press, New York.

Holdsworth, P. M., 1971, 'User's testing schedule for the Wallingford probe system', Rept. No. **10**, Inst. of Hydrology, Wallingford, England.

Hosegood, P. H., Sanderson, P. R. and Bridle, M. K., 1969, 'Manual of dilution gauging with sodium dichromate', *Water Res. Assoc.*, I.L.R. No. **34**.

Ingram, H. A. P., 1967, 'Problems of hydrology and plant distribution in mires', *J. Ecol*, **55**, 711–724.

Jones, A., 1970, 'Soil piping and stream channel initiation', *Water Res. Res.*, **7**, 602–610.

Kilpatrick, F. A., 1968, 'Flow calibration by dye dilution measurement', *Civil Engineering*, Am. Soc. Civil Engrs., February 1968, 74–76.

Kirkby, M. J., 1968, 'Infiltration, throughflow and overland flow', in Chorley, R. J., (Ed.), *Water, Earth and Man* Methuen, Ch. 5.1 588 pp.

Kirkby, M. J. and Chorley, R. J., 1967, 'Throughflow, overland flow and erosion', *Bull. Internat. Assoc. Sci. Hydrology*, **12**, 5–21.

Klute, A. and Gardner, W. R., 1962, 'Tensiometer response time,' *Soil Sci.*, **93**, 204–207.

Knapp, B. J., 1973, 'A system for the field measurement of soil water movement', *Tech Bull. Brit. Geomorph. Res. Gp.*, No. **9**, 26 pp.

Knapp, B. J., 1974, 'Hillslope throughflow observation and the problem of modelling', *Spec. Publ. Inst. Brit. Geogrs.*, No. **6**, 23–31.

Knutsson, G., 1968, 'Tracers for groundwater investigation', in Eriksson, E., Gustafsson, Y. and Nilsson, K., *Ground Water Problems (Proc. Stockholm Symposium*, 1966) Pergamon, pp. 123–152.

Østrem, G., 1964, 'A method of measuring water discharge in turbulent streams', *Geogr. Bull.* **21**, 21–43.

Parizek, R. and Lane, B. E., 1970, 'Soil water sampling using pan and deep-pressure-vacuum lysimeters', *J. Hydrology*, **11**, 1–21.

Parker, G. G., 1963, 'Piping, a geomorphic agent in landform development of the drylands', *Publ., Internat Assoc. Sci. Hydrology*, No. **65**, 103–113.

Pavlakis, G. and Burden, L., 1972, Hysteresis in the moisture characteristics of clay soil', *J. Soil Sci.*, **23**, 350–361.

Pilgrim, D. H., 1966, 'Radioactive tracing of storm runoff on a small catchment. I. Experimental technique, II. Discussion of results', *J. Hydrology*, **4**, 289–305 and 306–326.

Pond, S. F., 1971, 'Qualitative investigation into the nature and distribution of flow processes in Nant Gerig', *Subsurface Hydrology*, Rept. No. **28**, Inst. Hydrology, Wallingford, England.

Poulovassilis, A., 1962, 'Hysteresis of pore water, an application of the concept of independent domains', *Soil Sci.*, **93**, 405–412.

Poulovassilis, A., 1969, 'The effect of hysteresis of pore water on the hydraulic conductivity', *J. Soil Sci.*, **20**, 52–56.

Reynolds, E. R. C., 1966, 'The percolation of rainwater through soil demonstrated by fluorescent dyes', *J. Soil Sci.* **17**, 127–132.

Reynolds, S. G., 1970, 'The gravimetric method of soil moisture determination. I. A study of equipment and methodological problems. II. Typical required sample sizes and methods of reducing variability. III. An examination of factors influencing soil moisture variability', *J. Hydrology*, **11**, 258–300.

Richards, L. A., 1947, 'Pressure membrane apparatus, construction and use', *Agric. Engr.*, **28**, 451–454.

Richards, S. J., 1965, 'Soil suction measurements with tensiometers', in *Methods of Soil Analysis, Am. Soc. Agronomy*, Monograph No. **9**, 153–163.

Rodda, J. C., 1967, 'The rainfall measurement problem', *Publ. Internat. Assoc. Sci. Hydrology*, No. **78**, 215.

Scott, C. H., Norman, V. W. and Fields, F. K., 1969, 'Reduction of fluoresence of two tracer dyes by contact with a fine sediments', *US Geolog. Surv. Prof. Paper*, **650-B**, 164–168.

Smart, P. L. and Brown, M. C., 1973, in press, 'The use of activated carbon for the detection of the tracer dye rhodamine WT, *Proc. 6th Internat. Congress of WT, Olomouc, Czechoslovakia*, 1973.

Smith, D. B., 1972, 'Groundwater tracing with radioactive tracers', *Atom (Bull. UK Atomatic Energy Authority)*, No. **192**, October 1972, 178–180.

Smith, D. B., 1974. 'Tritium water tracing, *J. Soc. Water Treatment and Exam.*, **22**, 250–258.

Smith, D. B., Wearn, A., Richards, H. J. and Rowe, P. C., 1970, 'Water movement in the unsaturated zone of high and low permeability strata using natural tritium', in *Isotope Hydrology*, 1970, *Internat. Atomic Energy Authority*, Vienna, pp. 73–87.

Strangeways, I. C. and McCulloch, J. S. G., 1965, 'A low-priced automatic hydrometeorological station', *Bull. Internat. Assoc. Sci. Hydrology*, **10**, 57–62.

Tsukamota, Y., 1961, 'An experiment on subsurface flow', *J. Japanese Soc. Forestry*, **43**, 61–68.

United States Geological Survey, various dates, *Surface Water Techniques Book 1–Hydraulic Measurement and Computation*, published *seriatim* by chapters.

Van't Woudt, B. D., 1954, 'On factors governing sub-surface storm flow in volcanic ash soils, New Zealand', *Trans. Am. Geophys. Union*, **35**, 136–144.

Wagner, G. H., 1962, 'Use of porous ceramic cups to sample soil water within the profile', *Soil Sci.*, **94**, 379–386.

Webster, R., 1968, 'The measurement of soil water tension in the field', *New Phytology*, **65**, 249–258.

Weyman, D. R., 1970, 'Throughflow on hillslopes and its relation to the stream hydrograph', *Bull. Internat. Assoc. Sci. Hydrology*, **15**(2), 6/1970, 25–33.

Weyman, D. R., 1974, 'Runoff process, contributing area and streamflow in a small upland catchment', *Spec. Publ. Inst. Brit. Geogrs.*, No. **6**, 33–43.

Whipkey, R. Z., 1965, 'Subsurface stormflow from forested slopes', *Bull. Internat. Assoc. Sci. Hydrol.*, **10**(3), 74–85.

Whipkey, R. Z., 1969, 'Storm runoff from forested catchments by subsurface routes', *Publ. Internat. Assoc. Sci. Hydrol.* No. **85**, 773–779.

Wilson, J. F., 1968, 'Fluorometric procedures for dye tracing', *US Geological Survey, Techniques for Surface Water Investigations*, Book 3, Ch. A.12.

Wilson, C. M. and Smart, P. L., 1973, Personal communication.

CHAPTER 4

Flow within the soil

R. Z. Whipkey,

*Agricultural Research Service,
New England Watershed
Research Centre, South
Burlington, Vermont 05401,
USA*

and

M. J. Kirkby,

*School of Geography,
University of Leeds, UK*

4.1 INTRODUCTION

4.1.1 The importance of subsurface flow

A realization of the importance of subsurface flow for the generation of streamflow has arisen from two major conflicts between observations in small plots and the assumptions of traditional hydrological models of whole basins. The first conflict arises because whole-basin models associate a peaked stream flood hydrograph with widespread overland flow from its catchment, caused by rainfall intensities in excess of the soil-infiltration capacity. Field observations show that overland flow need not occur at all (Pierce, 1965; Rawitz *et al.*, 1970), or may be confined to only a small part of the catchment, and that infiltration capacities need not be exceeded (Betson, 1964; Dickinson and Whiteley, 1970). It is concluded that flood peaks may be generated, in at least some circumstances, by subsurface flow.

The second conflict between whole basins and small plots arises because low flows, long after rainfalls, are normally associated with groundwater. Long recession curves, with flows continuing for a month or more, may however occur in some cases without any identifiable groundwater body (Hewlett, 1961). Again, it must be concluded that subsurface flows may be adequate to supply these low streamflows.

4.1.2 Measurements of subsurface flow

The measurement of subsurface stormflow is a difficult task that has been attempted in several ways by a number of researchers. The undertaking is complicated by the necessity to identify the phenomenon for a unique hillside or catchment. When this has been accomplished, a technique or structure that adequately measures and represents this portion of total storm discharge must be designed. Historically, measurement techniques have been aimed at providing empirical determination of that portion of the hydrograph contributing direct runoff via underground flow routes. Discharge was measured from hillside plots in order to account for all outflow and this was a step in the right direction. But recognition of the reality of subsurface stormflow, throughflow, or quick interflow whetted hydrologists' imaginations and desire to learn more of the physical processes involved. Many new concepts were born, and it remained for physical experiments to verify the soundness of each.

Attempts to measure flow under field conditions have been judged by their success in catching the elusive direct runoff occurring through subsurface routes. Where measurable evidence of subsurface stormflow has been found, researchers have generally been inclined to accept the empirically-determined rates and quantities as 'what happens in the natural physical system'. When a researcher repeated the experiment under similar antecedent conditions with similar input and reproduced past output, his joy was unbounded.

Early plot attempts to study runoff from natural and artificial storms used techniques that primarily measured surface discharge. Earth dikes, metal cutoffs, and the like, served to exclude foreign water and prevent loss of plot water. However, there was little apparent concern for the infiltrated water which by-passed the plot by subsurface-flow processes. Prior to the 1930s, attempts to re-create subsurface stormflow under controlled plot conditions are not well documented in the literature. Allusions to subsurface stormflow as a mysterious but critical component of storm hydrograph were often hidden in the literature, tantalizingly so to those researchers hoping for a quantitative explanation. The past two decades have seen an increasing number of reports on subsurface-stormflow studies, some of them described in other chapters. It is however difficult to review even all the completed work on subsurface stormflow processes, and there are numerous experiments started but never completed. There are also continuing experiments not yet reported in the literature. It is not therefore intended to provide a complete summary of past results.

Researchers studying subsurface stormflow from open agricultural soils found evidence of surface runoff during both natural and man-made rainstorms (Amerman, 1956; Dunne and Black, 1970; Minshall and Jamison, 1965). Others (Pierce, 1965; Troendle, 1970; Tsukamoto, 1961; and Whipkey, 1965), studied subsurface stormflow on forested (or forest-derived) soils and found little or no infiltration excess overland flow. Subsurface stormflow appeared to be the primary source of direct runoff, provided that initial infiltration of rainfall was rapid and antecedent moisture conditions provided the condition for seepage outflow at some point downslope. Hewlett (1961; Hewlett and Hibbert, 1963) reported outflow (from a disturbed soil plot located on a hillside) derived from a seep zone which in turn was supplied by soil moisture moving in the unsaturated state. Weyman (1973) found similar seepage to steep channel banks from undisturbed soil horizons. Whipkey (1967), Knapp (1970) and others reported observations of flow occurring from natural soil structural cracks and channels formed by decayed tree roots. Except for some attempts (a) to relate subsurface stormflow to landform processes by Kirkby and Chorley (1967) and to piping (Institute of Hydrology, 1971), (b) to develop specific models by Hewlett and Troendle (1975), and Snyder and Asmussen (1972), and (c) to develop a general deterministic model by Freeze (1972), little generalization or quantitative analyses of the natural process have been done as an integral part of these experiments.

4.1.3 Empirical generalizations

Perhaps the most striking qualitative conclusion of the field experiments is the huge variations which are found from catchment to catchment, and to a lesser extent, between plots in a single catchment and from time to time, in a single plot. The variations are so large that it is difficult to say more than that subsurface flow may, at some sites, be rapid enough to generate flood peaks and may continue long enough at other sites to account for observed stream low flows.

At some sites, subsurface flow appears to be generated mainly above a single discontinuity in the soil and organic matter profile, below which the soil is relatively impermeable. At other sites, the main location of subsurface flow in the soil profile varies with antecedent moisture conditions. The development of soil layering, whether pedogenic or sedimentary, and the thickness of the relevant layers, is thus one important factor in determining the distribution of subsurface flow.

Within the soil, hydraulic erosion may enlarge pores and cracks into significant pipes which carry large amounts of subsurface flow where soil materials are suitable (Berry, 1970). In other circumstances, subsurface flow may follow and maintain, if not enlarge, channelways arising from other causes. Animal burrows, shrinkage cracks and root-holes are among the potentially important factors in forming such channels (Fletcher *et al.*, 1954; Bell, 1968). Particularly in areas of heavy rainfall organic soils may bridge over and enclose pre-existing open channels, to make their course at least intermittently subsurface.

These are among the mechanisms which may be responsible for allowing subsurface flow to divide into a continuum of types. At one extreme some flow may travel through distinct continuous pipes, and at the other extreme through textural capillary pores. Between these extreme end-members, flows may occur in a range of textural and structural voids which may be more or less continuous along the flow direction. The voids may be classified according to grain size and by structural or textural types. The flows may be classified as capillary (within water-filled pores) or non-capillary (in partly air-filled pores) and as bulk flows (travelling as a plug in discontinuous voids) or pipe flows (travelling separately in continuous voids). Where pipe flow occurs, most of the subsurface storm flow will be through large continuous voids, which are clearly capable of giving a rapid response to rainfall. They differ from open channels at the surface mainly on account of the time taken for rainfall to reach them from the surface.

Even in the absence of non-capillary pores or pipes, experiments differ in the importance assigned to saturated and unsaturated subsurface flow. All that may be stated with confidence is that the most nearly saturated soil layers tend to give the greatest flow, other things being equal. Thus, at sites where saturated flow normally occurs, unsaturated flow may largely be neglected and measurements may concentrate on collecting direct water-flow data. Where, however, mainly unsaturated flow is found, the unsaturated flow hydraulics are clearly crucial, so that measurements of moisture tension are as important as measurements of moisture content and flow.

Overland flow tends to be most common at the base of slopes, because soils are more nearly saturated there so that less rain can infiltrate into them. The greater saturation may be partly due to differences in soil characteristics, but is also due to accumulation of subsurface flow from upslope. If subsurface flow is rapid, it may be sufficient to raise soil-water levels at the slope-base during the course of a single storm. Even slow subsurface flow has been shown to raise downslope soil water levels between storms (Hewlett, 1961; Weyman, 1973), and lead to wetter antecedent moisture conditions at the slope base. These effects will lead to a more than linear response of stream flow to both antecedent moisture and storm rainfall; and the non-linearities will be greater in cases where the subsurface flow is more rapid. This is seen as a final important factor influencing the observed differences in subsurface flow response between sites, even within a single catchment.

4.2   THE PHYSICAL PROCESSES

4.2.1   Conditions for lateral diversion of flow

Flow within the matrix of the soil is mainly laminer, obeying Darcy's law. The hydraulic conductivity, however, increases with soil-moisture content as larger and larger pores are filled, and the majority of the flow occurs in the largest filled connecting-pores (Figure 4.1). Thus, a coarse-grained soil may carry a large and rapid flow when saturated or nearly so, while a fine-grained soil with no large pores carries a lesser flow. Under drier conditions at a higher soil-moisture tension however, the interconnected pores in the fine-grained soil may still be saturated, so that its flow is as high as before; scarcity of such fine pores in the coarser soil allow it to carry only a much smaller flow. These comparisons have been made for equal soil-moisture tensions, as it is *tensions* which equalize where differing soil materials are brought into contact. This topic is discussed at length in Chapter 2.

In layered soils, conditions for throughflow are readily met. If, for example, a coarse-grained layer overlies a finer soil at a gently sloping site, then at high rates of steady percolation from rainfall, the coarse soil readily allows infiltration at a low-moisture tension. Below the contact, even saturated infiltration in the fine layer is

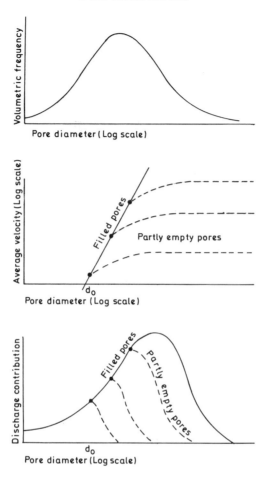

Figure 4.1 Schematic relationship between grain-size distribution and dominant pore size for subsurface flow

likely to be less rapid than the rainfall intensity, so that the fine material builds up to saturation, and a saturated layer backs up within the coarse material above (Figures 4.2A and 2.14). Under the gravity gradient of a gentle slope, there will be some downslope subsurface flow in all layers, but the great majority of the subsurface discharge will come from immediately above the contact, where the largest pores are water-filled because the soil is both saturated and coarse grained. Under dry conditions the same build-up may not occur because the finer layer may be able to carry more infiltration water than the coarser (Figure 4.2B), for the reasons described above. Where a fine layer overlies a coarser layer, no comparable concentration of subsurface flow occurs (Figure 4.2c and D). Since flow velocities

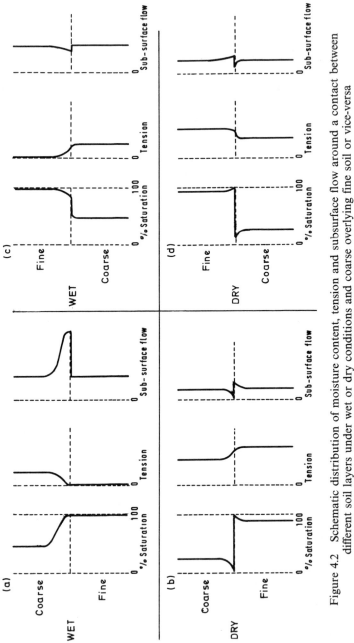

Figure 4.2   Schematic distribution of moisture content, tension and subsurface flow around a contact between different soil layers under wet or dry conditions and coarse overlying fine soil or vice-versa

increase with the square of the pore diameter, the quantitative predominance of a saturated subsurface flow layer, if one occurs, means that the unsaturated subsurface flow in other layers may be neglected for many purposes.

In real soils, sharp contacts may be unusual, but gradual changes in soil properties are a normal feature, and lead to high subsurface flow rates under suitable conditions. In coarse-textured soils, the normal sequence of soil horizons down from the surface, from litter to partly organic A horizons to illuviated B horizons, is typically a sequence of reducing porosity and increasing clay content. It therefore tends to assist in the diversion of throughflow, and most experiments which have measured appreciable subsurface flow are from forest and other soils which follow this normal pattern (e.g. Whipkey, 1965; Weyman, 1973). The transition from B to C horizons tends to be in the reverse direction, but a second zone of potential lateral flow diversion may occur at the base of the C horizon at the contact with an impermeable bedrock. Measurements to data however suggest that diversion closer to the surface, if present, tends to account for more of the subsurface flow, and with a shorter response time. In soils with a high silt–clay content (finer than a sandy loam), most of the subsurface flow is not in textural but in structural pores (Knapp, 1970), which also tend to be smaller at increasing depths.

Under conditions of uniform downward percolation in a horizonated soil, the percolation rate (Q) per unit cross-section is

$$Q = K - D\frac{\partial m}{\partial z} \qquad (4.1)$$

where

$K(m, z)$ = permeability

$D(m, z)$ = diffusivity = $-K\dfrac{\partial \phi}{\partial m}$

$\phi(m, z)$ = soil-moisture tension

$m(z)$ = volumetric moisture content

z = depth below soil surface.

In a soil which is becoming gradually less permeable at depth, moisture content must increase downwards for steady percolation through the soil column, and the maximum rate of percolation is K_s, the saturated permeability. At a given rate of percolation, Q, (perhaps determined by rainfall intensity), saturated conditions occur for all depths below that at which $K_s = Q$, and on a slope s, the percolating water which cannot penetrate the deeper layers is diverted as subsurface flow (Figure 4.3). In this case however the level of peak subsurface flow depends strongly on the percolation rate, and is not at a fixed level in the soil. Dominant depths for subsurface flow may however be empirically related to the distribution of rainfall intensities at the site.

In a sloping soil mass of uniform composition, the permeability term in equation

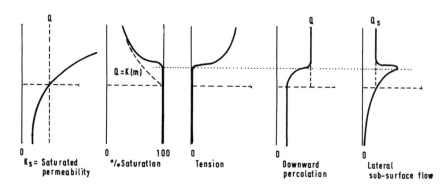

Figure 4.3 Schematic distribution of saturated permeability, soil moisture, tension, downward percolation and subsurface flow for a soil which gradually becomes more impermeable with depth

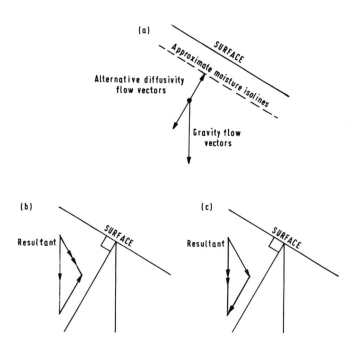

Figure 4.4 General conditions for downslope flow. (a) Directions of gravity and diffusivity flow; (b) triangle of vectors and resultant flow direction if soil is drier upwards (outward diffusivity flow); (c) triangle of vectors if soil is wetter upwards

4.1 acts vertically downwards in the direction of the gravity potential gradient, and so has some downslope component (Figure 4.4). Since the isolines of equal moisture tend to be almost parallel to the surface (except near a discontinuity such as a stream bank) in a uniform soil, the diffusivity term in equation 4.1 acts at right angles to the soil surface, in an outward direction if the soil is drier towards the surface, and vice-versa. If this diffusivity term is directed outwards, then the resultant flow vector lies between the downslope direction and the vertical: if inwards, between the vertical and the perpendicular *into* the soil. Under steady percolation, neither condition is expected. Following a rain event, surface layers are likely to be wetter than deeper layers, so that little lateral flow is normally expected. The only exception to this is where soil moisture is backed up above an impermeable or less-permeable layer beneath. Thus rapid downslope subsurface flow is seen essentially as a function of permeable overlying impermeable soil layers, either at a sharp contact or in a gradual transition. The experiments of Dunne and Black (1970) perhaps come closest to a vertically-uniform soil, and support this argument that subsurface flow is then minimal. This case is analysed mathematically in Chapter 6, and its implications are shown fully there.

4.2.2 The interaction of rainfall with subsurface flow

At the start of a rainstorm at least some of the water is entering the soil and percolating downwards, raising soil-moisture content as it goes. Unless rainfall exceeds the infiltration capacity of the soil, this percolating water does not saturate the soil, and does not produce any appreciable lateral flow. There may, however, be localized saturated flow down shrinkage cracks, root holes and animal burrows. Where percolation reaches a less-permeable layer which will not accept the full percolation flow, it will become saturated from its surface and a saturated wetting front will slowly penetrate into it. Meanwhile, a saturated layer will tend to back up above the interface and give rise to significant throughflow as described in the previous section.

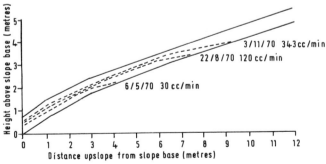

Figure 4.5   The saturated zone on a hillslope measured at different levels of basal outflow. The water table is shown as a broken line between the ground surface and bedrock (solid lines) (after Weyman, 1971)

This layer of saturated subsurface flow, if it exists, will also be receiving subsurface flow from upslope, and during a long wet period, the discharge in the layer will tend towards a steady state in which discharge is proportional to drainage area (or slope length), and so increases downslope. This increased flow, unless offset by increasing slope gradient or increasingly permeable soils downslope, will give rise in turn to an increasing depth of saturation downslope, acting effectively as a perched water table within the soil. An example of this is shown in Figure 4.5, for a saturated layer above an impermeable bedrock.

Rainwater percolating to a zone of saturated soil water raises the level of the saturated zone directly, as is shown schematically in Figure 4.6. If this process continues for long enough, then the saturated layer may build up to the surface, producing overland seepage of previously subsurface flow ('return flow'), and also

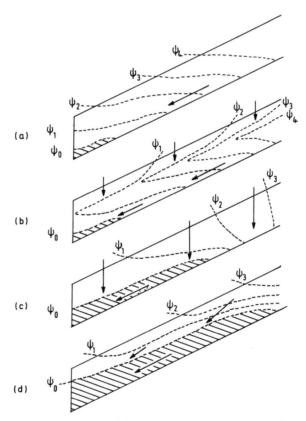

Figure 4.6   Generalized tension and flow patterns on an experimental slope during a storm. (a) Pre-storm state with lateral unsaturated flow; (b) after rain, unsaturated downward percolation added; (c) percolation reaching saturated zone which starts to grow; (d) peak discharge—maximum saturated lateral flow (from Weyman, 1973)

preventing the entry of further rainfall, which therefore runs off directly as 'saturation overland flow', even though the soil-infiltration capacity is not generally exceeded. At any moment, the area of saturated soils which will yield overland flow immediately in this way is termed the 'dynamic contributing area' (Weyman, 1974), which is continuously changing in size between the minimum area of low-flow stream-water surface, and the whole catchment.

The various physical components of the hillslope discharge in response to a rain event may thus be summarized as follows, in order of rapidity of response:

(1) *Infiltration excess overland flow.* Where rainfall intensity is greater than infiltration capacity, overland flow may occur over a large proportion of the catchment, and is likely to dominate the whole of the hydrograph peak.

(2) *Saturation overland flow.* This is of two types: (a), Within the area of previously saturated soils, all rainfall will be directed as overland flow. Both saturation and infiltration-excess overland flow respond without delay to rainfall, and the hydrograph shape they produce is formed by the times of travel over the catchment surface to the streams and along them. (b), Growth of the saturated area during the rainstorm may occur *either* from the rapid downslope flow of subsurface water *or* from the addition of vertically percolating water. After the delay required for saturation to reach the soil surface, further rainfall is added to other forms of overland flow. Most studies suggest that the latter mechanism is much more important than the former unless subsurface flow is exceptionally rapid.

(3) *Return flow.* Subsurface flow may return to the surface during a storm, in which case it is usually much less important than the saturation overland flow which accompanies it. Return flow may also occur however after rainfall has ceased, in which case it may be an important, and reasonably rapid, contribution to the hydrograph. This can happen if subsurface flow is forced to the surface by a downslope thinning of permeable soil layers; or if the subsurface flow thickness is increased by either convergence of flow-lines in a hollow or slowing of flow velocities on a concavity (Dunne and Black, 1970).

(4) *Saturated subsurface flow.* Flow which continues to the slope base within the soil is always slower than overland flow unless large pipes are involved. Where the flow is saturated however, it may take place fast enough to be considered as a contributor to the hydrograph peak, especially where flow in the categories above is small. The lags involved are first the time taken to percolate down to the saturated layer, and second, the travel times along it.

(5) *Unsaturated subsurface flow.* Even slower responses are associated with unsaturated flow which can only produce the hydrograph peak if nothing else is present. It may however continue for many weeks and be the dominant component of low flow discharge.

(6) *Groundwater flow.* The difference between saturated subsurface soil flow and groundwater flow is that percolation times, hydraulic gradients, and permeabilities all tend to be lower for the latter. If there is appreciable groundwater flow therefore, it tends to behave more like unsaturated than

saturated throughflow in terms of its response times. Where sub-soils and bedrock are permeable, groundwater flow then tends to replace unsaturated subsurface flow as the dominant supplier of low flow to streams.

There is thus a bewildering number of possibilities for the dominant processes and response times in forming a slope hydrograph. At very low rainfall intensities (relative to soil permeabilities), unsaturated subsurface flow or groundwater flow may be dominant, giving response times of many weeks. At moderate storm intensities, saturated subsurface flow often becomes more important in well-horizonated soils, with or without some saturation overland flow. At extreme storm intensities, infiltration-excess overland flow may dominate. It becomes easy to see how the wide range of observed responses can arise, and difficult to make detailed generalizations or models for more than a narrow range of soil conditions at a time. The result of both discussion and experiments however seem to converge on a 'normal' behaviour for vegetated areas with pronounced soil horizonation. During rainfall, the most important components are saturation-excess overland flow if present, and the *vertical* percolation of rainwater to increase the saturated area. Lateral flow is normally negligible compared to these. After rainfall has ceased, subsurface flow becomes important, and flow in the normally longer interval between rainstorms establishes the zones of saturation and near-saturation for the next rainstorm. It also locally produces return flow, and contributes to the low-flow regime of the stream. This topic is pursued more fully in Chapters 7 and 8.

4.3   CONTROLLING FACTORS

4.3.1   Soil and vegetation

The physical properties and depth of the soil are probably the most important controls on subsurface flow production at a site. If the texture is coarse (with predominant sand and stones), vertical flow usually dominates; and when this soil is deep, subsurface flow response may be delayed. If the texture is fine, resistance to vertical flow results and lateral or shallow subsurface flow sometimes occurs quickly. If the fine-textured soil is deep to the wetting zone (greater than 2-m deep), it may be physically impractical to construct open-face cut-offs with tile or trough collector systems. Where the soil is shallow to a wetting zone (less than 2 m), or distinctly layered with major portions of flow occurring from shallow horizons, it is less difficult to construct a satisfactory downslope cut-off seepage collection system.

Soil structure is also extremely important. Fissures, cracks, or channels are less likely to occur or to be of importance in coarse-textured soils. In fine-textured or layered soils, cracks, fissures and/or channels are more likely to occur providing possible routes for flow, and largely replacing textural voids as the main avenues for unsaturated and saturated flow. When and if large structural voids connect, pipe flow may also be initiated in low permeability strata. Ultimately, such flow can contribute to deep seepage to groundwater sources. Quick response of instrumented plots at the lowest zones overlying impermeable strata is thought to be due to water being routed in a pipe-like manner through otherwise slowly permeable discontinuities (Jones, 1971).

Vegetation cover is directly related to (a) the maintenance of infiltration capacity, and (b) the conditioning effect of organic material on soil structure, bulk density, and porosity. The effects of agricultural annual cover crops may however be counteracted by the mechanical effects of ploughing, discing, and harrowing. Subsurface stormflow has been shown over 'plough-pans' created by repeated ploughing compaction (Minshall and Jamison, 1965). Crop residue below the depth of mechanical disturbance (roots and incorporated plant materials), however, can markedly improve soil structure and macro-porosity, increasing the inherent permeability of deeper zones. Undisturbed agricultural fields (permanent pasture and hayfields) show the effect of constant plant cover from the surface down to the depths of significant root development. Where these soils are undisturbed by mechanical cultivation, the dead and decaying root material contributes to development of macro-porosity in the depths at which root penetration occurs.

The most noteworthy effects of cover are found in undisturbed forest stands (Gaiser, 1952; Van Dijk, 1958; Chamberlin, 1972). Here dead and decaying root systems (at depths sometimes greater than 2 m) can create important fissures or channels for free-water conduction, mainly in a vertical direction. Under old stands where root growth and root decay has affected the structure of the soil profile, patterns of subsurface stormflow may be completely different from the same type of soil not under forest cover. Fine-textured or well-layered soils may have their saturated permeability characteristics increased by the presence of root channels which intersect and penetrate soil layers or textural discontinuities. This is important for attempts to measure flow from successive horizons. A good soil texture also depends on the maintenance of a dense faunal population, particularly or earthworms. They are usually abundant in undisturbed soils with an adequate organic content where the soil pH is greater than 4.5–5.0. Soil fauna produce burrows and excreta, both of which tend to increase soil permeability in all directions (*see also* Chapter 2).

Land-use, while highly interrelated with vegetation cover, may have effects independent of cover. Adverse land-use practices commonly have the greatest effect on infiltration; such abuses or over-use include overgrazing by sheep and cattle, repeated burning of litter and humus layers on the forest floor, and topsoil loss by accelerated erosional processes. Studies attempting to represent a regional soil-cover-physiographic complex should therefore be conducted on sites characteristic of the region.

Lithology is indirectly significant, acting as parent material for soil development, but its rôle is usually lessened by the influence of pedological and organic processes. Lithology is most important where soil-forming processes have been least active, or have been active for only a short time. For example, soils developed on glacial or recent alluvial materials may be poorly developed, so that the properties of the parent material are dominant. In this case, layering is most dependent on the stratigraphy and structure of the material. A knowledge of the shape, composition and orientation of the material is then most important, together with information on the presence of natural channels, faults etc. in the plane of the impeding layer.

Climate also has an indirect effect, acting particularly through the development of soil organic matter. It also, however, acts directly through rainfall intensity and rates of evapotranspiration to determine the moisture budget of the soil and hillslope. Where deep soils have soil moisture thoroughly depleted by dense vegetative cover during the growing season subsurface stormflow seldom occurs until soil-moisture deficits are eliminated by dormant season recharge. Subsurface stormflow studies which take place only during the growing season may therefore lead to negative results as those are the periods of highest evapotranspiration. One such study, conducted on deep coarse-textured soils found little to no storm discharge—either surface or subsurface. This period, typifying the maximum evaporation season, is seldom an important one for subsurface stormflow in temperate areas; for example, in sections of the United States where the annual precipitation is less than about 1300 mm. Such effects may be minimized if artificial-rainmaker techniques are available.

⌐4.3.2　Topography

The position of measurement sites on a hillslope is also a major factor in determining the amount of subsurface flow observed. Distance from the divide is related to the amount of flow accumulated from upslope and slope gradient helps to determine the rate of flow. Both slope-profile convexity or concavity and the convergence or divergence of flow lines in plan also influence rates of flow. These topographic factors interact with the soil factors already discussed, and the two sets are related via the soil catena concept to produce complex areal patterns of subsurface flow and overland flow.

Any measurement site inevitably interrupts the natural flow patterns on the hillside, but a site cut into a natural channel bank has least influence, in that water is normally flowing out from the face in saturated form. At any other site, a pit interrupts the flow more drastically. Under wet conditions, water flowing into collectors will tend to produce a draw-down effect around the pit, so that the effective collecting area is greater than the pit width (Figure 4.7A), while under dry conditions the build-up to saturation required to drain water from the pit produces upslope tension gradients, which divert flow away from the pit (Figure 4.7B). This effect leads to an exaggeration of the flow differences between wet and dry conditions. It may be reduced by inserting plot boundaries (to the full depth of the pit) some distance upslope, or by making a longer pit along the contour. Knapp (1970) showed, in a very wet area of Wales, that a 1-m pit effectively drained 3 (high

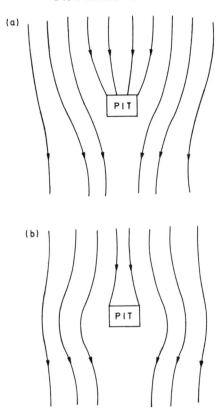

Figure 4.7 Schematic flow-lines in plan around a pit for measuring subsurface flow. (a) Under wet conditions; (b) under dry conditions

flows) to 4 (low flows) times its own width. Substantial draw-down effects also occurred in the vertical, so that discharges into a pit appeared to come from greater than their undisturbed flow depths. The pit size is also relevant to obtaining an adequate sample of the slope flow, which may be very laterally variable, particularly if pipes are present. Pits which intersect pipes or other large voids produce not only a disproportionate quantity of flow, but may also produce a faster-than-average response to rain. In most cases the behaviour of a pit is similar to that of an open ditch or shallow bore-hole, and there is an extensive literature in these contexts.

In progressing from the divide downslope, and assuming to begin with a constant gradient and uniform soils, the accumulated subsurface flow increases more or less linearly with distance (or slope drainage area). If a saturated layer is formed above an impeding horizon, this layer will also thicken downslope, so that the top of the saturated zone, where subsurface flow tends to be greatest (Figure 4.3), is progressively further above the impeding horizon. Downslope sites may therefore give greater measured flows (Jamison and Peters, 1967), but it may be more difficult to associate the flows with the soil layer responsible for producing them. This may be a severe practical problem, especially where soil properties are changing only gradually with depth.

The effect of slope gradient is a direct one, since laminar flow rates are directly proportional to gradient, other things being equal. In practice however, soils tend to vary consistently with gradient within a catchment. Perhaps the most usual relationship is for soils to be more permeable on steep slopes, so that subsurface flow velocities are increased more than proportionately with increasing gradient. On gentle slopes in humid areas the slow drainage may lead to development of peaty A horizons, which hold large quantities of water and also impede lateral subsurface flow because of their low permeability (Bay, 1969). Soils also tend to be thicker on gentler gradients, and this increases their water-storage capacity and partly counteracts their tendency to saturation.

Where slope gradient is changing downslope, flows will travel faster or slower in response. On a concavity, the slowing of a saturated subsurface flow causes the saturated layer to become thicker, and it may bring saturated conditions up to the surface. The area of return flow will also be a part of the dynamic contributing area. A hollow or area of flow convergence (in plan) will have a similar effect in raising the level of soil saturation. These areas, and areas of anomalously-thin soil above the impeding horizon (Betson and Marius, 1969) are the areas of a catchment which tend to be associated with return-flow seepage, and to form the nucleus of the dynamic contributing area. Since concave hollows are typically present around the head of valleys, these areas are generally the most dynamic in their response to both seasonal soil-moisture differences and storm rainfall. This theme is explored more fully in Chapters 7–9.

4.4 THE SOIL HYDROGRAPH

4.4.1 Flow produced by an impeding layer

Comparison of the process analysis above with the measured hydrographs of subsurface flow reported by Dunne and Black (1970), Knapp (1970), Weyman (1973), Whipkey (1965) and others allows some explanation of the way in which subsurface flow is normally produced. One or more impeding layers, or a progressive decrease in permeability with depth, appears to be a prerequisite for appreciable subsurface stormflow, and it is normal for this flow to be concentrated in a saturated layer.

In a many-layered soil, the first critical layer is that with the lowest saturated permeability. The second critical layer is the layer which has the next lowest permeability, and is *above* the first. The third critical layer is the layer which has the next lowest permeability, and is *above* the second, and so on. In this way a sequence of *critical* soil layers may be established. Each layer acts an an impeding layer if rainfall intensity exceeds its saturated permeability. The return period for that rainfall intensity may be roughly associated with the frequency with which the layer acts to impede percolation and direct subsurface flow. Where soil changes continuously with depth, the relationship between depth and frequency of impediment will also be a continuous one.

For clarity, a simple two-layer soil is mainly discussed here, but the principles apply equally to many-layered or gradually-changing soils. Rainfall commonly infiltrates at less than the maximum infiltration capacity (if not, the surface soil

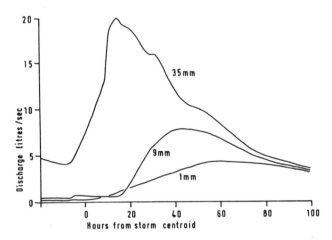

Figure 4.8   Variation in form of measured subsurface-flow hydrograph with antecedent moisture. Values refer to previous 5-days' rainfall in millimetres (from Weyman, 1971)

itself acts as an impeding layer) and percolates downwards to the top of the impeding contact or the top of the pre-existing saturated layer. The lag time between the beginning of rainfall and the beginning of the hydrograph rise depends on this percolation time, and so is less when the antecedent soil moisture is higher. This is illustrated in Figure 4.8 for a soil with gradually-changing properties, and in Figure 7.16A and B for a layered soil. Figure 4.8 shows the effect of 53–57-mm storms following 5-day periods with 1, 9, and 35 mm of antecedent rainfall. Figure 7.16A shows the response to a 2-hr simulated rainfall totalling 102 mm falling on an initially dry soil. Even with this large rainfall, there is a delay of over $2\frac{1}{2}$ hr before the subsurface hydrographs begin to rise, whereas in Figure 7.16B, for a 31-mm storm on an initially wet soil, the delay is only 50 mins.

Figure 7.16A also illustrates a second point about the percolation process. It seems that the percolating rainwater behaves as a pulse or 'plug' which takes an appreciable time to reach the saturated zone or impeding layer, but that it is then added to the saturated layer rather rapidly, so that the peak-flows from layers above the impeding contact occur almost simultaneously. In Figure 7.16A the main impeding contact is at 90-cm depth, and flow in the 0–56-cm layer occurs only after the 56–90-cm zone is saturated. Nevertheless, the responses of the two layers on the rising limb of the hydrograph are very closely in time with one another.

In a simple model of the soil hydrograph response, the subsurface flow begins to rise from its previous value towards a new peak as the pulse of percolating water reaches either the impeding layer or the top of the saturated layer, whichever comes first. In the former case, the time lag from the onset of rainfall is mainly dependent on the depth to the impeding layer, and hence on rainfall intensity. In the latter case, the time lag depends on the storage level in the soil, as well as on the rate of percolation (which is itself dependent on rainfall intensity). This is illustrated schematically in Figure 4.9 for a soil profile with properties which change gradually with depth.

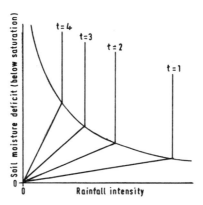

Figure 4.9   Schematic dependence of time-lag to subsurface stormflow peak on antecedent moisture and rainfall intensity

Within the impeding layer, a wetting front penetrates downwards, initially very rapidly, and then more slowly as the average diffusivity gradient becomes less. This behaviour is similar to that of infiltration at the soil surface. Figure 4.10 shows schematically the conditions in a two layer soil, which is subjected to rainfall downslope from an artificial plot boundary. The figure shows conditions after steady rainfall of long duration. Near the upper end of the plot, the small downslope discharge can be balanced by percolation into a shallow saturated layer within the impeding layer, where percolation is rapid owing to the high average diffusivity gradient. As discharge increases linearly downslope, the depth of saturation within the impeding layer needed to support the discharge also increases. This increase in depth in turn reduces the diffusivity gradient and hence

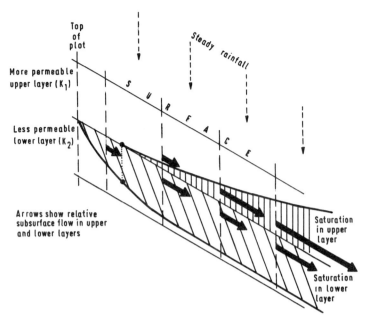

Figure 4.10   Saturated layers and subsurface flow above
and below the contact with an impeding layer, under
conditions of steady rainfall

the percolation rate into the impeding layer, until a point is reached where not all
the rainfall can enter the impeding layer. Downslope from this point a saturated
layer begins to back up in the upper, permeable layer. Figure 4.10 illustrates a case,
similar to that shown in Figure 7.16 in which most of the *moisture* percolates into
the impeding layer, but most of the *flow* occurs in the permeable layer. This flow
forms the main part of the subsurface stormflow hydrograph, and the upper
saturated zone may also reach up to the surface as return flow and, in doing so,
augment the contributing area. The steady state conditions of Figure 4.10 develop
progressively downslope from the top of the plot as rainfall continues.

Once rainfall has ceased to reach the top of the upper saturated layer, it loses
water both by downslope flow and further percolation into the impeding layer.
Flow from near the surface stops first (e.g. the 0–56-cm layer in Figure 7.16A), and
then flow from immediately above the impeding layer. For the soils studied by
Whipkey (1965), no more than 16 % of the storm rainfall flowed out within 24 hr.
The remainder is left, mainly within the impeding layer, where it can continue to
flow, at first in a saturated layer and then in entirely unsaturated form, for many
weeks.

Even with only two soil layers, different ratios between the two permeabilities
and the relative intensity of rainfall can lead to a wide range of possible soil-
hydrograph forms, which are illustrated schematically in Figure 4.11, notionally
for a constant storm duration on an initially dry soil. Figure 4.11A shows the

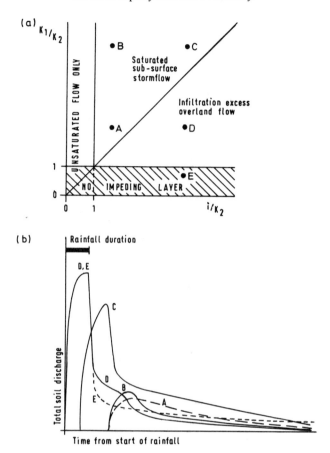

Figure 4.11   Regimes of soil flow for a two-layer soil in
relation to permeabilities and rainfall intensity. (a)
Conditions for different flow regimes; (b) schematic soil
hydrographs for each regime

possible subsurface flow regimes which may occur, and Figure 4.11B the broad
types of total hydrograph response, ignoring the additional possibility of return
flow because of its strong dependence on position down the slope. At point A, the
small difference between the soil layers requires a thick saturated zone in the
upper layer, and less water is available for subsurface flow. This build-up of water
however helps to maintain percolation into the impeding layer after rainfall has
ceased, so that flows are maintained for a long time. At B the higher upper-layer
permeability results in a higher peak, but less water is available to maintain the
recession limb. At C, more water builds up in both layers, and the rate of
percolation is higher at the higher rainfall intensity, so that an enlarged version of
the B hydrograph is produced, but with a shorter lag time. At D and E much of the

rainfall cannot infiltrate into the soil, and an initial overland flow peak dominates the hydrograph. As rainfall intensity increases at a site, it will tend to produce an increased and earlier version of the same hydrograph form provided that flow remains in the same regime, as from B to C; but radical changes may occur as boundaries in Figure 4.11a are crossed, as occurs between points A and D.

Some examples of the various types of flow regime may be seen from published subsurface flow hydrographs. Figure 4.8 shows types A to C; Figure 7.16b shows types A or B; Figure 7.16a and 7.24 shows types B or C, and Figure 7.26 shows types D or E. Factors other than the properties of two such simple soil layers are however operating in most cases.

Although there is a need for many more systematic measurements of subsurface flow, enough work has been done to suggest that there is a basis for setting up at least simplified models of the flow behaviour in a hillside soil. Perhaps the most important simplifications which have been discussed here relate to the need to treat the soil as a series of layers rather than as a uniform medium, and the possibility of confining attention mainly to flow in saturated zones, where differences in soil-moisture tension are only of secondary importance.

4.4.2 The soil hydrograph as part of the basin hydrograph

The soil hydrograph is important not only in terms of local conditions, but also as a major component of first the hillside, and second, the basin hydrographs. This topic is discussed further in Chapters 7 and 8, but some comments are relevant at this stage. Figure 4.12 shows some of the factors responsible for producing the soil, slope and basin hydrographs. In this chapter, the formation of the soil hydrograph has been discussed. The soil flows have one rôle as a direct component of the slope hydrograph. They also establish the soil-moisture conditions which determine where saturation will occur on a hillside, which in turn determines the areas where rainfall is directly diverted as saturation-excess overland flow, and where return flow will occur. This aspect of the slope hydrograph is discussed in Chapter 7. What is relevant to the present discussion is that where saturation-excess overland flow occurs, it dominates the slope hydrograph peak. The result is that in many humid–temperate drainage basins, the soil hydrograph determines the *timing* of the slope hydrograph peak, but its *magnitude* also depends on rainfall intensity at that time.

As water flows into the soil, out again into stream channels and along them, it may be considered as passing through a series of stores, each of which delays it according to its properties. In this kind of system, it has been shown that the overall hydrograph shape is mainly determined by the response times of the slowest stores (Wooding, 1965). Applying this argument to the formation of a basin hydrograph, it can be argued that only one aspect of the soil hydrograph is likely to have a direct impact on the shape of the basin hydrograph. Once subsurface flow has drained into the impeding soil layer, it drains very slowly, with a response time of several

weeks. This is slower than any other store, except possibly for channel travel time in
the world's largest river basins. As a result, the soil hydrograph is likely to be a
major determinant of drainage basin recession curves where a suitable impeding
layer is present in the soil, and where groundwater does not play a major rôle.

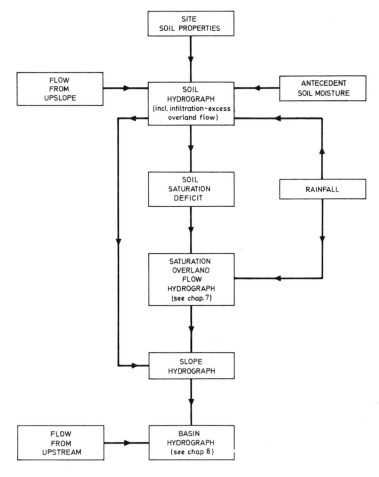

Figure 4.12   The rôle of the soil hydrograph in slope and basin
hydrograph formation

The normal rôles of subsurface flow in a drainage basin are thus seen to be first,
the establishment of soil moisture distribution over the area of the basin, and
second, the direct formation of recession curves and low flows in the basin. The
conditions under which subsurface flow acts in these rôles, are commonly though
not always present in humid areas, and are associated with permeable topsoils.

REFERENCES

Amerman, C. R., 1956, 'The use of unit-source watershed data for runoff prediction'. *Water Resources*, **1**, 499–508.

Bay, R. R., 1969, 'Runoff from small peatland watersheds', *J. Hydrology*, **9** (2), 90–102.

Bell, G. L., 1968, 'Piping in the Badlands of Dakota'. *Proc. 6th Annual Engineering Geology and Soils Engineering Symposium, Boise, Idaho*, 242–257.

Berry, L., 1970, 'Some erosional features due to piping and sub-surface wash with special reference to the Sudan'. *Geografiska Annaler*, **52A**(2), 113–119.

Betson, R. P., 1964, 'What is watershed runoff?', *J. Geophys. Res.*, **69**, 1541–1552.

Betson, R. P. and Marius, J. B., 1969, 'Source areas of storm runoff', *Water Res. Res.*, **5**(3), 574–582.

Bunting, B. T., 1961, 'The rôle of seepage moisture in soil formation, slope development, and stream initiation', *Am. J. Sci.*, **259**, 503–518.

Burykin, A. M., 1957, 'Seepage of water from soils in mountainous regions of the humid subtropics', (English translation), from *Pochvovedenie*, No. **12**, 90–97.

Chamberlin, T. W., 1972, 'Interflow in the mountainous forest soils of coastal British Columbia', *Mountain Geomorphology*, Ed. D. Slaymaker and H. J. McPherson, Tantalus Research, Vancouver, pp. 121–127.

Dickinson, W. T. and Whiteley, H., 1970, 'Watershed areas contributing to runoff', *Proc. of the Gen. Assembly, Int. Assoc. of Scientific Hydrology*, N.Z., pp. 12–16.

Dunne, T. and Black, R. D., 1970, 'An experimental investigation of runoff production in permeable soils', *Water Res. Res.*, **6** (2), 478–490.

Fletcher, J. E., Harris, K., Peterson, H. B., and Chandler, V. N., 1954 'Piping', *Trans. Am. Geophys. Union*, **35**, 258.

Freeze, R. A., 1972, 'Rôle of subsurface flow in generating surface runoff—1. Base flow contributions to channel flow'. *Water Res. Res.*, **8**(3), 609–623.

Hewlett, J. D., 1961, 'Soil moisture as a source of base-flow from steep mountain watersheds'. *USDA, FS, SEFES*, Station Paper **132**.

Hewlett, J. D. and Hibbert, A. R., 1963, 'Moisture and energy conditions within sloping soil mass during drainage', *J. Geophys. Res.*, **68**, 1081–1087.

Hewlett, J. D. and Troendle, C. A., 1975, Non-point and Diffused Water Sources: a Variable Source Area Problem, *Symposium on Watershed Management, Utah State Univ., Aug. 11–13, 1975*, ASCE, New York, pp. 21–46.

Hursh, C., 1944, 'Report of sub-committee on subsurface flow'. *Trans. Am. Geophys. Union*, **25**, 743–746.

Hursh, C. R. and Fletcher, P. W., 1942, 'The soil profile as a natural reservoir', *Proc. Soil Sci. Soc. Am.*, **7**, 480–486.

Hursh, C. R. and Hoover, M. D., 1941, 'Soil profile characteristics pertinent to hydrologic studies in the Southern Appalachians', *Proc. Soil Sci. Soc. Am.*, **6**, 414–422.

Institute of Hydrology, Research 1970–71, 'Runoff processes on Plynlimon', 26–30.

Jamison, V. C. and Peters, D. B., 1967, 'Slope length of claypan soil affects runoff', *Water Res. Res.*, **3**(2), 471–480.

Jones, A., 1971, 'Soil piping and stream channel initiation', *Water Res. Res.*, **7**(3), 602–610.

Kirkby, M. J. and Chorley, R. J., 1967, 'Throughflow, overland flow, and erosion' *Bull. Internat. Assoc. Sci. Hydrology*, XIIe Annee, 3, 5–21.

Knapp, B. J., 1970, *Patterns of Soil Water Movement on a Steep Upland Hillside, Central Wales*, Unpublished Ph.D. thesis, Reading Univ. 213 pp.

Minshall, N. W. and Jamison, V. C., 1965 'Interflow in claypan soils', *Water Res. Res.*, **1**, 381–390.

Pierce, R. S., 1965, 'Evidence of overland flow on forest watersheds', *Internat. Symposium on Forest Hydrology*, Pennsylvania State Univ.

Rawitz, E., Engman, E. T. and Cline, G. D., 1970, 'Use of the mass balance method for examining the rôle of soils in controlling watershed performance' *Water Res. Res.*, **6**(4) 1115–23.

Roessel, B. W. P., 1951, 'Hydrologic problems concerning runoff in headwater regions', *Trans. Am. Geophys. Union*, **31**, 431–442.

Synder, W. M. and Asmussen, L. E., 1972, 'Subsurface hydrograph analysis by convolution', *J. Irrigation and Drainage Division, ASCE*, 98 (IR3) Proc. Paper **9213**, 404–418.

Troendle, C. A., 1970, 'Water storage, movement and outflow from a forested slope under natural rainfall in West Virginia', abstract, EOS, *Trans. Am. Geophys. Union*, **51**, 279.

Tsukamoto, Y., 1961, 'An experiment on subsurface flow', *Jr. Japanese Soc. Forestry*, **43**, 61–68.

Van Dijk, D. C., 1958, 'Water seepage in relation to soil layering in the Canberra district', *CSIRO, Div. of Soils, Commonwealth of Australia*, Report 5/58, 13 pp.

Vasilyev, I. S., 1948, 'The experience of a study of surface and subsurface runoff in a Podzolic forest soil'. *Pochvovedenie*, No. **5**, 312–324.

Weyman, D. R., 1973. 'Measurements of the downslope flow of water in a soil', *J. Hydrology*, **20**, 267–288.

Weyman, D. R., 1974, 'Runoff process, contributing area and stream flow in a small upland catchment', *Inst. Brit. Geographers*, Special Publication, No. **6**, 33–43.

Whipkey, R. Z., 1965, 'Sub-surface stormflow from forested slopes', *Internat. Assoc. Sci. Hydrology*, Xth Annee. Bull., **2**, 74–85.

Whipkey, R. Z., 1967, 'Storm runoff from forested catchments by subsurface routes'. In *Proc. of the Leningrad Symp.: Floods and their computation, Gentbrugge, Belgium, Internat. Assoc. Sci. Hydrology*, pp. 773–779.

Wooding, R. A., 1965, 'A hydraulic model for the catchment-stream problem. I. Kinematic wave theory'. *J. Hydrology*, **3**(3/4), 254–282.

CHAPTER 5

Overland flow

W. W. Emmett,

United States Geological Survey, Denver, Colorado, USA

5.1 DEFINITIONS AND CONCEPTS

A area (ha or km^2)
D average depth (m)
D_d drainage density (kilometers of stream per square kilometer of area)
F_r Froude's number, $F_r = V/(gD)^{0.5}$
K coefficient of runoff
L_o average length of overland flow (m)
L_s stream length (km)
M exponent reflecting the degree of turbulence
Q discharge (m^3/sec)
R_e Reynolds' number, $R_e = 4VD/v$
S slope (expressed as the tangent of the slope angle)
V mean velocity (m/sec)
V_s surface velocity
f exponent of depth, $D \propto Q^f$
f_f Darcy–Weisbach friction factor, $f_f = 8gDS/V^2$
g acceleration due to gravity $= 9.81$ m/sec^2
m exponent of velocity, $V \propto Q^m$
n Manning resistance coefficient
q unit discharge (m^3/sec/m)
y exponent of friction, $f_f \propto Q^y$
v kinematic viscosity (m^2/sec)

It is important to be familiar with the terminology of those components of hydrology pertinent to a particular aspect being discussed. The italicized words in the paragraph which follows are key words in an understanding of overland flow. Cover-to-cover readers of this book will realize that almost all of the italicized words are discussed in detail in other sections or chapters of the book.

    A raindrop falls, but never reaches the earth's land surface. It has been *intercepted*, perhaps by a tree leaf or a rooftop. Its moisture is returned to the

145

F

atmosphere by *evaporation, transpiration,* or a combination of both, in which case it is termed *evapotranspiration.* A second raindrop falls and lands in a natural or artificial depression in the earth's surface. It is termed *depression storage* and as such, it never enters into the process of water removal known as *runoff.* Another raindrop falls but it penetrates the land surface, or *infiltrates* into the ground. The amount of infiltration, or *infiltration rate,* is variable among different soils and even for a given soil it is variable and dependent on the previous dampness, or *antecedent moisture,* condition of the soil. Infiltrated water may become part of the soil's *field moisture,* may continue its *percolation* through the soil to the *water table* and become part of the *groundwater* system, or may respond to gravity and flow down-gradient at shallow soil depths as *interflow* or *throughflow.* Water travelling as interflow or throughflow may remain at shallow depth until it discharges into a stream channel and becomes *channel flow,* may enter into the groundwater system, or may return to the land surface and is known as *return flow.* In this latter instance, it becomes that part of *surface runoff* which is termed *overland flow.* Other raindrops fall directly into stream channels and other bodies of water and thus contribute to surface runoff. Such water is termed channel precipitation but is not part of *overland flow* because, in the strict sense, it does not flow over the land surface en route to the stream channel. Still other raindrops fall on the land surface and respond to gravity by flowing down-gradient over the land surface. Such flow is termed *overland flow* and occurs when the number of raindrops, or amount of *precipitation,* exceeds the *infiltration capacity* of the soil and the *depression storage capacity* of the land surface. The consumptive utilization of water by evaporation and transpiration must also be included in the total amount of precipitation needed to generate overland flow, but considering the brief time span of overland flow, these effects have much less significance than the infiltration and depression storage capacities. Water en route downslope as overland flow is termed *detention storage.*

    In summary, overland flow is defined as the flow of water over the land surface toward a stream channel and is the initial phase of surface runoff. It is sometimes referred to as sheet flow because the water is envisioned as moving in a sheet downslope over a plane surface to the nearest concentration point or channel. The term 'sheet flow' is nearly synonymous with overland flow. The latter term is used preferentially in this chapter and the term 'sheet flow' is reserved as a special description of overland flow. Nearly all surface runoff starts as overland flow in the upper reaches of a watershed and travels at least a short distance in this manner before it reaches a rill or channel. Once a known flow rate reaches a defined channel, it becomes channel flow and its action may usually be characterized adequately by standard hydraulic procedures (Chow, 1959).

5.2 DIFFICULTIES IN A GENERAL DESCRIPTION

In contrast to channel flow, in overland flow the variables are more difficult to define precisely and the use of a simple hydraulic procedure for predicting overland flow and its characterizations is beset with many difficulties. Overland flow is both unsteady and spatially varied since it is supplied by rain and depleted by infiltration, neither of which is necessarily constant with respect to time and location. The flow may be either laminar or turbulent or a combination of these two conditions. Flow depths may be either below or above critical, or the depths may change from subcritical to supercritical. Under certain conditions the flow may become unstable and may give rise to the formation of roll waves, or rain waves as they are often called. The action of raindrop impact on the sheet of flowing water complicates further the overland flow problem.

On a plane surface, such as a paved parking area or a laboratory flume, a thin film of water may flow downslope with little variation of depth in the cross-slope direction. Even on natural slopes, it is likely that much of the ground is covered by surface detention during overland flow. However, topographic irregularities that occur on natural slopes are sufficient to direct most runoff water into lateral concentrations of flow. These concentrations of flow weave anastomosing paths downslope and often give the appearance of flow in a wide, shallow-braided channel. Over the length of slope for most overland flow, Reynolds' number, a measure of fluid turbulence, normally remains in the regime considered as laminar flow, but the flow is not truly laminar because of the disturbance by falling raindrops and the influence of topographic irregularities. Such a disturbed flow is capable of eroding and transporting sediments. The lateral concentrations of flow are conducive to the formation of rills and gullies, but rills and gullies do not form on all slopes.

The hydraulic characteristics of overland flow are dependent on many factors including the intensity and duration of precipitation (or melting snow and ice), the texture or type of soil as reflected by its infiltration capacity, the antecedent soil-moisture condition, the density and type of vegetation, and topographic features including the number and size of surface depressions and mounds, slope steepness, and length of slope. The geomorphic characteristics, or the capabilities of overland flow as a landscaping agent, are generally dependent on the hydraulic characteristics, but no simple description of the hydraulics of overland flow on natural hillslopes is possible because the hydraulic parameters vary rapidly over time and space.

It becomes necessary to make detailed observations of overland flow at several locations and use these detailed observations from which to draw generalizations about the hydraulic flow parameters. Reliance is also placed on laboratory experiments to provide generalizations for the effects of each variable despite the difficulties of direct transfer to the field case.

As more and more general observations of overland flow have become available, a recent trend has been towards considering the mathematical modelling of overland flow. These analytical and theoretical solutions to the hydraulics of overland flow are still partially incomplete, primarily due to the paucity of physical observations of overland flow. Real-life observations are necessary, not only to provide correct input information to the modelling attempts, but also to provide verification of the theoretical predictions. Chapter 6 provides a discussion of the mathematical treatment of overland flow. This chapter presents a description of the physical characteristics of the hydraulics of overland flow. Most of the discussion is based on the writer's published (Emmett, 1970) and unpublished studies of overland flow, and is not separately referenced. All other studies, generally in the discussion of results, are referenced where they appear in the text.

5.3   QUANTIFICATION OF OVERLAND FLOW

Few, if any, hydrologists have made contributions as significant to modern hydrology as those of the New England engineer Robert E. Horton, in the 1930s and 1940s. The first of these works described his theories of infiltration capacity and surface detention (Horton, 1933). In a study of the basic behaviour of laminar sheet flow (Horton, Leach and van Vliet, 1934), values of depth and velocity associated with the shallow flow of water over sloping surfaces were predicted. Horton (1936 and 1938) continued to work on a quantitative description of overland flow, and his efforts resulted in a classic work on geomorphology (Horton, 1945) which not only provided interpretation of the erosional development of streams and their drainage basins, but also provided the framework for many more recent studies.

Overland flow probably occurs in every watershed to some degree. Whether it is extensive enough, or the flow length great enough, to be of geomorphic and hydrologic significance needs to be determined. One method for estimating the average length of overland flow is that developed by Horton (1945). Horton shows that the average length of overland flow, L_o, can be estimated by the relation

$$L_o = \frac{1}{2D_d} \tag{5.1}$$

where D_d is the drainage density defined as

$$D_d = \frac{\sum L_s}{A} \tag{5.2}$$

where $\sum L_s$ is the sum of the stream lengths for the watershed and A is the drainage area of the watershed.

Thus, if the lengths of all channels that are fed directly by overland flow can be measured, it is possible to estimate the average length of overland flow. Such an attempt was made for a 0·83-km^2 grassed watershed operated by the Stillwater Hydraulics Laboratory in Oklahoma (Ree, 1963). Every waterway visible on an aerial photograph of the watershed was measured and the sum of the lengths of these drainageways was about 14·9 km. This value and the value of the drainage area were substituted into Equations 5.1 and 5.2 and the average length of overland flow was calculated to be 28 m. This value is comparable to results of laboratory experiments by Izzard (1946) who found that lengths of overland flow may be as much as 22 m.

For overland flow on natural slopes, Horton (1945) postulated that a condition of mixed flow exists; that is, areas of fully turbulent flow are interspersed with areas of laminar flow. The depth of overland flow for steady-state conditions can be estimated by the use of channel flow formulae. For turbulent flow, depth may be estimated by combining the continuity equation

$$q = DV \tag{5.3}$$

and the Manning equation

$$V = \frac{1}{n} D^{0 \cdot 67} S^{0 \cdot 50} \tag{5.4}$$

where

q = the unit discharge (m^3/sec/m)
D = the mean depth (m)
V = the mean velocity (m/sec)
n = the Manning resistance coefficient
S = the slope gradient (m/m)

In all calculations in this chapter, depth is assumed to be equivalent to hydraulic radius because the cross-section of overland flow is very wide and shallow. Equations 5.3 and 5.4 are combined to yield

$$q = \frac{1}{n} S^{0 \cdot 5} D^{1 \cdot 67} \tag{5.5}$$

which for a given slope may be written as

$$q = KD^{1 \cdot 67} \tag{5.6}$$

and, expressed as depth

$$D = \left(\frac{q}{K}\right)^{0 \cdot 60} \tag{5.7}$$

For laminar flow, a form of the Poiseuille formula may be used to estimate depth. This is expressed as

$$q = \frac{gSD^3}{3v} \tag{5.8}$$

where

g = the acceleration of gravity (9·81 m/sec^2)
v = the kinematic viscosity (m^2/sec)

For a particular slope, Equation 5.8 may be written as

$$q = KD^3 \tag{5.9}$$

or

$$D = \left(\frac{q}{K}\right)^{0\cdot33} \tag{5.10}$$

For either turbulent or laminar flow, depth can be expressed as

$$q = KD^M \tag{5.11}$$

in which M is the exponent for depth and reflects, in part, the degree of turbulence. The value of M for fully turbulent flow is $1\cdot67$ and is 3 for fully-laminar flow. Thus, with increases in discharge, depth increases more rapidly in turbulent flow than laminar flow. For mixed flow, as Horton postulates occurs in nature, values of M would range between these two extremes.

The depth calculated by Equations 5.7 and 5.10 is the depth at a selected downslope point, or for a given value of discharge. For various discharges, a theme analogous to the hydraulic geometry of streams introduced by Leopold and Maddock (1953) may be used to describe variations in depth. The general technique in hydraulic geometry is to relate changes in the hydraulic parameters of flow to changes in discharge. A further convenience afforded by the use of hydraulic-geometry analyses is that they allow the data of overland flow to be compared easily to the more voluminous collection of data for flow in river channels.

Because water temperature (and therefore values for the viscosity of water) is variable, Reynolds' number, R_e, has generally been substituted for discharge in the analysis of data. Reynolds' number, here proportional to the discharge per unit width, is a dimensionless parameter relating the effect of viscosity to inertia and is defined as

$$R_e = \frac{4VD}{v} \tag{5.12}$$

Use of Reynolds' number does not influence the comparison of data with other studies of hydraulic geometry and the use of Reynolds' number is convenient in visualizing whether the flow is laminar or turbulent. It should be noted that many authors use Equation 5.12 without the constant of 4. Thus values of Reynolds' number as used in this chapter may differ from the values used elsewhere by a factor of 4.

For uniform flows, the analysis of overland flow is similar to at-a-station hydraulic geometry because, for a given discharge, depths are constant downslope. With superposing of rainfall, and thus with increasing downslope discharge, the analysis is similar to the case of downstream hydraulic geometry. At-a-station relations may also be developed by considering only that data collected at the same downslope position.

The hydraulics of overland flow are determined most easily by laboratory measurements of uniform flow at shallow depths and of spatially-varied flow resulting from simulated rainfall on a plane surface. The results of these laboratory studies will then be compared to field studies of simulated rainfall on small runoff plots.

5.4 LABORATORY TESTS WITH UNIFORM FLOW

In the Hydraulics Laboratory of the US Geological Survey, Washington, DC, an impervious, smooth plane surface of adjustable slope was used to study uniform flow over a range of shallow depths from 0·9 to 3 mm (0·0029–0·0435 ft), and at slope values of 0·0033–0·0775. A uniform sand-grain roughness with a median grain diameter of 0·5 mm was next applied to the flume floor and similar tests were conducted. The principal measurements included depth, surface velocity, discharge, and water temperature. These data allow the computation of additional hydraulic parameters describing the characteristics of the flow.

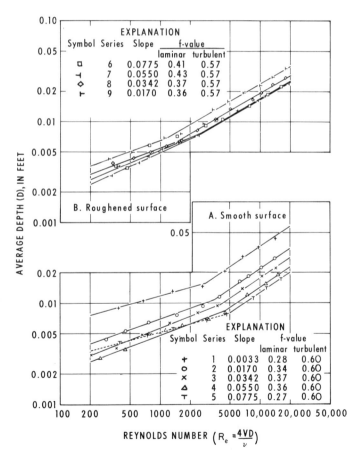

Figure 5.1 Depth of uniform flow as a function of Reynolds' number

Nine series of tests were conducted, five on the smooth surface and four on the roughened surface. Figure 5.1 illustrates the relation between the depth of uniform flow and Reynolds' number. The two most apparent observations are the break in the relation at a certain critical value of Reynolds' number between 1,500 and 6,000 and the increase in depth with decreasing slope. The critical Reynolds' number marks a change in regime from laminar flow at a smaller Reynolds' number to turbulent flow at greater Reynolds' numbers. For both smooth and roughened surfaces, the critical value of Reynolds' number increases with increased slope. This indicates that the shallower flows on the steeper slopes are somewhat more stable against change to turbulent flow.

In terms of hydraulic geometry, the depth may be expressed as:

$$D \propto Q^f \tag{5.13}$$

A similar expression utilizing Reynolds' number is

$$D \propto (R_e)^f \tag{5.14}$$

The values of f are the reciprocal of the M values (refer to Equation 5.11) used by Horton (1945). Values of f are tabulated in the explanation blocks in Figure 5.1. For turbulent flows over the smooth surface, the value of f is 0·60 for all slopes. This value is equal to the theoretical value (from Equation 5.7. For laminar flow over the smooth surface, values of f range from 0·27 to 0·37. These values tend to centre around the theoretical value of 0·33 from Equation 5.10. The scatter of values around a central value is attributed to the accuracy of data rather than any trend that might be suggested.

For the roughened surface, the value of f for turbulent flow is 0·57. Values of f less than 0·60 would indicate flow less than fully turbulent. The effect of a roughened surface is to retard the flow near the bed of the flume. In flows as shallow as those of overland flow, 'flow near the bed' may represent a considerable part of the entire depth. Thus, it is quite reasonable to expect extremely shallow flows over roughened surfaces to exhibit some tendencies of laminar flow.

The laminar flow regime for the roughened surface produces values of f ranging from 0·36 to 0·43. One explanation for the values slightly higher than the theoretical value of 0·33 is that no correction was applied to values of depth as measured from the top of the roughness element. At the extremely shallow depths for flows in the laminar region, generally less than 2 mm (0·007 ft), a correction that allows for the voids between bed roughness particles and is added to the measured depths would increase the shallowest depths proportionately most; the effect would be to decrease the value of f. For depths as large as those in the turbulent flow region, the effect would be negligible.

The effect of roughness on values of depth is both general and complex. Because of the additional resistance to flow, roughness increases the depth of flow for a given discharge. The maximum influence of roughness appears near the transition from laminar to turbulent flow. In this region, depths on the roughened surface are from 15 % greater (for the less steep slopes) to 30 % greater (for the steeper slopes) than

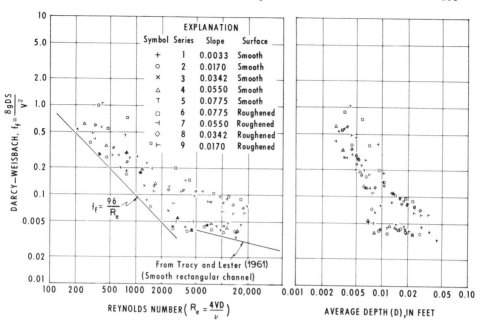

Figure 5.2   Darcy–Weisbach friction factor for uniform flow as a function of Reynolds' number and average depth

depths on the smooth surface. Differences are less pronounced at both low and high Reynolds' numbers.

The effect of flume slope on depth is to decrease depth for increasing slopes. The relation is hyperbolic; that is, as slope approaches zero, depth approaches infinity and as slope approaches high gradients, depths approach some minimum value. A roughened surface tends to dampen this effect. That is, the approach to some minimum depth regardless of further increases in slope occurs at a smaller gradient for the roughened surface. Thus, for flume slopes of 0.0775 and 0.0550, data on the roughened surface nearly describe a single curve while for the smooth surface, the data are still separated by a small distance.

A measure of the resistance to flow, the Darcy–Weisbach friction factor, is plotted in Figure 5.2 as a function of Reynolds' number and average depth. The Darcy–Weisbach friction factor, f_f, is defined as

$$f_f = \frac{8gDS}{V^2} \tag{5.15}$$

The expression for velocity in terms of hydraulic geometry is

$$V \propto Q^m \tag{5.16}$$

or

$$V \propto (R_e)^m \qquad (5.17)$$

The friction factor, in terms of hydraulic geometry, may be written as

$$f_f \propto Q^y \qquad (5.18)$$

or

$$f_f \propto (R_e)^y \qquad (5.19)$$

When slope, S, is constant, equation 5.15 may also be written as:

$$f_f \propto \frac{D}{V^2} \qquad (5.20)$$

or

$$f_f \propto \frac{(R_e)^f}{(R_e)^{2m}} \qquad (5.21)$$

Thus for a constant gradient, the slope, y, of the friction line as a function of Reynolds' number is $f - 2m$. Also, since $VD = q$, then $f + m = 1$ or $m = 1 - f$. For turbulent flow, the slope of the line is $0.60 - 2(0.40) = -0.20$ and for laminar flow it is $0.33 - 2(0.67) = -1.0$. Negative values of y indicate a decrease in resistance to flow with increasing Reynolds' number. For laminar flow in a smooth rectangular channels, the equation of the line with a slope of -1.0 is

$$f_f = \frac{96}{R_e} \qquad (5.22)$$

For rough channels, values of the friction factor would be higher and would plot above the lower limits defined by Equation 5.22.

For turbulent flows, the friction factor–Reynolds' number relation has a slope of -0.20. For a smooth surface flume, this value is indicated in Figure 5.2 with a line representing data from experiments by Tracy and Lester (1961).

The data of Figure 5.2 plot higher than the limits indicated by the equations applicable to a smooth surface. This illustrates the pronounced effect of channel roughness on the friction factor. The smooth flume in the overland flow study was, in fact, not completely smooth as indicated by the data plotting higher than the equations for a smooth surface.

Using the average value of the friction factor at a Reynolds' number of 20,000, reference to a Stanton or Moody diagram gives a relative roughness of 0·03 for the roughened surface and 0·002 for the smooth(er) surface. At an average depth 9.1 mm (0·03 ft) for a Reynolds' number of 20,000, a computed absolute roughness is 0·27 mm (0·0009 ft) for the roughened surface and 0·018 mm (0·00006 ft) for the smooth surface. One-half of the diameter of the 0·5-mm grain roughness used on the flume is 0·25 mm (0·0082 ft). This is very close to the computed roughness for the roughened surface. The computed value for the smooth surface is not unreasonable for sanded plywood with a paint finish.

Only two roughnesses are involved with the two surfaces but because each plotted point in Figure 5.2 has a different depth, the data include many values of relative roughnesses. For a Reynolds' number of 300, depths are approximately 0·9 mm (0·003 ft) and the relative roughness is 0·3 for the roughened surface. This is 10 times rougher than at a Reynolds' number of 20,000. If points representing equal depths were connected, one would find the beginnings of a family of curves, each representing a given relative roughness. The curves are not drawn in Figure 5.2 because the data are too sparse. However, it is interesting to note that most of the data lie in a range of friction factors and relative roughnesses much greater than those included on conventional Moody diagrams. This again illustrates the tremendous influence of even small surface roughnesses on flows as shallow as those that occur in overland flow.

The right half of Figure 5.2 again illustrates that plotted points represent differing depths. To maintain geometric similarity between the depth of flow and scale of roughness, friction factors are plotted against depth. The tendency to converge into a single curve for laminar flow and into two curves, one for each roughness, at higher Reynolds' numbers is apparent. The remaining scatter in data is most likely within the accuracy of the experiment. From Equations 5.3 and 5.15, the friction factor may be expressed as

$$f_f = \frac{8gD^3S}{q^2} \tag{5.23}$$

from which it can be seen that the percentage error in friction factor is three times the percentage error in depth. Using a nominal depth of 1·5 mm (0·005 ft)—typical of depths at low Reynolds' numbers—an error of only 0·15 mm in the measurement of depth yields a 30% error in friction factor.

Perhaps the most widely used open-channel formula is the Manning equation (Equation 5.4) used by Horton (1945). Computed values of Manning's n are plotted in Figure 5.3 as a function of Reynolds' number and average depth. In terms of hydraulic geometry, Manning's n can be expressed by:

$$n \propto \frac{D^{0·67}}{V} \propto \frac{Q^{0·67f}}{Q^m} \propto \frac{(R_e)^{0·67f}}{(R_e)^m} \tag{5.24}$$

Thus the slope of the Manning's relation as a function of Reynolds' number is $0·67f - m$ or $-0·45$ for laminar flow and $0·0$ for turbulent flow. The data closely follow these values. The previous discussion concerning the Darcy–Weisbach friction factor is applicable to both plots of Figure 5.3.

The results of the laboratory tests of shallow uniform flow at low Reynolds' numbers are generally confirmed by the results of other investigations, for example, Parsons (1949), Straub (1939), and Owens (1954). Small differences do arise because of small differences in equipment and experimental techniques, but the behaviour as shown in Figures 5.1–5.3 is both general and adequate for further description of overland flow.

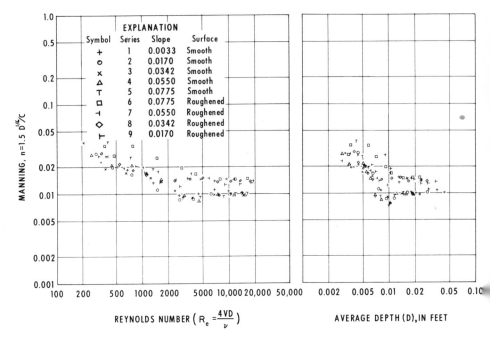

Figure 5.3   Manning's *n* for uniform flow as a function of Reynolds' number and average depth

5.5   LABORATORY TESTS WITH SIMULATED RAINFALL

At each slope position of the uniform flow tests, spatially-varied flow resulting from the uniform application of artificial rainfall was studied at five intensities of rainfall. These intensities were about 90, 120, 155, 215 and 290 mm/hr (3·5, 4·7, 6·1, 8·5 and 11·5 in/hr), respectively. No attempt was made to determine the impact velocity or drop size of the falling rain droplets. Natural rainstorms at a given intensity include a wide range of drop sizes and the drop-size distribution by volume for different rainfall intensities shows mean drop sizes of 1, 2, and 3 mm for intensities of 0·25, 25, and 100 mm/hr, respectively (Laws and Parsons, 1943). However, the percentage by volume contributed by drop sizes within 0·12 mm of the sizes listed above are, respectively, only 28, 12 and 9 % of the total rain, the remainder being about equally contributed between larger and smaller drops. Water drops falling through the air approach a terminal velocity which varies with drop size. The relation of distance of fall to drop-fall velocity was studied by Laws (1941) and Gunn and Kinser (1949). For drop sizes of 1·25, 2·0, and 3·0 mm, a respective fall distance of about 4·5, 9 and 12 m was needed to obtain terminal velocity of 4·8, 6·6 and 8·1 m/sec. The average drop size in the studies discussed here was about 0·5 mm, somewhat smaller than natural raindrops for the intensities of rainfall simulated, but the impact velocity approximated the terminal fall velocity for the drop size generated.

Keulegan (1944) has reported that the retarding effect of falling rain on shallow flows is small. However, other investigators (Izzard, 1944; Parsons, 1949; Woo and Brater, 1962; Yoon and Wenzel, 1971; Shen and Li, 1973) have reported that the effects may be significant. These effects will be discussed in more detail later. Apparently, the geomorphic or erosional effects of falling raindrops are related in part to the depth of flow into which they fall. Palmer (1965) investigated the soil loss by waterdrop impact forces for three drop sizes and with various depths of a water layer over the soil surface. Water-layer depths were varied from 0 to 30 mm. Maximum soil losses for drop sizes of 2·9, 4·7 and 5·9 mm occurred at a critical depth of the water layer of 2, 4 and 6 mm, respectively. Thus the critical depth occurs in a region where a 1:1 relation exists between the drop diameter and the depth of the water layer.

For tests with artificial rainfall, depths and velocities were measured to determine the downslope changes, and the total discharge measured was distributed over the runoff surface to determine the average intensity of rainfall. The maximum Reynolds' numbers which occurred with flows from artificial rainfall were less than 1,500. The data from uniform flow tests indicate this is entirely within the region of laminar flow. The effect of falling rain sufficiently disturbed the flow of water that injections of dye were rapidly dispersed. Although this flow has some characteristics of turbulent flow, it exhibits most of the properties of laminar flow. This type of flow does not belong to any of the classifications of laminar, transitional, or turbulent flow. In this chapter, runoff from artificial rainfall will be called disturbed flow.

As discharge increases downslope, values of Reynolds' number increase. Thus, increasing Reynolds' numbers indicate both increasing downslope distance and discharge. The downslope increase in depth of non-uniform flows resulting from a uniform increase in discharge is illustrated as a function of Reynolds' number in the lower half of Figure 5.4. Data illustrated in Figure 5.4 are for the flume conditions of test series 3 and 8 for uniform flow (*see* Figure 5.1). All five rainfall intensities are represented by the plotting symbols; the higher-numbered tests are the lower intensity rainfalls. The effect of plotting values of depth against Reynolds' number rather than downslope distance is to eliminate the influence of increasing rainfall intensities. The single line drawn through the data of Figure 5.4 represents the average downslope increases in depth due to increasing discharge.

The most important effect of the falling raindrops is to retard the flow and increase the depth for a given discharge. Consideration of the momentum exchange between the mass of falling water and the mass of water as surface flow would predict this increase in depth. The momentum of the falling rain has little downslope component compared to the surface runoff. One would expect that the increase in depth would be least for the lowest intensities of rainfall and the highest rates of surface runoff. The present data do no entirely confirm this hypothesis. The effects of rainfall intensities are masked because lowest intensities are accompanied by lowest runoff rates, highest intensities by the highest runoff rates, and the overall effect is a balancing of the momentum exchange so that the percentage effect is

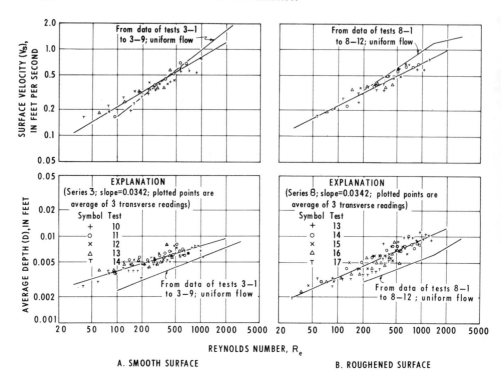

Figure 5.4   Depth and surface velocity as a function of Reynolds' number for sprinkled tests

roughly the same for all intensities of rainfall. For a constant intensity of rainfall, greater depths downslope should be less affected by raindrop impact than shallower upslope depths. The depth profile for the smooth surface in Figure 5.4 illustrates this by a convergence of the plotted data to the line representing depths from the uniform flow test. That is, the f value for the runoff from rainfall is less than the f value for uniform flow. The data for test series 3 (smooth surface) plotted in Figure 5.4 indicate that the percentage increase in depth over uniform flow depth is about 60% at a Reynolds' number of 100 and decreases to about 35% at a Reynolds' number of 1,000. This increase in depth due to rainfall impact is considerably greater than the average of 17% reported by Parsons (1949).

The data for test series 8 plotted in Figure 5.6 do not show a reduction in the increase of depth with an increase in Reynolds' number. However, if a correction in depth to allow for the voids between grains in the roughness was added to the measured depths, the data would more closely conform to that from the smooth surface. For the uncorrected data from the rough surface plotted in Figure 5.4, the increase in depth is about 50% at a Reynolds' number of 200 and about 65% at a Reynolds' number of 1,000.

Measurements of the downslope changes in surface velocity are shown in the upper parts of Figure 5.4. For all laboratory tests, there appears to be some reduction in surface velocity in the upstream reaches of the flume. The apparent upslope reduction in velocity is attributed to the greater retarding effects of high rainfall intensities in the region where surface velocities and momentum of flow are initially small.

Summary curves of the Darcy–Weisbach friction factor are plotted in Figure 5.5 as a function of Reynolds' number. The curves average a value of resistance to flow about four times greater than the theoretical value of $f_f = 96/R_e$ for uniform flow on a smooth surface. However, with the present set of data, friction factors for uniform flow tests (*see* Figure 5.2) were also greater than the theoretical value ($f_f = 96/R_e$) by a factor of about 2. Therefore, the isolated effect of the artificial rainfall used in the investigation was to about double the friction factor over that for flows without rainfall.

The recent work by Woo and Brater (1962), Yoon and Wenzel (1971), Li (1972), and Shen and Li (1973) more clearly define the effects on the friction factor by various intensities of rainfall. They show that the increase in values of the friction factor by falling rain is dependent on rainfall intensity. The higher intensity rainfalls have the greater effect, but the average increase in friction factor is of the

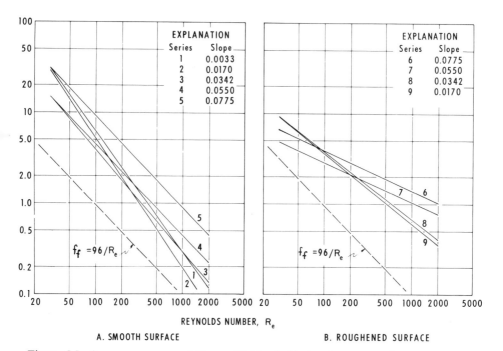

Figure 5.5   Summary curves of Darcy–Weisbach friction factor as a function of Reynolds' number for sprinkled tests

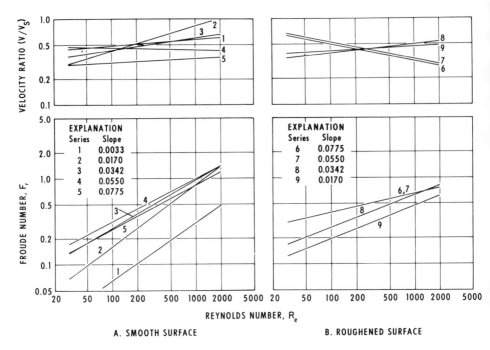

Figure 5.6 Summary curves of Froude number and velocity ratio as a function of
Reynolds' number for sprinkled tests

order presented in Figure 5.5. Most investigators also report an increase in value of friction factor with an increase in flume slope. This is the trend shown in Figure 5.5, but the effect is often insignificant compared to experimental accuracy.

The summary curves in Figure 5.6 show the relation of Froude's number and the ratio of mean velocity to surface velocity as functions of Reynolds' number. Froude's number is defined as the ratio of inertial forces to gravitational forces; flows with Froude's number greater than 1 are supercritical, flows with Froude's number less than 1 are subcritical. Within the range of conditions investigated, most of the flows observed in the smooth channel and all of those observed in the roughened channel were in subcritical flow. Over longer slopes, however, supercritical flows could be expected. Theoretically, the slope of the line relating Froude's number to Reynolds' number is 0·5 for laminar flow. The present data for the smooth-surface tests indicate a value slightly higher than 0·5 and the roughened-surface tests have a value somewhat less than 0·5. This difference is related to the difference in f values. (Lest the reader jump to early conclusions, it will be shown in the next section that Froude's number for natural slopes is nearly constant with increasing discharge and generally stays below a value of 0·2.)

The upper parts of Figure 5.6 show values of the ratio of mean velocity to surface velocity. Nearly all values of the ratio are smaller than 0·67, the theoretical value for laminar uniform flow. This indicates that mean velocity is retarded more than surface velocity and this influence is greatest at the lower Reynolds' numbers. For seven of the nine test series, the slope of the line relating the velocity ratio to Reynolds' number is positive.

5.6 FIELD INVESTIGATION OF OVERLAND FLOW

Seven field sites in west-central Wyoming were selected to determine the transfer value of the laboratory data to natural conditions. Runoff plots at the field sites were 2·1 m (7 ft) wide, about 14 m (45 ft) long, and approximately represented four groundslope angles. These slope gradients were about 0·003, 0·10, 0·20 and 0·33. To adequately describe the field sites, detailed topographic maps were prepared, surficial soil samples were analysed for particle-size distribution, and estimates were made of overall vegetation density and overstory cover. Table 5.1 lists the name and some of the characteristics of each of the sites. Figure 5.7 shows the slope profiles of the sites.

Table 5.1 Characteristics of overland flow field sites

Site name	Average elevation (m)	Slope aspect	Ground slope		Estimated vegetation cover (%)	Median soil-particle diameter (mm)
			(m/m)	(degrees)		
New Fork River, Site 2	2,180	S60°E	0·0290	1°40''	8	0·09
Pole Creek, Site 1	2,200	N75°W	0·0960	5°31''	20	0·38
New Fork River, Site 1	2,180	S50°E	0·1000	5°44''	10	0·15
Boulder Lake Site, 1	2,230	N30°W	0·1880	10°53''	28	0·16
Pole Creek, Site 3	2,230	N40°W	0·2080	12°00''	35	0·10
Boulder Lake Site, 2	2,230	N30°W	0·3315	19°21''	22	0·32
Pole Creek, Site 2	2,210	N05°E	0·3320	19°23''	28	0·48

Each field site was sprinkled with artificial rain at an intensity of approximately 200 mm/hr (8 in/hr) and, generally, runoff of about 100 mm/hr (4 in/hr) was measured at the lower end of the plot. The infiltration capacity of all sites was thus about 100 mm/hr. This value of infiltration may appear high, but, as examples, investigations by Smith and Leopold (1942) and Hadley and McQueen (1961) report values of infiltration comparable to those found in the overland flow study. Beginning with the initial runoff, flow rate was recorded to determine the rising hydrograph and the infiltration characteristics. Periodic sampling of runoff water

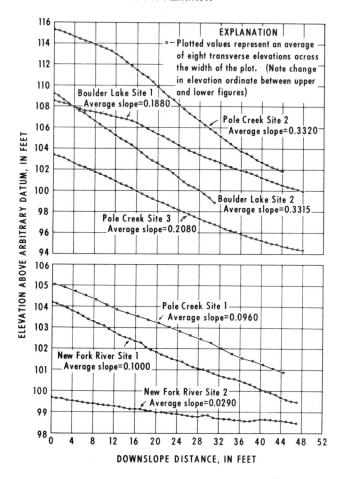

Figure 5.7   Hillslope profiles of field sites

provided data on the sediment concentration of the flow. After the infiltration rate became constant, values of depth and surface velocity, and their downslope changes, were measured. The flow rarely occurred as a uniform sheet of water and the majority of water travelled downslope in several lateral concentrations of flow; however, these concentrations were not considered rill flow. The flow concentrations were mapped by dye tracings to show the general pattern of flow. Each site exhibited a unique flow pattern dependent mostly on the physical characteristics of the slope. Runoff from some of the sites was also characterized by surface detention in a series of puddles formed by barrier dams of organic debris. Surface runoff occurs, in part, by a succession of failures of these barriers.

Figures 5.8–5.11 are illustrative of the microtopography of the field sites and the resulting flow patterns. It is emphasized that not all of the runoff occurs in the concentrations of flow shown in Figures 5.8–5.11. The general appearance of runoff at most sites was one of omnipresent surface detention, easily detected by the glistening of the sheet of water in sunlight. Dye tracings showed that this sheet of water moved slowly downslope and often moved laterally to join the concentrated areas of flow. Few areas approached stagnation because continuing rainfall forced runoff. Therefore, the flow patterns shown represent only the concentrations of flow and not all of the flow.

Pole Creek Site 1 is shown in Figure 5.8. This site was relatively free of topographic irregularities and surface runoff was essentially downslope. Note in

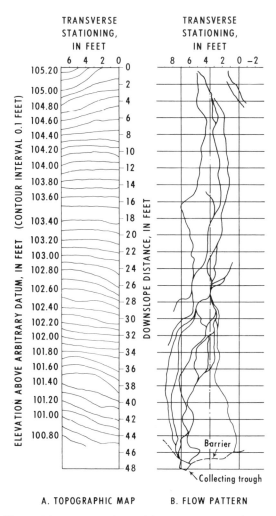

A. TOPOGRAPHIC MAP   B. FLOW PATTERN

Figure 5.8   Topography and flow pattern at Pole Creek
Site 1

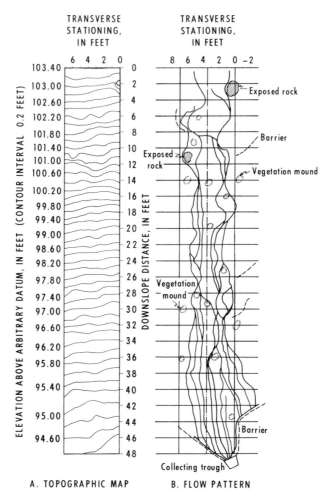

Figure 5.9   Topography and flow pattern at Pole Creek
Site 3

Figure 5.8 the trend of the flow pattern in the lower half of the plot in response to the curvature of the contour lines.

   The slope of Pole Creek Site 2 was great enough to override the influence of minor topographic irregularities. The general pattern of flow at this site was directly downslope with little anastomosing of the flow concentrations. A similar flow pattern, as shown in Figure 5.9, was traced for Pole Creek Site 3. However, the less steep slope of Pole Creek Site 3 begins to show the influence of small topographic features and the curve of flow lines around topographic highs. In the lower quarter of the runoff plot at Pole Creek Site 3, depths of flow are sufficiently great and are evenly-enough distributed so that flow was nearly uniform across the plot. Note also the shift in direction of the flow pattern in this area in response to a curvature in slope direction.

The ground slope at New Fork River Site 1 is flat enough that small topographic features are obvious in the topographic map in Figure 5.10 and are visible in the flow pattern responding to the topography. The gradient at the lower end of the plot was such that water ponded in the lower 0·6 m (2 ft) of slope. In this ponded area, sediment was deposited as a delta and indicates the effectiveness of overland flow to erode and transport sediments. The eroded sediments apparently were derived as sheetwash since no rilling was observed. The ponding had no effects on

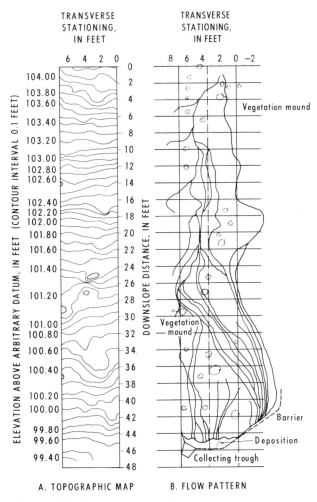

Figure 5.10   Topography and flow pattern at New Fork
River Site 1

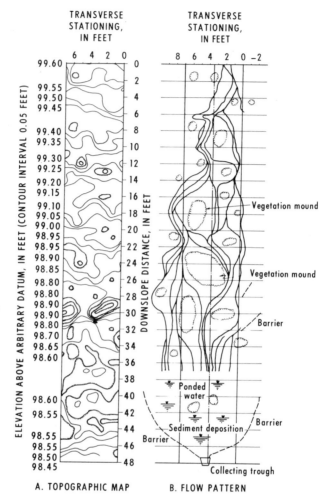

Figure 5.11   Topography and flow pattern at New Fork
River Site 2

the upslope hydraulics of flow and affected only the analysis of sediment concentrations in the runoff water.

The flattest slope of the sites investigated was for New Fork River Site 2. Irregular surface features are unmistakable in the topographic map and the flow pattern in Figure 5.11. The flow is definitely directed around the topographic highs and follows the micro-valleys indicated by the contour lines. Water in the lower eight feet of this site was also ponded and there was some deposition of sediment in this area.

The topography and resulting flow patterns for the two Boulder Lake sites are not unlike those just described for the other two sites.

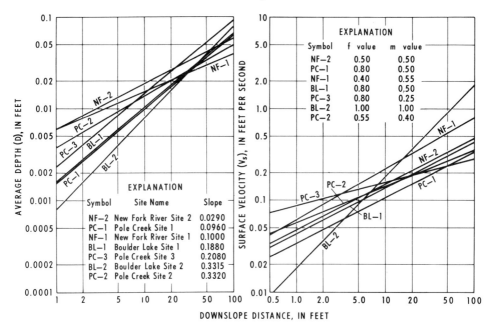

Figure 5.12 Downslope profiles of depth and surface velocity at field sites

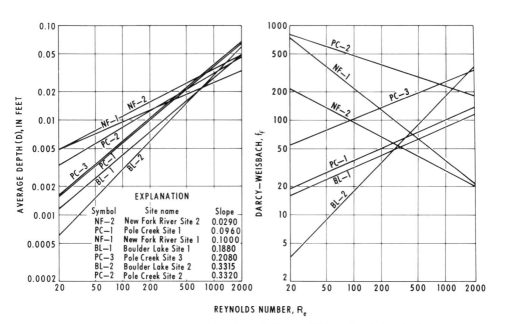

Figure 5.13 Summary curves of depth and Darcy–Weisbach friction factor as a function of Reynolds' number, field sites

Summary curves of downslope profiles of depth and surface velocity are shown in Figure 5.17 for all field sites. Values of f for the exponent of depth range from 0·40 and 1·00 and m exponents for velocity range from 0·25 to 1·00. These values of m and f will be discussed in more detail later. Summary curves of depth as a function of Reynolds' number are shown in the left graph in Figure 5.13.

Summary curves of the Darcy–Weisbach friction factor–Reynolds' number relation are shown in the right half of Figure 5.13 for all field sites. The straight-line relations are computed lines of best fit taken from the opposite graph in the figure. The slope of the straight-line relation, as determined from Equation 5.21, is equal to $f - 2m$ or $f - 2(1 - f)$. For a value of f equal to 0·67, the slope of the friction factor relation is 0; smaller values of f yield negative slopes, and greater values of f give positive slopes. The wide range in values of slope for the friction factor relation is due to the range in values of f for the depth relations.

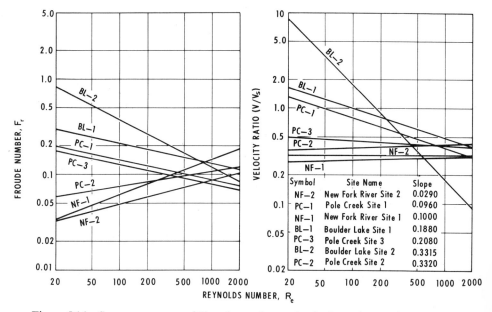

Figure 5.14   Summary curves of Froude number and velocity ratio as a function of Reynolds' number, field sites

Summary curves for Froude's number as a function of Reynolds' number are shown in the left graph in Figure 5.14. Positive or negative slopes for the straight line relation depend on whether the f value is greater or less than 0·67.

Summary curves of the ratio of mean velocity to surface velocity are shown as a function of Reynolds' number in the right graph in Figure 5.14. The near horizontal-to-negative range in slopes of the lines indicates a relatively more rapid downslope increase in surface velocity than mean velocity.

5.7 COMPARISON OF FIELD RESULTS WITH LABORATORY DATA

Probably the most important data presented in this chapter are those from the field sites. Few studies have replicated the detailed field measurements of depth and velocity occurring in overland flow. And, the most important analysis of those data is a comparison of it with the laboratory and theoretical data. If as appears likely, the more rapid advances in hillslope hydrology will be in mathematical modelling, it is imperative that these model be predictive of the scant amount of empirical data that exist. Thus, the comparison of field and laboratory data in this section is of significance, especially as it relates to the geomorphic implications of overland flow as a landscaping agent.

The most apparent difference in field and laboratory data is the greater depths occurring in the field runoff. This is not an unexpected observation. The laboratory data indicated that the effect of roughness is to retard the flow and increase the depth; the sand-grain roughness in the laboratory increased the depths up to 30 % over those on a smooth surface. Surface roughness of the field sites is difficult to estimate. The roughness of field sites consists of both particle roughness (sand-grain roughness in bare areas and plant sprouts in vegetated area) and form roughness (topographic irregularities). Mean grain sizes of the soil particles at the field sites are smaller than those in the laboratory roughness so that the increase in depths is due primarily to vegetation and topographic characteristics of the field sites. Depths at the upslope end of the field plots were comparable to depths at the upstream end of the flume in the laboratory. Thus the downslope rate of increase in depths is greater for the field sites. Figure 5.12 shows f values for field sites range from 0·40 to 1·00 (average 0·69) and compared to an average f value of 0·48 for the rough-surface tests in the laboratory.

Higher f values are related both to the magnitude of relative roughness (degree of overall retardance) and the character of the runoff (for example, ponding as at Pole Creek Site 1). The f values for the field sites show no correlation with ground slope, or, at least, the data indicate that other characteristics of the field plots override the influence of slope. Since no correlation could be established between percentage vegetation cover (or type of vegetation) and f value, the increase in f values at field sites is attributed to topographic form and, dependent on form, the character of runoff. No two field sites are identical in form and no two depth profiles are the same.

Values of f are related to downslope changes in resistance to flow. Resistance to flow, expressed in this chapter as the Darcy–Weisbach friction factor, f_f and Manning's n, describe bulk resistance to flow rather than resistance attributable to grain roughness alone. With increasing downslope discharge, one would expect resistance to flow to decrease downslope as relative roughness decreased (as in all laboratory cases). Overriding influences, such as ponding, are analogous to tremendous retarding forces. Thus, depending on topographic form, relatively smoother surfaces may show higher resistance to flow and different rates in downslope changes in roughness.

The average of the downslope changes in roughness at the field sites is approximately zero; this indicates a downslope increase in relative roughness. Absolute roughness is probably not increasing, so the apparent increase in roughness is due to a decrease in runoff efficiency and is related to the microtopography of the site. As a first approximation to overland flow on natural ground surfaces, a *f* value of 0·67 and no downslope change in roughness may be used to estimate the hydraulic parameters. Absolute values of depth and the resistance term may be approximated from Figure 5.13. Values of relative roughness corresponding to values of the Darcy–Weisbach friction factor are beyond those shown on conventional Moody diagrams, but they would appear in some instances to have a value greater than 1. This is not unreasonable comparing the shallow depths of flow to the magnitude of vegetation and topographic barriers.

It is difficult to compare values of the friction factor at field sites to values from the laboratory tests. However, in general, the field data indicate a ten-fold increase in resistance on the natural field plots compared to the laboratory surfaces.

Values of *m* for surface velocity are generally less for the field sites than for the laboratory flume. As depth enlarges its rôle in absorbing downslope increases in discharge, mean velocity must absorb less of the change. Thus, *m* values for surface velocity in the field must be lower than laboratory values. An average value of the ratio of mean velocity to surface velocity for field data is about 0·4 to 0·5. One reason for this low value of the velocity ratio is that a maximum velocity was measured in the field. The leading edge of the dye trace was measured and as the dye merged into concentrated area of flow, the velocity timed was greater than an average surface velocity over the width of the plot.

The downslope change in Froude's number for field data varied, but on the average, Froude's number was nearly constant down the slope and was considerably lower than values from laboratory data. This behaviour is related to differences in depths of field runoff compared to laboratory tests. An average value of Froude's number is 0·1, which is well within the regime of subcritical flow.

5.8 EROSION AND SEDIMENT TRANSPORT

One of the most ubiquitous processes occurring on hillslopes is the erosion, transportation, and deposition of debris by running water. The formation of rills is one consequence of the flow of water. However, some slopes may show no rills and may be undergoing uniform degradation by sheet erosion. Both rills and sheet erosion are the products of overland flow. However widespread overland flow may be, it is one of the most elusive processes to observe and measure. This fact has made difficult the collection of quantitative data to help resolve the questions of why and how rills develop. In fact, little is known of the general mechanics of slope erosion by overland flow.

The author, together with Leopold and Myrick (1966) has measured hillslope erosion for nearly 10 years in a semi-arid area of New Mexico. Despite efforts to observe overland flow from thunderstorms occurring during the several weeks of residence at the project area each year, overland flow was never observed in the field. Yet, during the period of measurement, surface erosion on unrilled slopes yielded 13,600 tons per square mile per year or 98 % of the sediment production from all sources. Obviously, surface erosion on these unrilled slopes must be the work of unconcentrated overland flow, but without the detailed measurements of hillslope erosion, the full importance of overland flow was not apparent in the field. The question therefore remains: why do slopes degrade by sheet erosion rather than develop rills?

The presence of rills was not observed at any of the field sites reported in this chapter and rilling is not common in the general area. Flow concentrations occurring at the sites were dictated by microtopographic features, but the paths followed by concentrations of flow were not in discernible rills. Nor, during the course of sprinkling at each site (generally about 6 hr or longer), were rills observed to be formed by flow concentrations. Rilling is generally considered to be evidence of more accelerated erosion than sheet erosion. Sediment concentrations at New Fork River Site 1 were the highest observed at that time in the investigation and rilling was considered most likely to occur at this site. However, after nearly 10 hr of sprinkling at an intensity of 8·5 in/hr, no observable rills had been formed. The sprinkling intensity was raised to 10·5 in/hr and continued for over 6 hr. Still no rills were formed by the increased runoff.

The appearance of rills on a soil surface during overland flow as influenced by slope steepness, runoff rate, and presence or absence of rainfall was reported by Meyer and Monke (1965) for a laboratory investigation using glass spheres as a non-cohesive bed material. They reported that erosion occurs predominantly by rilling and the intensity of erosion increases with increasing slope steepness and runoff rates. Rainfall tended to level the bed surface, thereby smoothing its rill-roughened surface. For a 10 % slope, the same as the slope steepness at New Fork River Site 1, Meyer and Monke reported that erosion was rapid and rilling was pronounced. Rills were long narrow chutes and were directed predominantly downslope. As erosion rates increased with increased runoff rates, they reported that erosion tended to be uniform since potential rills were filled by the great rates of soil movement before they could fully develop.

Rills may or may not develop on unrilled surfaces if some threshold is exceeded which causes a change in the degradation rate. Such a threshold may be exceeded, for example, because of climatic change, but equilibrium could be maintained by equally altering the erosion rate throughout the drainage system. It is interesting to note that even at the high intensity of rainfall applied to the test plots, this theshold was not exceeded. It follows that the potential for increased erosion was absorbed by increased depths of flow rather than by higher velocities and accelerated erosion by rilling. To maintain this equilibrium, each slope had developed both a form and a resistance to flow, manifested in a complex interaction of vegetation and microtopographic form, to which the depth and velocity components of overland flow must adjust.

Schumm (1962) discussed the development of resistance to flow to maintain equilibrium between pediments and hillslopes in an analysis of miniature pediments developed on badland topography in South Dakota. Using the Manning formula to estimate velocity, Schumm applied a value of Manning's n to the rougher but more steep hillslopes which was three times greater than the value of n for the smoother pediment surfaces. The actual values of n used by Schumm were probably low, but the relative order of magnitude appears reasonable. The pediment slopes were about eight times less steep than the hillslopes. Assuming that the depth of the sheet of water moving over the hillslope was the same as that moving over the pediment, the computed values of velocity was the same for the hillslopes as for the pediments. Thus, in the case of the pediment, the decrease in roughness apparently compensates for the decrease in slope angle. Such mutual adjustment of the component variables is the key to equilibrium.

This example by Schumm and the data from Figure 5.14 showing values of Froude's number and its downslope change should serve to discredit the existence of a hydraulic jump as overland flow on natural hillslopes passes from a steep slope to a moderate or flat slope. A hydraulic jump occurs as flow goes from supercritical to subcritical, but Schumm's example shows no decrease in velocity at the base of the steep slope because changes in resistance to flow compensate for the smaller value of slope. And the data of Figure 5.14 show that the value of the Froude number for natural slopes, even steep slopes, is on the order of 0·1 and is essentially constant over the length of the slope. Although it is possible to have supercritical flow in river channels, especially with flash flooding in ephemeral channels, it is unlikely that overland flow is ever supercritical and thus never offers the opportunity for a hydraulic jump.

The sediment concentrations observed in runoff samples of overland flow illustrate the ability of overland flow to erode and transport sediments. The analytical results of the several sediment samples from each site are summarized as averages for each site in Table 5.2. The average values of sediment and organic content from Table 5.2 show a positive correlation with ground slope, and a negative correlation with vegetation cover.

One important observation is the higher concentrations of sediment in the initial runoff and the relatively rapid decrease in concentrations during the remainder of the runoff. This observation was illustrated by data of New Fork River Site 1. From a sediment concentration of 228 mg/litre after 24 min of runoff, the concentration decreases to 184 mg/litre at 35 min, 41 mg/litre at 82 min, and 36 mg/litre at 119 min. Similar results were found by Lowdermilk and Sundling (1950). Their studies indicate that the erosion rate decreases throughout a simulated rainstorm as the finest particles were removed in surficial flow. Their removal led to the domination of the soil surface by larger particles until ultimately an erosion pavement was formed. Similar results were also found by Swanson, Dedrick and Weakly (1965). However, data collected by the author do not strictly support the

Table 5.2 Average values of sediment sample analyses compared to ground slope and vegetation cover at overland flow field sites

Site name	Ground slope (m/m)	Estimated vegetation cover (%)	Total sediment and organic content (mg/litre)	Sediment content of sample (mg/litre)	Organic content of sample (mg/litre)
New Fork River, Site 2	0·0290	8	22·1	19·3	2·8
Pole Creek, Site 1	0·0960	20	67·6	21·3	46·2
New Fork River, Site 1	0·1000	10	126·3	115·8	10·5
Boulder Lake, Site 1	0·1880	28	8·8	5·8	3·0
Pole Creek, Site 3	0·2080	35	26·0	4·0	22·0
Boulder Lake, Site 2	0·3315	22	87·3	75·3	11·7
Pole Creek, Site 2	0·3320	28	44·7	38·3	6·4

pavement theory. Comparison of sediment concentrations observed early in the runoff from a second day of sprinkling to the concentrations at the end of the previous day's sprinkling shows the second day's initial concentration to be considerably higher than the preceding day's final concentration. Since a number of new fine-grained particles could not be produced in the short interval between sprinklings, higher initial sediment concentrations appear to be related to some process making soil particles ready for transport. Over a single night, as in the present investigation, the responsible process is most likely a wetting–drying effect on the soil. Between natural storms, processes making soil ready for transport would include wetting–drying, wind, frost action, churning by animals, and even weathering where intervals are long.

Splash erosion by raindrop impact before a protective layer of surface detention is build up is also important in the initial high-sediment concentrations (Borst and Woodburn, 1942). However, the present data do not fully confirm the conclusion of Borst and Woodburn that raindrop splash, not runoff, is responsible for soil loss. A number of other investigators have shown the importance of raindrop impact on erosion (for a summary, *see* Smith and Wischmeier, 1962). However, raindrop impact with little runoff is not likely to be an effective agent of erosion. As surface detention builds up everywhere and depths increase downslope, the effect of waterdrop impact lessens. The data of Palmer (1965) indicate that for the size of waterdrops and depths of flow in the present investigation, there was probably little splash erosion due to raindrop impact.

One of the most detailed studies of overland flow and erosion is being conducted as part of the Public Lands Hydrology Program of the US Geological Survey. These studies are being conducted at Badger Wash in Western Colorado, an area underlain by Mancos Shale. The general area is one for which the US Geological Survey has previously published considerable hydrologic, geologic, and biologic information. Overland flow was generated by simulated rainfall over land surfaces which were representative of small, but complete basins with watershed divides. Erosion was measured both by detailed transects of ground surface elevations and by the sediment content of runoff water. Data from these studies are just beginning to be analysed, but preliminary results (R. F. Hadley, in a written communication) are worthy of mention here. On each of two basins (Hadley Basin and Lusby Basin), rainfall was simulated on two consecutive days. On the first day, termed a 'dry run', rainfall was applied to the slopes without any pre-wetting and the antecedent soil moisture was low. A similar rainfall was applied on the day following a dry run and is termed a 'wet run'. The hillslope profile of both basins is convex, straight, concave, and the average ground slope is about 0·2.

On Hadley Basin, a dry run was made with a rainfall intensity of 31 mm/hr until 22 mm of water was applied. Measured runoff was 4·4 mm, or 20 % of the rainfall applied. During the wet run, 27 mm of water was applied and runoff was 11 mm, or 41 % of the rainfall applied. Over the 400 m^2 (0·10 acre) basin, sediment transported past the exit from the basin was 51·5 kg (113·2 lb; 1275 kg/ha) during the dry run, and 157 kg (346 lb; 3890 kg/ha) during the wet run. At Lusby Basin, 28 mm of rainfall was applied at an intensity of 37 mm/hr, and 7·8 mm of runoff (28 %) was recorded for the dry run and 13·3 mm of runoff (48 %) was recorded for the wet run. The sediment yield for the 485 m^2 (0·12 acre) basin was 89 kg (196 lb; 1840 kg/ha) for the dry run and 184 kg (405 lb; 3800 kg/ha) for the wet run. The data are conclusive that for a given rainfall rate, the runoff rate is dependent on infiltration (antecedent soil moisture) and that the greatest erosion rates are related to highest runoff rates. Thus, despite any eroding effects of raindrop impact, there must be sufficient overland flow for transport eroded sediment downslope. An analysis of variance for net erosion on convex, straight, and concave segments of the Badger Wash hillslopes shows significant differences at the 5 % level for erosion between different hillslope segments. Straight segments of slope have the greatest erosion and there was no significant differences between convex and concave segments.

The values of sediment concentrations in Table 5.2 and the data from Badger Wash are adequate proof that overland flow can be effective as an eroding and transporting agent. It is interesting to note that the values of Reynolds numbers' from all field tests were well within the regime of laminar flow as defined in Figure 5.1. As previously mentioned, overland flow is disturbed by rainfall, the actual characteristics of flow being somewhere between laminar and turbulent (primarily laminar in upslope reaches but more turbulent as slope length increases). Regardless of the exact characteristics of overland flow, sediment was being eroded and transported.

Sediment transport occurring at the low values of Reynolds' number are in agreement with Bagnold's (1955) observation that turbulence is not an essential requisite of sediment transport. The present data tend to invalidate King's (1953) canon 27 of landscape evolution that laminar flow is non-erosive.

Prediction of soil erosion requires some knowledge of the characteristic of overland flow. A popular soil erosion formula in the United States is the 'universal soil-loss equation' proposed by Wischmeier and Smith (1960). However, this equation needs as input data some measure of the kinetic energy (the product of mass and square of the velocity) of overland flow. Without measurements of the velocity of overland flow, it would be difficult to accurately use the universal soil-loss equation. The importance of various slope segments (convex, straight or concave) cannot be overlooked. It has been proposed (Wischmeier and Smith, 1965) that the slope gradient at the lower end of each slope segment be used in the universal soil-loss equation, but this would contradict the results found at Badger Wash. Soil erosion by overland flow is discussed in detail in a later chapter. It is mentioned here only to indicate that the results of one study can often be contradicted by the results of another study. And, much of the controversy will not be resolved until we gain more knowledge of the hydraulic and geomorphic implications of overland flow.

REFERENCES

Bagnold, R. A., 1955, 'Some flume experiments on large grains but a little denser than the transporting fluid, and their implications', London, England, *Proc. Inst. Civil Engs.*, Paper **6041**, 174–205.

Borst, H. L. and Woodburn, R., 1942, 'The effect of mulching and methods of cultivation on runoff and erosion from Muskingum silt loam: St. Joseph, Michigan', *Agric. Engr.*, **23**, 19–22.

Chow, V. T., 1959, *Open Channel Hydraulics*, McGraw-Hill, New York, 680 pp.

Emmett, W. W., 1970, 'The hydraulics of overland flow on hillslopes', *US Geolog. Surv. Prof. Paper* **662A**, 68 pp.

Gunn, R. and Kinser, G. D., 1949, 'Terminal velocity of water droplets in stagnant air', *J. Meteorology*, **6**, 243–248.

Hadley, R. F. and McQueen, I. S., 1961, 'Hydrologic effects of water spreading in Box Creek basin, Wyoming', *US Geolog. Surv. Water-Supply Paper* **1532-A**, 48 pp.

Horton, R. E., 1933, 'The rôle of infiltration in the hydrologic cycle', *Trans. Am. Geophys. Union*, **14**, 446–460.

Horton, R. E., 1936, 'Hydrologic interrelations of water and soils', *Proc. Soil Sci. Soc. Am.*, **1**, 401–437.

Horton, R. E., 1938, 'The interpretation and application of runoff plot experiments with reference to soil erosion problems', *Proc. Soil Sci. Soc. Am.*, **3**, 340–349.

Horton, R. E., 1945, 'Erosional development of streams and their drainage basins; hydrophysical approach to quantitative morphology', *Bull. Geol. Soc. Am.*, **56**, 275–370.

Horton, R. E., Leach, H. R. and van Vliet, R., 1934, 'Laminar sheet flow', **15**, *Trans. Am. Geophys. Union*, Part 2, 393–404.

Izzard, C. F., 1944, 'The surface profile of overland flow', *Trans. Am. Geophys. Union*, **25**, 959–968.

Izzard, C. F., 1946, 'Hydraulics of runoff from developed surfaces', Washington, D.C., Highway Research Board, Nat. Res. Council, Natl. Acad. Sci., *Proc. 26th Annual Meeting*, 17 pp. (reprint).

Keulegan, G. H., 1944, 'Spatially variable discharge over a sloping plane', *Trans. Am. Geophys. Union*, **25**, 956–958.

King, L. C., 1953, 'Canons of landscape evolution', *Bull. Geolog. Soc. Am.*, **64**, 721–752.

Laws, J. O., 1941, 'Measurements of fall velocity of water droplets and raindrops', *Trans. Am. Geophys. Union*, **22**, 709–721.

Laws, J. O. and Parsons, D. A., 1943, 'Relation of raindrop size to intensity', *Trans. Am. Geophys. Union*, **24**, 452–460.

Leopold, L. B., Emmett, W. W. and Myrick, R. M., 1966, 'Channel and hillslope processes in a semiarid area, New Mexico', *US Geolog. Surv. Prof. Paper* **352G**, 193–253.

Leopold, L. B. and Maddock, Thomas, Jr., 1953, 'The hydraulic geometry of stream channels and some physiographic implications', *US Geolog. Surv. Prof. Paper* **252**, 57 pp.

Li, R. -M., 1972, *Sheet Flow Under Simulated Rainfall*, M.S. thesis, Fort Collins, Colorado, Colorado State Univ., 88 pp.

Lowdermilk, W. C. and Sundling, H. L., 1950, 'Erosion pavement formation and significance', *Trans. Am. Geophys. Union*, **31**, 96–100.

Meyer, L. D. and Monke, E. J., 1965, 'Mechanics of soil erosion by rainfall and overland flow', *Trans. Am. Soc. Agr. Engrs.*, **8**, 572–577 and 580.

Owens, W. M., 1954, 'Laminar to turbulent flow in a wide open channel', *Trans. Am. Soc. Civil Engrs.*, **119**, 1157–1175.

Palmer, R. S., 1965, 'Waterdrop impact forces', *Trans. Am. Soc. Agric. Engrs.*, **8**,(1) 70–72.

Parsons, D. A., 1949, 'Depths of overland flow', *Soil Conserv. Ser. Tech Paper* **82**, 33 pp.

Ree, W. O., 1963, 'A progress report on overland flow studies', *Soil Conserv. Ser. Hydraulic Engrs. Meeting, August 12–16, 1963, New York*, 18 pp. (mimeo).

Schumm, S. A., 1962, 'Erosion on miniature pediments in Badlands National Monument, South Dakota', *Bull. Geolog. Soc. Am.*, **73**, 719–724.

Smith, D. D. and Wischmeier, W. H., 1962, 'Rainfall erosion', *Adv. Agronomy*, **14**, 109–148.

Smith, H. L. and Leopold, L. B., 1942, 'Infiltration studies in the Pecos River watershed, New Mexico and Texas', *Soil Sci.*, **53**, 195–204.

Shen, H. W. and Li, R. -M., 1973, 'Rainfall effect on sheet flow over smooth surface', *Proc. Am. Soc. Civil Engrs.*, **99**, (HY5), 771–792.

Straub, L. G., 1939, 'Studies of the transition region between laminar and turbulent flow in open channels', *Trans. Am. Geophys. Union*, **20**, 649–653.

Swanson, N. P., Dedrick, A. R. and Weakly, H. E., 1965, 'Soil particles and aggregates transported in runoff from simulated rainfall', *Trans. Am. Soc. Agric. Engrs.* **8**, 437–440.

Tracy, H. J. and Lester, C. M., 1961, 'Resistance coefficients and velocity distribution—smooth rectangular channel', *US Geol. Survey Water-Supply Paper* **1592A**, 18 pp.

Wischmeier, W. H. and Smith, D. D., 1960, 'A universal soil-loss equation to guide conservation farm planning', *Trans. Seventh Internat. Congress Soil Sci.*, **1**, 418–425.

Wischmeier, W. H. and Smith, D. D., 1965, 'Predicting rainfall-erosion losses from cropland east of the Rocky Mountains', *US Dept. Agric. Handbook*, No. 282, 47 pp.

Woo, D.-C. and Brater, E. F., 1962, 'Spatially varied flow from controlled rainfall', *Proc. Am. Soc. Civil Engrs.*, **88**(HY6), 31–56.

Yoon, Y. N. and Wenzel, H. G., Jr., 1971, 'Mechanics of sheet flow under simulated rainfall', *Proc. Am. Soc. Civil Engrs.*, **97**(HY9), 1367–1386.

CHAPTER 6

Mathematical models of hillslope hydrology

R. A. Freeze

Department of Geological Sciences, University of British Colombia, Vancouver, B.C., Canada

> *Nature's indifferent toward the difficulties*
> *it causes a mathematician.*
>
> FOURIER

6.1 INTRODUCTION

In the earlier chapters of this book the authors have provided qualitative descriptions of the basic hillslope hydrologic processes that provide delivery of rainfall from hillslope to stream. They have reviewed the mechanisms of surface and subsurface flow, and they have introduced and defined the various hydraulic, hydrologic and hydrogeologic parameters that control the different components of the flow system. In this chapter we will presuppose an understanding of the groundwork laid in the earlier chapters and will proceed to examine the role that mathematical models can play in improving our understanding of hillslope hydrologic processes.

6.6.1 Mathematical models

The technique of analysis inferred by the term *mathematical model* may not be clear to every reader. It is a four-step process involving (1) an examination of the physical problem, (2) replacement of the physical problem by an equivalent mathematical problem, (3) solution of the mathematical problem with the accepted techniques of mathematics, and (4) interpreting the mathematical results in terms of the physical problem.

The nature of the mathematical model that may be chosen to solve a specific physical problem is far from unique. Very early in the process there are several major decisions that must be reached with regard to the type and sophistication of the mathematical model that will be employed, These decisions can be aided by

G

reference to the classic paper by Amorocho and Hart (1964) and more recently by Clarke (1973). These papers provide an introduction to mathematical modelling concepts, and a classification of techniques, as they are applied in hydrology.

It is worth summarizing Clarke's lucid analysis of the terminology of hydrologic modelling. He defined a *system* as a set of physical processes that convert an input variable or variables into an output variable or variables. A *variable* is understood to be a characteristic of the system that can be measured, and that assumes different numerical values at different times. It is differentiated from a *parameter*, which is a quantity characterizing the system that does not change with time. Hydrologic models are concerned with the relationships between hydrologic variables that describe those aspects of the system's behaviour that interest us. The general form of a hydrologic model can be set down as

$$y_t = f(x_{t-1}, x_{t-2}, \dots; y_{t-1}, y_{t-2}, \dots; a_1, a_2, \dots) + \varepsilon_t$$

The vector variable $\{x_t\}$ ($t = \dots -1, 0, 1, 2 \dots$) is the input, and the vector variable $\{y_t\}$ ($t = \dots -1, 0, 1, 2, \dots$) is the output. The a's are the system parameters. The function defines the nature of the model and the error ε_t expresses the lack-of-fit with reality.

A model is termed *stochastic* or *deterministic* according to whether or not it contains random variables. If any of the variables x_t, y_t or ε_t is thought of as having a distribution in probability, then the model is stochastic; if not it is deterministic. Clarke (1973) further defines the terms *conceptual* and *empirical* according to whether or not the function is suggested by consideration of physical processes acting on the input variables. The terms conceptual and empirical thus refer to the function f, whereas the terms stochastic and deterministic refer to the presence or otherwise of a probabilistic structure for any of the variables.

The majority of hydrologic models fit into the category stochastic–empirical. Examples are the predictive use of rainfall-runoff correlations or the time-series analysis of streamflow sequences. This black-box approach is widely-used in applied hydrology and has great power in satisfying the hydrological needs of engineering design. It does not, however, provide any insight into the internal mechanisms of the hydrologic cycle. For the hillslope hydrology problem and the purposes of this text, we need a fully-illuminated white-box approach selected from the class of conceptual models. As noted by Clarke, a conceptual model need not be deterministic, and the ultimate modelling goal may well be conceptual models with stochastic variables that correctly reflect the data uncertainties that always exist. Nevertheless, to date, almost all conceptual models have been conceptual–deterministic, and those described in this review will be so, too. In the earlier chapters we have already seen at least one such model. The theoretical infiltration model of Philip (1957) discussed in Chapter 2 is based on the physics of flow in unsaturated porous media and invokes Darcy's law, the equation of continuity, and an understanding of the interrelationships between the various flow parameters to simulate the physical process of infiltration.

6.1.2 Boundary-value problems

Conceptual–deterministic models, based as they are on the physics of flow, usually take the form of mathematical boundary-value problems of the type pioneered by the developers of potential-field theory and as applied in physics to such problems as the conduction of heat through solids (Carslaw and Jaeger, 1959). To fully define a transient boundary-value problem for subsurface flow we need to know:

(i)   the size and shape of the region of flow
(ii) the equation of flow within the region
(iii) the boundary conditions around the boundaries of the region, and their spatial and temporal distribution
(iv) the initial conditions and their spatial distribution
(v)   the spatial and temporal distribution of the hydraulic or hydrogeologic parameters that control the flow
(vi) a mathematical method of solution.

If the model is for a steady-state system, then requirement (iv) is removed and the temporal distributions referred to in items (iii) and (v) are no longer needed. The hillslope hydrology problem, however, is inherently time-dependent and steady-state analyses would have little value. The models discussed in this chapter are all transient.

Boundary-value problems may be set up for one of two purposes: either to try to improve our understanding of the mechanisms of flow, through simulation of hypothetical systems, or to analyse the hydrologic system at a specific field location. In this article we will present examples of both types. In field simulations it is necessary to itemize a seventh requirement of the boundary-value problem, namely; some past-measured records of output which can be used to check and calibrate the model results. The subject of model calibration is worthy of a chapter unto itself, but for the purposes of this text a few comments in the section on limitations of the mathematical modelling approach will have to suffice as an introduction to this thorny subject.

Having committed ourselves to a conceptual–deterministic approach utilizing a boundary-value problem as the mathematical model, we are once again faced with several decisions. These will probably take on more meaning for the reader if we now introduce the boundary-value problem that fully describes the processes of hillslope hydrology. As we shall see, this complete mathematical model has not yet been solved but it is instructive to examine this presumed ultimate goal before discussing our actual attainments.

6.1.3 The complete, but unsolved, boundary-value problem for hillslope hydrology

Figure 6.1 shows the physical setting of a hillslope feeding a stream channel. There may be overland flow across the land surface, subsurface stormflow (or interflow) in the near-surface soil layers, and deeper groundwater flow. Our aim is to use the time- and space-dependent rainfall inputs in a mathematical model whose output will be (i) the outflow streamflow hydrograph, *and* (ii) the time- and space-configurations of such internal variables as the water-table height, soil-moisture content, seepage-face height, depth and velocity of streamflow, and location of partial-area contributions. The region of flow for this boundary-value problem would encompass the entire subsurface regime, the stream channel, and the land surface. The complete mathematical model would simulate streamflow on the

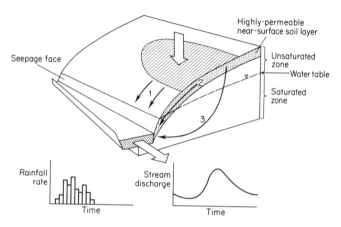

Figure 6.1 The mechanisms of delivery of rainfall to a stream channel from a hillslope in a small tributary watershed. 1, Overland flow; 2, subsurface stormflow; 3, groundwater flow (after Freeze, 1972a)

basis of the integrated lateral inflows from the surface and subsurface flow systems. We know, however, that these two contributing processes also interact with one another, in that the subsurface flow is dependent on the rates of infiltration from any overland flow running across the land surface. A complete model would have to allow treatment of such complex interactions as overland flow from an uphill partial area that subsequently becomes infiltration farther downslope where the unsaturated flow domain permits its entry into the soil. This completely integrated model has not yet been solved.

What can immediately be recognized, of course, is that this complete model is made up of a set of three component models, one for overland flow, one for subsurface flow and one for channel flow. Solutions *are* available for each of these component systems. In fact, recent research has led one step further. Models of each of these components have been presented which have boundary conditions that are compatible with the adjoining component model. In this way, two or more of these models can be *coupled* even though they have not been *fully integrated*. A decision has thus been forced on us by the state of the art. We will have to be satisfied with coupled component boundary-value problems rather than a single fully-integrated analysis. In this context, it is worth noting that there are very few theoretical problems associated with the fully-integrated approach; the limitations are imposed by computer capacities and data availability. These limitations will be discussed in more detail near the end of this chapter.

We have already noted the need for a transient analysis and this is true for all three components. In each case, we require transient equations of flow, a set of initial conditions, and the time-dependent as well as space-dependent properties of the input and output variables and boundary conditions.

We must also decide on the number of dimensions which we will specify to analyse the problem. The limitations of available computer capacity constrain us to using the lowest number of dimensions compatible with our aims. For the surface-flow components, on the slope and in the channel, a one-dimensional analysis suffices. For the subsurface-flow model we need a two-dimensional vertical cross-section taken in a direction parallel to the downslope flow, or roughly perpendicular to the stream.

For the subsurface model, we must now make two more decisions. First, it is clear that we need a model that will include both the *saturated* and *unsaturated* zones and our equation of flow must reflect this fact. Second, we must decide whether the model is to be limited to *homogeneous, isotropic* hydrogeological materials or whether we will allow more realistic cases of *heterogeneity* and *anisotropy*. The model that is presented in the following section allows very general heterogeneity but it does not allow anisotropy. The reasons for this latter limitation are presented there. It is also clear that our subsurface model must include whatever complications are caused on the existence of a *seepage face* on the stream bank.

Having noted the essential properties of the component boundary-value problems, we come to the last decision. What mathematical techniques will be used to obtain the solutions? Once again there are two broad categories: *analytical methods* and *numerical methods*, and once again we are forced in one direction without real choice by the properties of our problem. The classic techniques of analytical mathematical solution can be applied only to boundary-value problems with very regular properties, for example, square or circular regions with symmetrical boundary conditions and homogeneous, isotropic internal properties. For the hillslope-hydrology problem to be treated in any realistic sense, we must make use of the powerful numerical mathematical techniques that remove the limitations of regularity, symmetry and homogeneity. The choice of numerical simulation brings with it the automatic marriage of the model and the digital computer.

6.1.4 The three component models of hillslope hydrology

The three component mathematical models of hillslope hydrology can be summarized as follows:

(1) Transient, saturated–unsaturated, *subsurface* flow in a two-dimensional

cross-section with heterogeneous, isotropic media; boundary conditions that allow time- and space-dependent arrival of rainfall on the upper surface, and outflow to the stream through a transient seepage face.

(2) One-dimensional, transient, *channel flow* with boundary conditions that allow time- and space-dependent arrival of lateral inflow.

(3) One-dimensional treatment of the sheetflow representation of transient *overland flow;* boundary conditions that allow time- and space-dependent arrival of rainfall, infiltration to the subsurface system, and outflow to the stream.

In the three sections to follow, mathematical models representing these three components will be developed in detail, and in the subsequent section, two coupled models will be described: (i), the subsurface flow–channel flow model of Freeze (1972a and 1972b); and (ii), the overland flow–infiltration model of Smith and Woolhiser (1971a and 1971b).

All three models are solved with numerical techniques. It is beyond the scope of this chapter to provide a review of the various available numerical techniques, but references are provided, and the techniques actually used in the component models are described in some detail.

6.2   MATHEMATICAL MODEL OF SUBSURFACE FLOW ON A HILLSIDE

For the subsurface-flow model we will follow the development of Freeze (1971a) as it was applied to the hillslope hydrology problem (Freeze, 1972a and 1972b). The saturated–unsaturated analysis owes much to the work of Rubin and Steinhardt (1963), Rubin (1968), Hornberger, Remson, and Fungaroli (1969), and Verma and Brutsaert (1970).

6.2.1   Region of flow

Let us choose as the region of flow, the two-dimensional vertical cross-section ABCDEFGHA shown in Figure 6.2. We will assume this section to be in a plane parallel to the direction of subsurface delivery of water toward a stream. The

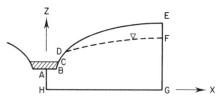

Figure 6.2   Region of flow for transient, two-dimensional, saturated–unsaturated model of subsurface flow

stream bounds the section at ABC and flows roughly perpendicular to it. The region of flow is bounded along CDE by the ground surface, with the portion CD adjacent to the stream denoting a possible seepage face. The basal boundary is a geological one separating the permeable near-surface soils from less-permeable underlying ones. In some cases this boundary may occur where the developed A, B and C soil horizons blend into the parent material; in other cases it may separate unconsolidated geological deposits from bedrock. If the permeability contrast across this boundary is large enough (say 2–3 orders of magnitude or more) we are justified in taking the boundary as impermeable and disregarding the very small contributions to the flow system that occur below it. The right-hand boundary EFG is in the plane separating our region of flow from the adjacent hillslope that feeds the adjacent tributary stream. The configuration of the flow system decrees that there is no flow across this plane and in our two-dimensional section it becomes an imaginary impermeable boundary. The left-hand boundary GH is an impermeable boundary in the same sense. It separates the fields of flow of the two hillslope sections that deliver water to the stream under analysis.

The region of flow includes both the saturated zone ABCDFGHA and the unsaturated zone DEF. The upper and lower boundaries may be highly irregular, and the region may harbour a complex, heterogeneous configuration of soil layers and geologic formations.

6.2.2   Equation of flow

The equation of flow is developed on the basis of the equation of continuity for transient flow through a saturated–unsaturated porous medium, and it is put into its usual form with the aid of Darcy's law.

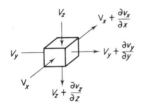

Figure 6.3 Continuity relationship in elemental volume of porous media

The equation of continuity is a statement of the conservation of mass during fluid flow through an elemental volume of the porous media. It states that the net rate of fluid mass flow into any elemental control volume within the porous media must equal the time rate of change of fluid mass storage within the element. Referring to the elemental volume shown in Figure 6.3, the equation of continuity can be written (Freeze, 1971a)

$$\frac{\partial(\Theta n\rho v_x)}{\partial x} + \frac{\partial(\Theta n\rho v_y)}{\partial y} + \frac{\partial\Theta n\rho v_z}{\partial z} = \frac{\partial}{\partial t}(\Theta n\rho) \qquad (6.1)$$

where

Θ = moisture content of the soil (decimal fraction)
n = porosity of the soil (decimal fraction)
ρ = density of water $[M/L^3]$
x, y, z = coordinate directions [L]
$\bar{v} = (v_x, v_y, v_z)$ = velocity of water [L/T]
t = time [T]

The right-hand side of (6.1) can be expanded to produce three terms

$$\frac{\partial}{\partial t}(\Theta n\rho) = \rho n\frac{\partial\Theta}{\partial t} + \rho\Theta\frac{\partial n}{\partial t} + n\Theta\frac{\partial\rho}{\partial t} \qquad (6.2)$$

that refer respectively to changes in the mass storage within the elemental volume due to (i) changes in the moisture content Θ (ii) changes in the porosity n, and (iii) changes in the fluid density ρ. The first of these effects is limited to the unsaturated zone, the second to the saturated zone. Changes in porosity are related to the compressibility of the porous medium; changes in the density to the compressibility of the fluid. For flow in a saturated–unsaturated system near the ground surface, the first term of (6.2) is much greater than the second and third. That is, for all practical purposes we can consider the compressibility of media and fluid to be negligible, and the porosity and fluid density to be constant in time and space. With this simplification, and reducing the equation from three dimensions to two, (6.2) becomes

$$\frac{\partial v_x}{\partial x} + \frac{\partial v_z}{\partial z} = \frac{\partial\Theta}{\partial t} \qquad (6.3)$$

For saturated flow in a two-dimensional homogeneous isotropic medium, Darcy's law takes the form

$$\bar{v} = (v_x, v_y) \qquad (6.4)$$

$$v_x = K\frac{\partial\Phi}{\partial x} \qquad (6.5)$$

$$v_z = K\frac{\partial\Phi}{\partial z} \qquad (6.6)$$

where

K = hydraulic conductivity of the soil [L/T]

Φ = hydraulic head [L]

For this case K is a constant in space and time. For *saturated* flow in a *heterogeneous* medium, K becomes a function of space due to the spatial heterogeneity, that is $K = K(F) = K(x, y, z)$ where F denotes a specific geologic formation or soil type. For *unsaturated* flow in a homogeneous medium, K is a function of the pressure head Ψ (Figure 6.4) and since this changes with time, so too does K, that is $K = K(\Psi) = K(t)$. For *unsaturated* flow in a *heterogeneous* media, then, $K = K(F, \Psi) = K(x, y, z, t)$. If we allow the medium to be anisotropic, we

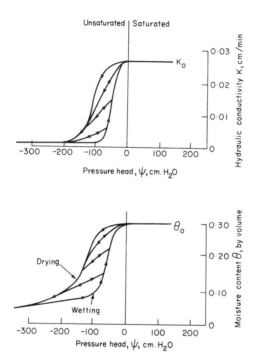

Figure 6.4 Functional relationships between hydraulic conductivity (K), moisture content (θ) and pressure head (ψ) for a naturally-occurring sand

introduce even further complications. In fact these complications are severe enough that their ramifications in the unsaturated flow domain have not yet been fully explored. In light of this, we will limit ourselves to heterogeneous, but isotropic, media. Darcy's law can then be written

$$v_x = K(F, \Psi)\frac{\partial \Phi}{\partial x} \tag{6.7}$$

$$v_z = K(F, \Psi)\frac{\partial \Phi}{\partial z} \tag{6.8}$$

Inserting (6.7) and (6.8) in (6.3) yields

$$\frac{\partial}{\partial x}\left[K(F,\Psi)\frac{\partial\Phi}{\partial x}\right] + \frac{\partial}{\partial z}\left[K(F,\Psi)\frac{\partial\Phi}{\partial z}\right] = \frac{\partial\Theta}{\partial t} \qquad (6.9)$$

Within any of the homogeneous formations F, that make up the heterogeneous system, we also know that the moisture content Θ is a characteristic function of the pressure head (Figure 6.4), that is, $\Theta = \Theta(\Psi)$. Soil physicists have denoted the slope of this curve, which is also a function of Ψ, the specific moisture capacity C

$$C(\Psi) = \frac{\partial\Theta(\Psi)}{\partial\Psi} \qquad (6.10)$$

For a heterogeneous media $\Theta = \Theta(F,\Psi)$ and $C = C(F,\Psi)$. Relating (6.10) to the right-hand side of (6.9) gives

$$\frac{\partial}{\partial x}\left[K(F,\Psi)\frac{\partial\Phi}{\partial x}\right] + \frac{\partial}{\partial z}\left[K(F,\Psi)\frac{\partial\Phi}{\partial z}\right] = C(F,\Psi)\frac{\partial\Psi}{\partial t} \qquad (6.11)$$

Recognizing that the total hydraulic head Φ is related to the pressure head Ψ by

$$\Phi = \Psi + z \qquad (6.12)$$

allows us to put (6.11) in its following final form

$$\frac{\partial}{\partial x}\left[K(F,\Psi)\frac{\partial\Psi}{\partial x}\right] + \frac{\partial}{\partial z}\left[K(F,\Psi)\left(\frac{\partial\Psi}{\partial z}+1\right)\right] = C(F,\Psi)\frac{\partial\Psi}{\partial t} \qquad (6.13)$$

If we know the shape of the functions $K(\Psi)$ and $C(\Psi)$ for each soil type F, we can solve Equation (6.13) for $\Psi(x,z,t)$. Knowing $\Psi(x,z.t)$ we can easily calculate $\Phi(x,z,t)$ from (6.12); and knowing the shape of the function $\Theta(\Psi)$ we can convert the $\Psi(x,z,t)$ results to $\Theta(x,z,t)$. Solution of the boundary-value problem will thus provide us with two-dimensional patterns of the moisture content, pressure head and hydraulic head throughout the region of flow at any time. The transient moisture-content patterns will provide us with the most-easily grasped diagrammatic description of the flow system; the pressure-head patterns will allow us to follow the transient behaviour of the water table (which will show up as the specific $\Psi = 0$ contour); and the hydraulic-head patterns will allow us to calculate the hydraulic gradients at various points in the system and, with the aid of Darcy's law, to calculate rates of infiltration at the ground surface and subsurface outflow to the stream.

6.2.3   Boundary conditions

We have written the equation of flow (6.13) with Ψ, the pressure head, as the independent variable. We could as easily have used Φ, the total hydraulic head, as the independent variable. We must regard the two formulations as equivalent and interchangeable. Having made this point we can proceed to outline the boundary conditions for our problem, first in the Φ-mode, in which they are conceptually clearer, and then in the Ψ-mode as required to correspond with our equation of flow.

Referring to Figure 6.2, along the basal impermeable boundary GH:

$$\frac{\partial \Phi}{\partial z} = 0 \qquad (6.14)$$

On the imaginary, vertical, impermeable boundaries AH and EG

$$\frac{\partial \Phi}{\partial x} = 0 \qquad (6.15)$$

On the stream bottom ABC:

$$\Phi = z_c \qquad (6.16)$$

where z_c is the elevation above datum of the surface of the stream (i.e. the elevation of point C). In our analysis we will assume z_c to be a constant with time; that is, we will not take into account the influence on the subsurface flow system of a rise in stream levels created by the subsurface outflows to the stream. This is a good assumption in wide, shallow streams or for the small first-order tributary streams that are our primary interest in this study.

On the upper boundary DE we allow a time- and space-dependent rainfall or evaporation rate $r(x, t)$ (with dimensions [L/T]), where r positive is a rainfall rate and r negative is an evaporation rate. If we restrict ourselves to rainfall rates less than the saturated hydraulic conductivity of the surface soils, then ponding will not occur on the surface, and there is no possibility of overland flow (Rubin and Steinhardt, 1963; Freeze, 1972b). Under these circumstances, all precipitation $r(x, t)$ becomes infiltration $I(x, t)$, and the boundary condition can be stated as

$$I(x, t) = K(x, t)\frac{\partial \Phi(x, t)}{\partial z} \qquad (6.17)$$

The value of $K(x, t)$ will depend on the value of $\Psi(x, t)$ at the point x at time t and on the nature of the $K(\Psi)$ curve for the surface soil at x.

Along the seepage face CD, the pressure head $\Psi = 0$, so, from (6.12):

$$\Phi = z \qquad (6,18)$$

The upper boundary of the seepage face, D is known as the exit point. For a transient flow system, its position is time-dependent. This moving exit point requires special attention in the numerical solution.

If we convert boundary conditions (6.14)–(6.18) to the Ψ-mode, they take the following form

On GH:

$$\frac{\partial \Psi}{\partial z} = -1 \qquad (6.14a)$$

On AH and EG:

$$\frac{\partial \Psi}{\partial x} = 0 \tag{6.15a}$$

On ABC

$$\Psi = z_c - z \tag{6.16a}$$

On DE:

$$I(x, t) = K(x, t)\left[\frac{\partial \Psi(x, t)}{\partial z} + 1\right] \tag{6.17a}$$

On CD:

$$\Psi = 0 \tag{6.18a}$$

6.2.4 Initial conditions

There are two sets of initial conditions that are hydrologically meaningful and mathematically tractable: (i) static conditions, and (ii) steady-state flow.

Under static initial conditions, it is assumed that there is no flow through the system. The hydraulic head $\Phi(x, z, 0)$ is constant for all (x, z), and the water-table ($\Psi = 0$) is horizontal and at an elevation level with the stream surface. Above the water table there will be an equilibrium configuration of pressure heads and moisture contents.

For steady-state flow, the time-dependent term is removed from the right-hand side of (6.13). The steady-state equation of flow becomes:

$$\frac{\partial}{\partial x}\left[K(F, \Psi)\frac{\partial \Psi}{\partial x}\right] + \frac{\partial}{\partial z}\left[K(F, \Psi)\left(\frac{\partial \Psi}{\partial z} + 1\right)\right] = 0 \tag{6.19}$$

The initial flow regime is determined by solving this reduced flow equation for $\Psi(x, z)$, given $K(F, \Psi)$ for each soil type F, and a set of steady-state boundary conditions around the region of flow (Figure 6.2). The allowable boundary conditions are (i) impermeable boundaries such as (6.14) and (6.15), and (ii) constant head boundaries along which Φ or Ψ are specified. Along the seepage face, $\Psi = 0$. The steady-state position of the exit point must be located by a series of trial-and-error steady-state solutions. The solution is considered satisfactory when the simulated position of the water table ($\Psi = 0$) coincides at the boundary with the pre-specified exit point (Freeze, 1971b).

6.2.5 Soil parameters

In the unsaturated domain, the hydraulic conductivity K and moisture content Θ are both functions of the pressure head Ψ for any soil type F. The functional relationships are hysteretic in that the curves differ depending on whether the soil is wetting or drying. To illustrate, Figure 6.4 shows the characteristic curves for a naturally-occurring soil known as Del Monte sand (Liakopoulos, 1965). The scanning curves between the main wetting and drying curves provide the necessary data for cases where the soil changes from wetting to drying or vice versa at some intermediate condition of saturation.

As noted in Equation 6.10 the specific moisture capacity C that appears in Equation 6.13 is simply the slope of the $\Theta(\Psi)$ curve. For saturated flow ($\Psi > 0$, Figure 6.4), the soil parameters are constant, and $K = K_0$, $\Theta = n$, and $C = 0$.

While hysteresis can be important in some applications and the model of Freeze (1971a) allows for its consideration, it seems likely that the uncertainties as to the exact form of the basic curves for field soils will more than outweigh the secondary influences of hysteresis in hillslope hydrology simulations. It is therefore common to utilize single $K(\Psi)$ and $\Theta(\Psi)$ curves to represent the unsaturated hydrologic properties of a soil.

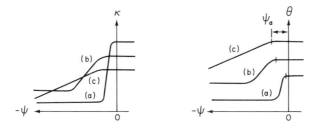

Figure 6.5 Single-valued functional relationships between K, θ and ψ for three hypothetical soils. (a), Uniform sand; (b) silty sand; (c), silty clay

The $K(\Psi)$ and $\Theta(\Psi)$ curves are strongly dependent on soil texture. Figure 6.5 shows curves for three hypothetical soil types: (a) a uniform sand, (b) a silty sand, and (c) a silty clay. The uniform sand shows a high saturated hydraulic conductivity, a low porosity, a high capacity over a narrow range of pressure heads, and low moisture contents at high tension heads. Such a soil would show a sharp gradient in moisture content and permeability across the water table. As a soil becomes less uniform and less permeable (curves b and c), the porosity increases and the capacity becomes more uniform. Moisture contents and permeabilities would tend to show more gradual changes in the vicinity of the water table in such soils.

The Ψ_a value shown in Figure 6.5 represents the air-entry pressure head. Over the range $0-\Psi_a$, conditions remain saturated even though the pressure heads are less than atmospheric. This gives rise to the tension-saturated zone above the water table, better known as the capillary fringe.

The characteristic curves can be determined in the laboratory using techniques that are well developed in the soil physics field (Black, 1965). Data on naturally occurring soils abound in the soil physics literature.

6.2.6 Numerical method of solution

Equation 6.13 is a non-linear parabolic partial-differential equation. There are many available methods of numerical solution and the reader is referred to the recent text by Remson, Hornberger and Molz (1971) for a systematic and detailed presentation of these various methods. In general, the schemes can be iterative or direct, and the finite-difference formulations can be implicit or explicit. An iterative–implicit approach should minimize the computer time for the solution and it also guarantees unconditional stability for all size time steps (although the accuracy is reduced by the use of larger time steps). Iterative techniques have an added advantage over direct techniques for our problem in that they allow for the re-calculation at each iteration of the terms that are functions of the dependent variable. In this section, I will briefly outline the line-successive over-relaxation (LSOR) scheme used by Freeze (1971a, 1971b and 1972a). It falls into the iterative–implicit category.

Consider the block-centered nodal grid in the x–z plan shown in Figure 6.6. In the x-direction the nodes are labelled $i = 1, 2, \ldots N$ in the z-direction they are labelled $j = 1, 2, \ldots M$. At time t, the time step is Δt^t with the superscript $t = 1, 2, \ldots$. The value of Ψ at x_i, y_j and t is denoted by Ψ^t_{ij}. All nodal spacings can be variable and the region can take on any shape that does not break the continuity of any vertical nodal column. Each node is specified as being within some formation F, so that $K^t_{ij} = K(F, \Psi^t_{ij})$ and $C^t_{ij} = C(F, \Psi^t_{ij})$.

The finite-difference approximations used for the first term of (6.13) are of the form

$$\frac{\partial}{\partial x}\left[K(F, \Psi)\frac{\partial \Psi}{\partial x}\right]^{t-\frac{1}{2}}_{ij} = \frac{1}{\Delta x_i}\left[K^{t-\frac{1}{2}}_{i+\frac{1}{2},j}\left(\frac{\partial \Psi}{\partial x}\right)^{t-\frac{1}{2}}_{i+\frac{1}{2},j} - K^{t-\frac{1}{2}}_{i-\frac{1}{2},j}\left(\frac{\partial \Psi}{\partial x}\right)^{t-\frac{1}{2}}_{i-\frac{1}{2},j}\right] \quad (6.20)$$

$$\left(\frac{\partial \Psi}{\partial x}\right)^{t-\frac{1}{2}}_{i-\frac{1}{2},j} = \frac{\Psi^t_{ij} + \Psi^{t-1}_{ij} - \Psi^t_{i-1,j} - \Psi^{t-1}_{i-1,j}}{\Delta x_i + \Delta x_{i-1}} \quad (6.21)$$

$$\left(\frac{\partial \Psi}{\partial x}\right)^{t-\frac{1}{2}}_{i+\frac{1}{2},j} = \frac{\Psi^t_{i+1,j} + \Psi^{t-1}_{i+1,j} - \Psi^t_{ij} - \Psi^{t-1}_{ij}}{\Delta x_i + \Delta x_{i+1}} \quad (6.22)$$

and those for the second term are analogous. The right-hand side of (6.13) is approximated as

$$\left[C(F, \Psi)\frac{\partial \Psi}{\partial t}\right]^{t-\frac{1}{2}}_{ij} = C^{t-\frac{1}{2}}_{ij}\left[\frac{\Psi^t_{ij} - \Psi^{t-1}_{ij}}{\Delta t^t}\right] \quad (6.23)$$

The full finite-difference approximation to (6.13) is then

$$
\left\{ \frac{1}{\Delta x_i} \left[K(\Psi_{\mathrm{I}}) \left(\frac{\Psi_{i+1}^t + \Psi_{i+1}^{t-1} - \Psi_i^t - \Psi_i^{t-1}}{\Delta x_i + \Delta x_{i+1}} \right) \right. \right.
$$

$$
\left. \left. - K(\Psi_{\mathrm{II}}) \left(\frac{\Psi_i^t + \Psi_i^{t-1} + \Psi_{i-1}^t + \Psi_{i-1}^{t-1}}{\Delta x_i + \Delta x_{i-1}} \right) \right] \right\}_j
$$

$$
+ \left\{ \frac{1}{\Delta x_j} \left[K(\Psi_{\mathrm{III}}) \left(\frac{\Psi_{j+1}^t + \Psi_{j+1}^{t-1} - \Psi_j^t - \Psi_j^{t-1}}{\Delta z_j + \Delta z_{j+1}} + 1 \right) \right. \right.
$$

$$
\left. \left. - K(\Psi_{\mathrm{IV}}) \left(\frac{\Psi_j^t + \Psi_j^{t-1} - \Psi_{j-1}^t - \Psi_{j-1}^{t-1}}{\Delta z_j + \Delta z_{j-1}} + 1 \right) \right] \right\}_i
$$

$$
= \left\{ C(\Psi_{\mathrm{V}}) \left(\frac{\Psi^t - \Psi^{t-1}}{\Delta t^t} \right) \right\}_{ij} \quad (6.24)
$$

For vertical LSOR, the terms can be grouped as

$$
- A_j \Psi_{i,j+1}^t + B_j \Psi_{ij}^t - C_j \Psi_{i,j-1}^t = D_j \quad (6.25)
$$

where A_j, B_j, C_j and D_j are developed from the groupings of the coefficients of (6.24). During the scanning of each vertical line on the x–z mesh (Figure 6.6), values of Ψ^t on adjacent lines, as calculated from the most recent iteration, are considered as known values, as are all values of Ψ^{t-1}. The set of equations (6.25) for a line scan,

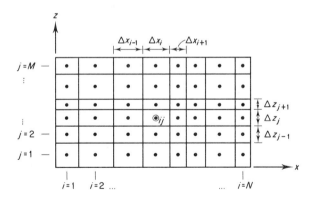

Figure 6.6   Nodal network used for development of finite-difference equations in numerical solution of subsurface flow equations

form a tri-diagonal matrix equation that can be solved by the well-known triangularization scheme embodied in the following recurrence relation

$$\Psi_{ij}^t = E_j \Psi_{i,j+1}^t + F_j \quad \text{for } j < M$$
$$\Psi_{ij}^t = F_j \qquad\qquad \text{for } j = M \tag{6.26}$$

where

$$E_j = \frac{A_j}{B_j - C_j E_{j-1}} \quad \text{for } j > 1; \quad E_1 = \frac{A_1}{B_1} \tag{6.27}$$

$$F_j = \frac{D_j + C_j F_{j-1}}{B_j - C_j E_{j-1}} \quad \text{for } j > 1; \quad F_1 = \frac{D_1}{B_1} \tag{6.28}$$

The E and F coefficients are calculated from $j = 1$ to $j = L$ using (6.27) and (6.28) and the Ψs are back calculated from $j = L$ to $j = 1$ using (6.26). After each iteration (superscript it) the calculated values of Ψ at each node are over-relaxed in the usual fashion:

$$\Psi^{t,it} = \omega \Psi_{(\text{calc})}^{t,it} + (1 - \omega)\Psi^{t,it-1}, \qquad 1 \leqslant \omega \leqslant 2 \tag{6.29}$$

At each iteration it is necessary to predict a pressure head value $\Psi_{(\text{pred})}$ at each node, from which the current estimates of K and C can be calculated. For the first iteration of the first time step:

$$\Psi_{(\text{pred})ij}^t = \Psi_{ij}^{t-1} \tag{6.30}$$

where $t = 0$ implies initial conditions. For the first iteration of later time steps

$$\Psi_{(\text{pred})ij}^t = (T^t + 1)\Psi_{ij}^{t-1} - T^t \Psi_{ij}^{t-2} \tag{6.31}$$

where

$$T^t = \Delta t^t / 2\Delta t^{t-1} \tag{6.32}$$

For later iterations of all time steps:

$$\Psi_{(\text{pred})ij}^{t,it} = \Psi_{(\text{pred})ij}^{t,it-1} + \lambda(\Psi_{ij}^{t,it-1} - \Psi_{(\text{pred})ij}^{t,it-1}), \qquad 0 \leqslant \lambda \leqslant 1 \tag{6.33}$$

The value of Ψ_I that appears in (6.24) is the predicted value of Ψ at the boundary between two nodal blocks. It is determined by

$$\Psi_I = \tfrac{1}{2}(\Psi_{(\text{pred})ij} + \Psi_{(\text{pred})i+1,j}) \tag{6.34}$$

The values of Ψ_{II} through Ψ_{IV} are determined analogously. Ψ_V is equal to $\Psi_{(\text{pred})ij}$. If the nodal boundary at which Ψ_I is calculated is also a geologic boundary between formations or soils with conductivity K_1 and K_2, then

$$K(\Psi_I) = \tfrac{1}{2}[K_1(\Psi_I) + K_2(\Psi_I)] \tag{6.35}$$

At any boundary node, boundary conditions can be imposed that specify the flux, the head, or no-flow conditions. For flux across the upper z-boundary of a nodal block, the first term in the finite-difference representation of flow in the z-direction in Equation (6.24) is replaced by $I/\Delta z$ where I-positive is an inflow rate and I-negative is an outflow rate. For a no-flow boundary, $I = 0$. At a head boundary, the head is specified at the node, so that Ψ_{ij}^t is known and can be used directly in the finite-difference scheme. Time-dependent flux or head conditions on the boundaries can be handled with programming adaptations that do not reflect any added mathematical complications to the underlying development.

In transient problems the position of the exit point D (Figure 6.2) that separates boundary conditions (6.17) and (6.18) may be time-dependent. To handle this complication, a test is made on the boundary nodes above and below D prior to each time step by using the predictor Equation (6.31) to see whether the seepage face is rising or falling. If it is rising, the time step is simulated without change, but on its completion the Ψ value at the node above D is checked. If it is less than zero, the seepage face rise during the time step was apparently not sufficient to reach the node. If it is greater than zero, the boundary condition at the node is changed from Equation 6.17 to Equation 6.18 and the time step is re-calculated. For a falling exit point, the boundary condition the node below D is changed from Equation 6.18 to Equation 6.17 prior to the time step, the calculations are carried out, and then the node just above the new exit point is checked in the same manner as for the rising face. There are thus many more re-calculated time steps under falling conditions than under rising conditions.

The single-valued or hysteretic functional relationships for the soil properties can be built into the computer program in the form of a table of coordinate values of K^* and C^* versus Ψ^* (where the asterisk denotes coordinate values of the input data). Linear interpolation is used between the coordinate points. This line-segment approach requires that the slopes of successive segments be within a factor of about 2 of one another. In other words, many Ψ^* coordinates are needed at the locations of major curvature. The smallest $\Delta\Psi$ coordinate spacing, which can be called $\Delta\Psi_{\text{critical}}^*$, is a kind of measure of the smoothness of the soil curves.

The LSOR technique is theoretically stable for all size time steps. In practice, however, a time-step limitation is introduced by the non-linearity of the K and C coefficients. If the Ψ value at a particular node is in the range of $\Delta\Psi_{\text{critical}}^*$, and $\Delta\Psi = (\Psi^t - \Psi^{t-1})$ exceeds $\Delta\Psi_{\text{critical}}^*$, an instability results. This instability takes the form of an oscillation in the succeeding Ψ values at the offending node. The oscillations, of course, quickly spread through the mesh. In that $\Delta\Psi_{\text{critical}}^*$ often occurs in the vicinity of $\Psi = 0$, the instability usually appears at the moving saturation front, which is, of course, a major focus of our transient solutions. Freeze (1971b) reports that the following criterion is sufficient to avoid problems although it is probably unduly restrictive. After each time step, his program checks all values of $\Delta\Psi$ and if any exceed $0.75\Delta\Theta\Psi_{\text{critical}}^*$, the Δt value is reduced accordingly and the step recalculated. The Δt values can be further optimized by using a geometric progression to increase the size of the time steps with time. This optimization is warranted by the nature of the transient solutions, which decay more or less exponentially.

For initial conditions of steady state flow the reduced equation 6.19 can be solved with the same LSOR scheme but with the finite-difference equations adjusted to remove the terms involving t.

For steady flow the optimum value of the over-relaxation parameter ω in (6.29) is around 1·88. In the transient simulations when (6.31) is used as a predictor, it replaces the need for over-relaxation and $\omega = 1·00$ can be used. Solution times are not very sensitive to the values of λ in (6.33).

Output from the computer model can usually be obtained in the form of print-outs and plots of pressure head, total head, and moisture-content fields at any desired time step. From the pressure-head diagram one can locate the position of the water table. From the total-head diagram one can determine the saturated–unsaturated flow pattern and determine quantitative values of the infiltration rates and subsurface outflow rates as a function of time and space.

6.2.7 Sample results

A simulation taken from Freeze (1972a) provides a suitable set of sample results. Consider the two-dimensional vertical cross-section shown in Figure 6.7(a). One might consider it to be the end slice from Figure 6.1. As actually simulated, however, the cross-section measures only 6 m wide by 3 m deep; a computer model equivalent of a large laboratory sand-tank experiment. The porous medium is homogeneous, isotropic, and highly permeable, $K_0 = 0·0044$ cm/sec. The porosity is 0·30, and the unsaturated characteristic curves are those of Figure 6.4. The cross-section is assumed to be uniform in the third dimension, and a stream flows perpendicular to the page across the width AF. The inflow function along the strip ED is one of constant flux with $I = 0·005$ cm/sec until surface saturation occurs; then constant head thereafter with $\Psi = 0$ until the specified end of the inflow. The rest of the boundaries are impermeable, but development of the seepage face is allowed at the base of the slope. The initial conditions are static.

This small permeable system exhibits a rapid response to the imposed inflows. The dashed hydraulic head contours of Figures 6.7(b), (c), and (d) illustrate the growth of the system with time. The growth of the zone of saturation is shown in these figures, and the composition of this water-table rise is summarized in Figure 6.7(e). As noted there, the water table remains horizontal for 40 min (while soil-moisture storage deficits are being satisfied above), and then begins to rise. A steady-state configuration is reached after 218 min. For a cessation of inflow at 277 min, the subsequent decline of the water table is shown in Figure 6.7(f). The recession is considerably slower than the rise.

Figure 6.7(g) shows the composition of the inflow function. The total inflow entering the flow system across ED remains constant until saturation occurs at the surface at $t_3 = 84$ min. Inflow declines rapidly until the steady state flow is established at $t_4 = 218$ min. The three small diagrams at the right of Figure 7(g)

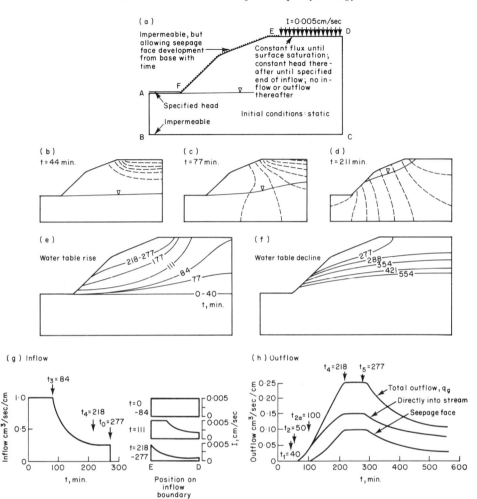

Figure 6.7 Saturated–unsaturated subsurface flow in a small idealized two-dimensional flow system (after Freeze, 1972a). (a) Boundary and initial conditions; (b), (c) and (d), transient hydraulic head contours (broken line) and water-table positions (solid line); (e) and (f), water table rise and decline; (g), inflow as a function of time and position; (h) outflow hydrograph

display the areal variations in the inflow rate I at various times. The lowest of these diagrams is the final steady-state inflow profile.

The outflow from the subsurface system is charted in Figure 6.7(h). Although the initial water-table rise occurs at $t_1 = 40$ min (at the right-hand side of the system), the outflow to the stream is not initiated until $t_2 = 50\,\text{min}$ and the build-up of the seepage face does not occur until $t_{2a} = 100\,\text{min}$.

Some further examples of subsurface flow simulation are included in the later section on coupled models. These later examples reflect conditions in hillslope cross-sections of a more realistic size and shape.

6.3   MATHEMATICAL MODEL OF FLOW IN A STREAM CHANNEL

In the preceeding section we have seen how numerical simulations of subsurface flow in hillslope cross-sections can be used to provide estimates of subsurface lateral inflow into streams. In a later section we will look into overland flow contributions to lateral inflow. In this section we will develop the mathematical modelling approach that can be used to analyse the flow in stream channels, when that flow is generated by lateral inflow from either source.

On the face of it, it would be more logical to investigate overland flow models first, but the development of such models is very similar to that for streamflow models, and it is more straightforward to present the basic analysis in terms of finite stream channels and then proceed to the adaptation to overland flow planes.

Flow in stream channels can be analysed with the classical methodology of open-channel hydraulics. The standard text is Chow (1959). Strelkoff (1969) provides an excellent recent review of the equations of flow and their development. Some numerical results of Ragan (1966) will be used in this chapter as sample solutions.

6.3.1   Region of flow

Let us first assume that flow is unidirectional, that is, that secondary transverse flow components within the stream cross-section are minimal. We can then proceed with a one-dimensional analysis, which we will take to be in the y-direction, perpendicular to the x–z plane of the subsurface cross-section in Figure 6.2. Let us further assume that at any given point the stream cross-section can be reasonably idealized as a *rectangular* open channel. If this is not the case, and some more complex idealization is required, the algebra used in the development of the flow equation may differ but the concept remains unchanged. For open-channel flow, a rectangular channel is completely specified by its width B.

The region of flow, if such it can be called, is a one-dimensional line representing the reach of stream from $y = 0$ to $y = L$ (Figure 6.8A). Within this reach the channel width B, the channel slope S_0, the resistance parameter n, and the lateral inflow q are allowed to vary with position; that is $B = B(y)$, $S_0 = S_0(y)$, $n = n(y)$ and $q = q(y)$. In fact, since q may also vary with time, $q = q(y, t)$. At any given point in the reach, the lateral inflow (Figure 6.8B), considered as a discharge per unit channel length, is given by:

$$q = rB + q_s + q_g \tag{6.36}$$

where

r = rainfall rate $[L/T]$

q_s = surface lateral inflow $[L^2/T]$

q_g = subsurface lateral inflow $[L^2/T]$

Transient flow in the reach is controlled by the flows at the end-points $Q(0, t)$ and $Q(L, t)$, by the nature of the lateral inflow function $q(y, t)$, and by the channel

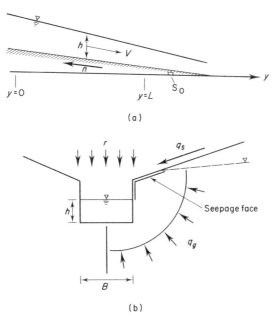

Figure 6.8   One-dimensional channel-flow model. (a), Longitudinal section; (b), cross section showing lateral inflow components (after Freeze, 1972a)

parameters $B(y)$, $S_0(y)$ and $n(y)$. The independent variables are y and t, and the dependent variables are the stream depth $h(y, t)$ and velocity $v(y, t)$. Knowing B, v and h one can calculate the stream discharge $Q(y, t)$ from

$$Q = Bvh \tag{6.37}$$

6.3.2   Equations of flow

Flow in open channels can be described by the shallow water equations. This pair of equations consists of an equation of continuity and an equation of motion. We will develop them for the special case of transient *gradually-varied*, *turbulent*, *subcritical* flow in a rectangular open channel of variable width. Before proceeding, however, we ought to examine the meaning of the italicized adjectives introduced in the previous sentence.

'Gradually-varied flow' refers to cases where the cross-section of flow varies gradually along the channel. Changes in depth and velocity are continuous along the reach and with time; accelerative effects are negligible. Flow in natural channels, without man-made constrictions, is usually gradually varied.

Turbulence is usually defined with reference to Reynolds' number (R_e):

$$R_e = \frac{vL}{v}$$
(6.38)

where

v = flow velocity $[L/T]$
L = a characteristic length $[L]$
v = kinematic viscosity $[L^2/T]$

Reynolds' number relates the forces of inertia to viscous forces during flow. When R_e is small, viscous forces are predominant, and flow is *laminar*. When R_e is large, inertial forces are predominant, and flow is *turbulent*. Flow in natural stream is almost always turbulent.

Froude's number

$$F_r = \frac{v}{\sqrt{gh}}$$
(6.39)

relates the inertial forces to gravitational forces. On mild slopes, when $F_r < 1$, inertial forces are predominant, and the flow is *subcritical* (or tranquil). On steep slopes, when $F_r > 1$, gravitational forces are predominant and the flow is *supercritical* (or rapid). By limiting our analysis to subcritical flow we are thus limiting ourselves to mild slopes and we may be removing from consideration certain steep first-order tributary streams that are of interest to us. Mathematically, however, the limitation is well worthwhile as it removes the non-uniqueness problems associated with the analysis of the full subcritical-supercritical range, and it ensures that hydraulically-complex hydraulic jumps and breaking surges will not haunt our analyses.

The equation of continuity for flow in an open channel states that the rate of fluid mass flow into any elemental reach must equal the time rate of change of fluid mass storage within the reach. Referring to Figure 6.9(a), the equation of continuity can be written as

$$\frac{\partial Q}{\partial y} - q = -\frac{\partial(Bh)}{\partial t}$$
(6.40)

Inserting (6.36) and (6.37) in (6.40) leads to

$$h\frac{\partial v}{\partial y} + v\frac{\partial h}{\partial y} + \frac{vh}{B}\frac{\partial B}{\partial y} - \frac{rB + q_s + q_g}{B} + \frac{\partial h}{\partial t} = 0$$
(6.41)

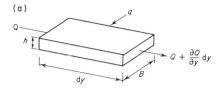

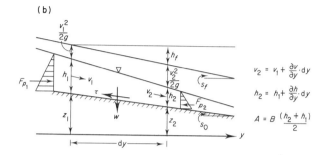

Figure 6.9 (a), Continuity relationship in elemental reach of open channel; (b), force and energy relationships in a reach of open channel

The equation of motion can be developed with reference to Figure 6.9(b). We will apply Newton's second law of motion in the form that states that the vector sum of the applied forces acting on the fluids in a reach must equal the rate of change of momentum within the reach. The total applied force (F_T) is given by

$$F_T = F_p + F_g + F_f \qquad (6.42)$$

where

F_p = unbalanced hydrostatic pressure force (acting in the upstream direction)
F_g = gravity force (in the downstream direction)
F_f = friction force (acting in the upstream direction).

We can expand these terms as

$$F_p = F_{p_2} - F_{p_1} = -\rho g A \frac{\partial h}{\partial y} dy \qquad (6.43)$$

$$F_g = W \sin S_0 = \rho g A S_0 \, dy \qquad (6.44)$$

$$F_f = TP \, dy = -\rho g A h_f = -\rho g A S_f \, dy \qquad (6.45)$$

where

P = the wetted perimeter of the channel
A = the average cross-sectional area of flow

All other symbols are identified on Figure 6.9(b).

The second equality in 6.44 assumes that S_0 is small. The second equality in (6.45) rests on the realization that the head loss h_f reflects the work done by the shear stress T. The third equality in (6.45) assumes that S_f is small.

The rate of change of momentum is given by

$$\frac{dM}{dt} = m_c\frac{dv}{dt} + m_s\frac{dv_s}{dt} + m_g\frac{dv_g}{dt} + m_r\frac{dv_r}{dt} \tag{6.46}$$

where

m_c = mass of fluid in the channel

m_s, m_g, m_r = mass of fluid entering the channel as surface lateral inflow, subsurface lateral inflow and rainfall, respectively

v = velocity of flow in the channel

v_s, v_g, v_r = velocity of inflow of surface lateral inflow, subsurface lateral inflow, and rainfall, respectively.

Each of the four terms of (6.46) can be expanded

$$m_c\frac{dv}{dt} = \rho A\,dy\left(\frac{\partial v}{\partial t} + v\frac{\partial v}{\partial y}\right) \tag{6.47}$$

$$m_s\frac{dv_s}{dt} = \rho q_s\,dy\,dt\left(\frac{v - v_s}{dt}\right) = \rho q_s(v - v_s)\,dy \tag{6.48}$$

$$m_g\frac{dv_g}{dt} = \rho q_g\,dy\,dt\left(\frac{v - v_g}{dt}\right) = \rho q_g(v - v_g)\,dy \tag{6.49}$$

$$m_r\frac{dv_r}{dt} = \rho r B\,dy\,dt\left(\frac{v - v_r}{dt}\right) = \rho r B(v - v_r)\,dy \tag{6.50}$$

Recognizing that v_s, v_g and v_r are all much smaller than v allows removing these terms from the right-hand side of (6.48), (6.49) and (6.50), and equating the sum of the expansions (6.43), (6.44) and (6.45) with the sum of the expansions (6.47)–(6.50), cancellation of common terms, we obtain

$$-g\frac{\partial h}{\partial y} + gS_0 - gS_f = \frac{\partial v}{\partial t} + v\frac{\partial v}{\partial y} + \frac{v}{A}(rB + q_s + q_g) \tag{6.51}$$

Gathering terms and noting that $A = Bh$ yields the equation of motion

$$v\frac{\partial v}{\partial y} + \frac{\partial v}{\partial t} + g\frac{\partial h}{\partial y} + g(S_f - S_0) + \frac{v}{Bh}(rB + q_s + q_g) = 0 \tag{6.52}$$

The slope S_f that appears in (6.52), and is defined on Figure 6.9(b), is called the energy slope. It can be related to the resistance parameter n by means of the empirical Manning relation

$$S_f = \frac{v^2 n^2}{N R^{4/3}} \tag{6.53}$$

where

R = hydraulic radius $[L]$
  = $Bh/(B + 2h)$

and N is a dummy variable that takes the value $1 \cdot 0$ in the SI system of units. The resistance parameter n is known as Manning's friction factor. It is best considered as a dimensionless number, with the resulting dimensional inconsistencies in (6.53) absorbed in the variable N. One should note that n has the same value in the SI system as in the Imperial system as long as the value of N is adjusted accordingly. There is no suitable theoretical foundation for the Manning formula, and this fact identifies the friction factor n as the weakest link in the combination of physical parameters that underlie our conceptual–deterministic approach. The relation (6.53) was initially proposed for steady flow but it is now widely used in transient analyses. Many workers have noted that n is not a true constant but is a function of the depth of flow. In this analysis, however, n, is taken as constant for all flow depths and at all times for any point on the reach. However, the value of n is allowed to vary along the length of the reach as it undoubtedly does in most natural channels.

Equations (6.41) and (6.52) constitute the shallow-water equations. Given the initial and boundary conditions, the Manning relation (6.53), and the input parameters $B(y)$, $S_0(y)$, $n(y)$ and $q(y, t)$, they can be solved to produce the depths and velocities of flow $h(y, t)$, and $v(y, t)$, and the stream discharge $Q(y, t)$.

6.3.3   Boundary conditions

For subcritical flow we must know one boundary condition at each end of the system. At the upstream boundary, $y = 0$, we will specify $Q(0, t)$, the time-dependent discharge entering the reach under study. Since $Q = Bvh$, and knowing B, we are in effect specifying the product value of vh, but we are not specifying their separate values. These values will be controlled by the boundary condition at the downstream boundary, $y = L$. There, we cannot specify the $Q(L, t)$ condition directly, as its magnitude is dependent on the nature of the lateral inflow function, $q(y, t)$. Rather, it is more reasonable to require that the flow at the downstream boundary be either *critical* or *normal*.

Critical flow is defined by equation (6.39) with the Froude number $F_r = 1 \cdot 0$. It is often specified for a free overfall at the downstream boundary and is thus well-suited to the analysis of flow in a reach controlled by a man-made constriction at its lower end. For flow in natural channels, a *normal* downstream boundary condition seems more representative, and Freeze (1972a) utilized this condition in his study. Normal flow is defined by equation (6.53) with $S_f = S_o$, so that:

$$v = \frac{N^{\frac{1}{2}} S_o^{\frac{1}{2}}}{n} \left[\frac{Bh}{B + 2h} \right]^{\frac{2}{3}} \tag{6.54}$$

If we generate $Q(L)$ on the basis of $Q(0)$ and q, and the equations of flow, then (6.54) will separate out the component values of v and h.

6.3.4 Initial conditions

As in the groundwater flow model, two sets of initial conditions are hydrologically reasonable: no flow, and steady-state flow. The latter is the simpler to analyse. For steady, spatially-varied flow, the time-dependent terms must be removed from Equations (6.41) and (6.52) and the partial derivatives become full. The steady-state flow equations then become

$$h\frac{dv}{dy} + v\frac{dh}{dy} + \frac{vh}{B}\frac{dB}{dy} - \frac{vB + q_s + q_g}{B} = 0 \tag{6.55}$$

$$v\frac{dv}{dy} + g\frac{dh}{dy} + g(S_f - S_o) + \frac{v}{Bh}(rB + q_s + q_g) = 0 \tag{6.56}$$

The initial flow regime is determined by solving this reduced set of flow equations for $h(y)$ and $v(y)$, given $B(y)$, $S_o(y)$, $n(y)$, $q(y, o)$, $Q_o(0)$ and relationship (54) at the downstream boundary.

6.3.5 Channel parameters

The only non-geometrical channel parameter needed as input to the mathematical model of channel flow is Manning's friction factor n. It is an empirical parameter that presumably depends on the stream-bottom vegetation and the nature of the stream-bed sediments. It can be measured in the field by an inverse analysis of streamflow records at a time and place when v, h and Q are known at both ends of a test reach. It is more usual, however, to estimate n by comparing the channel. under study with sets of type channels for which n is known. Chow (1959) provides many photographs of type channels for a wide range of friction factors.

6.3.6 Numerical method of solution

Equations 6.41 and 6.52 form a non-linear set of hyperbolic partial differential equations. A considerable number of numerical schemes have been proposed for their solution. Liggett and Woolhiser (1967) and Strelkoff (1970 provide detailed analyses of these solution methodologies and classify the various finite-difference schemes into the three generally-recognized categories: explicit, implicit, and characteristic. Amein and Fang (1969) provide an excellent operational review.

Difference schemes, based on the method of characteristics have the most secure theoretical foundation, and their convergence and stability properties are well understood. Explicit and implicit techniques, on the other hand, are simpler to work with.

For the purposes of this chapter, I will limit the presentation to a single explicit technique. This technique, known as the 'single-step Lax–Wendroff method', was recommended by Liggett and Woolhiser (1969) in the closure to their earlier article. Their empirical-stability investigations show that this scheme is stable and convergent, whereas many of the other commonly-used explicit schemes are not. As with all explicit schemes, stability requires the satisfaction of the Courant condition:

$$\Delta t \leqslant \Delta y / [|v| + (gh)^{\frac{1}{2}}] \tag{6.57}$$

where

$\Delta t = $ the time step

$\Delta y = $ the nodal distance on the discretized nodal representation of the y–t plane.

Although the Lax–Wendroff scheme is very fast at each time step, the requirement of (6.57) dictates a large number of time steps. An implicit scheme would probably involve less computational work, but the stability and convergence properties are less well documented.

The following development of the finite-difference equations for (6.41) and (6.52) in their direct form differs from that of Liggett and Woolhiser (1969), who used a dimensionless form of (6.41) and (6.52), but the approach is identical.

Consider the nodal network in the y–t plane shown in Figure 6.10. In the y-direction the nodes are labelled $j = 1, 2, \ldots M$; in the t direction they ae labelled $t = 0, 1, 2, \ldots$. The values of h and v at y_j and t^t are h_j^t and v_j^t. The nodal spacing in the y-direction must allow a variable nodal spacing, so that uneven reach lengths with given sets of properties can be modelled directly, and so that concentrated lateral inflows can be specified more efficiently. The nodal spacing at node j is denoted by Δy_j. The time step Δt chosen to carry the solution through time must also be variable. Its value should be as large as possible within the constraints laid down by (6.57).

With the nodal network shown in Figure 6.10, we can define three subsidiary nodal distances that will be used to complement the basic nodal spacing Δy_j. For any interior node

$$\Delta' y_j = \tfrac{1}{2}(\Delta y_j + \Delta y_{j+1}) \tag{6.58}$$

$$\Delta'' y_j = \tfrac{1}{2}(\Delta y_j + \Delta y_{j-1}) \tag{6.59}$$

$$\Delta''' y_j = \tfrac{1}{2}(\Delta y_j + \tfrac{1}{2}[\Delta y_{j+1} + \Delta y_{j-1}]) \tag{6.60}$$

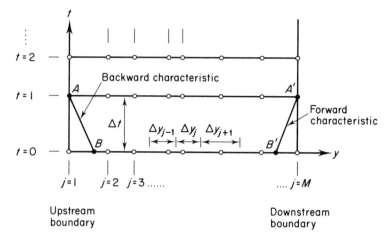

Figure 6.10   Nodal network for development of finite-
difference equations in numerical solution of shallow-
water equations

To develop the finite-difference equations, we will follow the lead of Liggett and
Woolhiser (1967) and define the variable

$$m = vh \tag{6.61}$$

Equations 6.41 and 6.52 can then be converted to the form

$$\frac{\partial h}{\partial t} + \frac{\partial m}{\partial y} + \frac{m}{B}\frac{\partial B}{\partial y} - P = 0 \tag{6.62}$$

$$\frac{\partial m}{\partial t} + \frac{\partial D}{\partial y} + G\frac{\partial B}{\partial y} + C = 0 \tag{6.63}$$

where

$$P = \frac{rB + q_s + q_g}{B} \tag{6.64}$$

$$C = gh(S_f - S_o) \tag{6.65}$$

$$D = \frac{m^2}{h} + \frac{gh^2}{2} \tag{6.66}$$

$$G = \frac{m^2}{Bh} \tag{6.67}$$

and the dependent variables are now m and h, rather than v and h. If we solve for m
and h, v can be determined from (6.61).

The single-step Lax–Wendroff scheme is developed by expanding h and m in Taylor's series about the point $(y, t + \Delta t)$. Equations 6.62 and 6.63 are used to expand the derivatives, and finite-difference approximations of the form

$$\left(\frac{\partial m}{\partial y}\right)_j \approx \frac{m_{j+1} - m_{j-1}}{2\Delta''' y_j} \tag{6.68}$$

$$\left(\frac{\partial m}{\partial y}\right)_{j+\frac{1}{2}} \approx \frac{m_{j+1} - m_j}{\Delta' y_j} \tag{6.69}$$

$$\left(\frac{\partial m}{\partial y}\right)_{j-\frac{1}{2}} \approx \frac{m_j - m_{j-1}}{\Delta'' y_j} \tag{6.70}$$

are used for m, B and D. For P we use:

$$\left(\frac{\partial P}{\partial t}\right) \approx \frac{P^{t+1} - P^t}{\Delta t} \tag{6.71}$$

The resulting explicit expression for h at any interior node is

$$
\begin{aligned}
h_j^{t+1} = h_j^t &- \frac{\Delta t}{2\Delta''' y_j}\left[m_{j+1} - m_{j-1} + \frac{m_j}{B_j}(B_{j+1} - B_{j-1}) - 2\Delta''' y_j P_j\right] \\
&+ \frac{(\Delta t)^2}{2\Delta y_j \Delta' y_j}\left[D_{j+1} - D_j + \left(\frac{G_{j+1} + G_j}{2}\right)(B_{j+1} - B_j) + \frac{\Delta' y_j}{2}(C_{j+1} + C_j)\right] \\
&- \frac{(\Delta t)^2}{2\Delta y_j \Delta'' y_j}\left[D_j - D_{j-1} + \left(\frac{G_j + G_{j-1}}{2}\right)(B_j - B_{j-1}) + \frac{\Delta'' y_j}{2}(C_j + C_{j-1})\right] \\
&+ \frac{(\Delta t)^2}{8B_j(\Delta'' y_j)^2}\Big\{[B_{j+1} - B_{j-1}][(D_{j+1} - D_{j-1}) + G_j(B_{j+1} - B_{j-1}) \\
&\quad + 2\Delta''' y_j C_j]\Big\} + \frac{\Delta t}{2}(P_j^{t+1} - P_j^t)
\end{aligned} \tag{6.72}
$$

where all non-superscripted variables on the right-hand side are evaluated at t.

Development of the corresponding expression for m requires the following definitions

$$E = \frac{\partial D}{\partial m} = \frac{2m}{h} \tag{6.73}$$

$$F = \frac{\partial D}{\partial h} = gh - \frac{m^2}{h^2} \tag{6.74}$$

$$J = \frac{\partial G}{\partial m} = \frac{2m}{Bh} \tag{6.75}$$

$$K = \frac{\partial G}{\partial h} = -\frac{m^2}{Bh^2} \tag{6.76}$$

$$H = \frac{\partial D}{\partial y} + G\frac{\partial B}{\partial y} + C \tag{6.77}$$

$$L = \frac{\partial m}{\partial y} + \frac{m}{B}\frac{\partial B}{\partial y} - P \tag{6.78}$$

The explicit finite-difference for m at any interior node can then be denoted as

$$m_j^{t+1} = m_j^t - \frac{\Delta t}{2\Delta''' y_j}[D_{j+1} - D_{j-1} + G_j(B_{j+1} - B_{j-1})]$$

$$+ 2\Delta''' y_j C_j] + \frac{(\Delta t)^2}{4\Delta y_j \Delta' y_j}\left\{(E_{j+1} + E_j)\left[D_{j+1} - D_j\right.\right.$$

$$+ \left(\frac{G_{j+1} + G_j}{2}\right)(B_{j+1} - B_j)$$

$$+ \frac{\Delta' y_j}{2}(C_{j+1} + C_j)\right] + (F_{j+1} + F_j)\left[m_{j+1} - m_j\right.$$

$$+ \frac{(m_{j+1} + m_j)(B_{j+1} - B_j)}{(B_{j+1} + B_j)}$$

$$- \frac{\Delta' y_j}{2}(P_{j+1} + P_j)\right]\right\} - \frac{(\Delta t)^2}{4\Delta y_j \Delta'' y_j}\left\{(E_j + E_{j-1})\left[D_j - D_{j-1}\right.\right.$$

$$+ \left(\frac{G_j + G_{j-1}}{2}\right)(B_j - B_{j-1})$$

$$+ \frac{\Delta'' y_j}{2}(C_j + C_{j-1})\right] + (F_j + F_{j-1})\left[m_j - m_{j-1}\right.$$

$$+ \frac{(m_j + m_{j-1})(B_j - B_{j-1})}{(B_j + B_{j-1})}$$

$$- \frac{\Delta'' y_j}{2}(P_j + P_{j-1})\right]\right\} + \frac{(\Delta t)^2}{4\Delta''' y_j}(B_{j+1} - B_{j-1})(J_j H_j + K_j L_j)$$

$$- \frac{(\Delta t)^2}{2}\left\{gS_o L_j - \left(\frac{2gn^2 m}{hR^{4/3}}\right)_j\left[H_j - \left(\frac{m}{2h} + \frac{2mR}{3h^2}\right)_j L_j\right]\right\} \tag{6.79}$$

While equations 6.72 and 6.79 may appear formidable, they are easily programmed. If we know the values of all the parameters and variables at all nodes along the line $t = 0$ (Figure 6.10), these equations allow the straightforward calculation of m and v at all the interior nodes along the line $t = 1$. They do not, however, provide the values at the boundaries (i.e. at $j = 0$ and $j = M$). An independent evaluation of these values is needed, and it is best carried out with the single-characteristic boundary solution outlined by Liggett and Woolhiser (1967). This technique makes use of the characteristic solution to Equations 6.41 and 6.52. The reader is referred to Amein and Fang (1969) for a lucid presentation of the underlying theory and to Liggett and Woolhiser (1967) for the details.

The slope of the characteristics is given by

$$\frac{dy}{dt} = v \pm \sqrt{gh} \tag{6.80}$$

and the equation of the characteristic lines by

$$d(v + 2\sqrt{gh}) = -g(S_f - S_o)\,dt - \left(\frac{v}{h} \pm \sqrt{\frac{g}{h}}\right)P\,dt \mp \sqrt{gh}\frac{v}{B}\frac{\partial B}{\partial y}\,dt \tag{6.81}$$

The finite-difference equivalent of Equation 6.80 can be used to determine the distances α and α' (Figure 6.10)

$$\alpha = \mp(v_A \mp \sqrt{gh_A})\Delta t \tag{6.82}$$

and the finite-difference analogue to Equation 6.81 can be developed for the specific characteristics AB and A′B′:

$$\frac{Q_A}{B_A h_A} \mp 2\sqrt{gh_A} = V_B \mp 2\sqrt{gh_B} + \frac{g\Delta t}{2}(S_{o_B} + S_{o_A})$$

$$- \frac{g\Delta t}{2}\left[S_{f_B} + \frac{(Q_A/B_A h_A)^2 n_A^2}{R_A^{4/3}}\right]$$

$$\tag{83}$$

$$- \left[\frac{(Q_A/B_A h_A) + v_B}{h_A + h_B} \pm \left(\frac{2g}{h_A + h_B}\right)^{\frac{1}{2}}\right]\left(\frac{P_A + P_B}{2}\right)\Delta t$$

$$\pm \left[\frac{g(h_A + h_B)}{2}\right]^{\frac{1}{2}}\left[\frac{(Q_A/B_A h_A) + v_B}{B_A + B_B}\right]\left(\frac{B_B - B_A}{\alpha}\right)\Delta t$$

where the upper signs are for the upstream boundary and the lower signs for the downstream boundary. For the latter case A becomes A′, B becomes B′, and α becomes α'. At the upstream boundary Q_A is given directly; at the downstream boundary Q_A is available as a function of h_A through the normal flow relationship.

Parameter values at B and B' are interpolated from the values at the adjacent nodes using the parameters α and α'. Equation 6.83 can be solved for h_A by the method of successive approximations. In that v_A in Equation 6.82 is initially unknown, it must be approximated by the previous value of v at the boundary node or its neighbour. When h_A is calculated, it can be used to correct the v_A-value and hence the α-value. This iterative process must be continued until the value of α converges within a satisfactory tolerance.

For the initial conditions of steady-state flow it is advantageous to apply yet another numerical technique. We now wish to solve the steady equations of flow (6.55) and (6.56) and as with the transient approach it is worthwhile defining m by (6.61) and converting (6.55) and (6.56) to an m–n base. The reduced equations will be identical to (6.62) and (6.63) but with the terms involving t dropped and the derivatives full rather than partial. Knowing $Q(0, 0)$ at $j = 1$ allows calculation of m, and a straightforward finite-difference analogue to the reduced form of (6.62) yields the remaining values of m_j for $j = 2$ to M. Using the normal flow relationship at $j = M$, one can calculate h_M with the successive approximations technique. To calculate the remaining values of h, the reduced form of (6.63) can be put in the form:

$$\frac{dh}{dy} = \frac{1}{\left(\dfrac{m^2}{h^2} - gh\right)}\left[\frac{2m}{h}\frac{dm}{dy} + gh(S_f - S_o) + \frac{m^2}{Bh}\frac{dB}{dy}\right] \tag{6.84}$$

Denoting the right-hand side by $f(y, h)$ and using finite-difference equations of the form of (6.68) for dm/dx and dB/dx, we can solve for h, using a second order Runge-Kutta predictor-corrector technique (McCracken and Dorn, 1964). The predictor at $j = M - 1$ is

$$h^{(0)}_{M-1} = h_M - \frac{\Delta'' y_M}{2}[f(y_M, h_M) + f\{y_{M-1}, h_M - \Delta'' y_M f(y_M, h_M)\}] \tag{6.85}$$

and for all other values of j, is

$$h^{(0)}_j = h_{j+2} - 2\Delta''' y_{j+1} f(y_{j+1}, h_{j+1}) \tag{6.86}$$

The corrector at all nodes is

$$h^{(i)}_j = h_{j+1} - \frac{\Delta'' y_{j+1}}{2}[f(y_{j+1}, h_{j+1}) + f(y_j, h^{(i-1)}_j)] \tag{6.87}$$

The mathematical model of flow in a natural stream channel is embodied in a computer program that calculates the initial flow regime from Equations 6.85 through 6.87, and carries the solution through time with Equations 6.72 and 6.79 for interior nodes, and (82) and (83) for boundary nodes. Output from the channel flow model provides the values of the depth, velocity and discharge at all nodes, at all time steps.

6.3.7 Sample results

The primary application of solutions to the shallow-water equations has been in the routing of major flood waves over considerable distances in large rivers. In such cases the discharge of the flood wave overwhelms the influence of the lateral inflow function. Our interest, on the other hand, lies in the generation of streamflow in small tributaries in upstream portions of watersheds where the lateral inflow is the primary source of flow. We are primarily interested in the nature of the lateral inflow function and its rôle in generating source hydrographs. With reference to Figure 6.11 for example, we would expect very different lateral-inflow functions

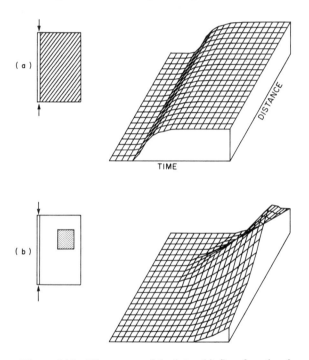

Figure 6.11   The nature of the lateral inflow function for
(a) areally-uniform rainfall and (b) rainfall of limited areal
extent (after Freeze, 1972a)

from a hillslope on which there has been an areally-uniform rainfall and one on which the rainfall has been areally limited. While the time-scale on this diagram would differ, depending on whether lateral inflows were mainly overland flow or mainly subsurface stormflow; the nature of the function would probably be similar.

Ragan (1966) has examined the influences of lateral inflow functions on rising hydrographs using both numerical simulations and a hydraulic experiment with a

H

60-ft tilting flume. His numerical technique was similar but not identical to the Lax–Wendroff scheme described above. His results were duplicated by Freeze (1972) using the Lax–Wendroff method during the initial testing phase of the latter study. Figure 6.12 shows the outflow hydrograph and some flow-depth profiles for a case with a lateral-inflow function that is triangular with time and evenly distributed along the channel. Figure 6.13 compares two cases in which the lateral inflow differs in its spatial distribution. In Figure 6.13, in both cases, the inflow function is triangular with time, but the time base is only half as long as the upper diagram in Figure 6.12.

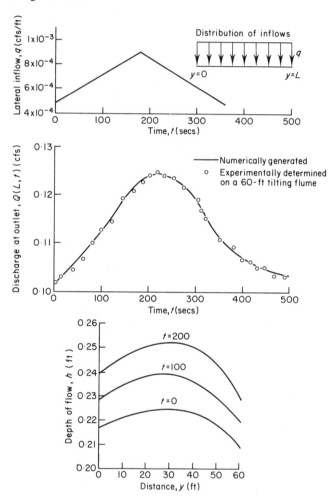

Figure 6.12   Numerical simulation of the outflow hydrograph and flow–depth profiles for a case with flow generated by a lateral inflow function that is triangular with time and evenly distributed along the channel (after Ragan, 1966)

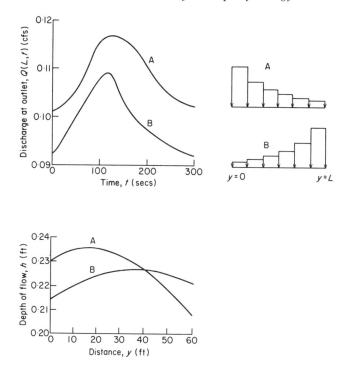

Figure 6.13   Influence of the spatial distribution of lateral inflow on outflow hydrographs and flow–depth profiles (after Ragan, 1966)

6.4   MATHEMATICAL MODEL OF OVERLAND FLOW ON A HILLSIDE

The detailed analysis of streamflow presented in the previous section will now stand us in good stead for a much briefer presentation of the approach to overland flow modelling. We will make use of the sheetflow representation of overland flow and a form of the shallow water equations that describes this form of 'open-channel' flow. The application of the shallow-water equations, or their kinematic approximation, to overland flow generation was pioneered by Henderson and Wooding (1964), Morgali and Linsley (1965) and Brakensiek (1966). The most complete and rigorous analysis is that of Woolhiser and Liggett (1967).

6.4.1   The full equations on a single plane

Let us take as the region of flow a one-dimensional line up and down the hillslope in a direction perpendicular to the topographic contours and to the stream in Figure 6.1. It is the line CDE in Figure 6.2. The geometric relationships are identical to those shown on Figure 6.8(a) except that the direction of flow is now x rather than y, and flow occurs as sheetflow rather than in a rectangular channel. The one-dimensional treatment will be taken as representative of flow in a unit width on the sheetflow plane, so that $B = 1$ in (6.37) and $Q = vh$. The hydraulic radius R, defined in connection with (6.53), is calculated for $B \gg h$, so that $R = h$. In the simplest analysis, the slope S_o is taken as constant from the ridge at $x = 0$ to the stream at $x = L$.

For an impermeable overland flow plane, the 'lateral inflow' is simply the rainfall $r(x, t)$, and in most analyses the function r has been taken as a single-step function ($r = r_1$ for $t < t_1$; $r = 0$ for $t > t_1$), constant over x. Kruger and Bassett (1965), Wooding (1965a) and Foster, Huggins and Meyer (1968) have interpreted this constant step-function as being the rainfall rate less an infiltration rate I that is also constant with time and space. In the following section we will examine some results of Smith and Woolhiser's (1971a and 1971b) coupled model where $r = r(t)$ and $I = I(x, t)$. For now, however, let us be satisfied with an impermeable overland flow plane of length L, slope S_o and roughness n, fed by a step function rainfall r.

For this case, the shallow-water Equations (6.41) and (6.52) become

$$h\frac{\partial v}{\partial x} + v\frac{\partial h}{\partial x} + \frac{\partial h}{\partial t} - r = 0 \qquad (6.88)$$

$$\frac{\partial v}{\partial t} + v\frac{\partial v}{\partial x} + g\frac{\partial h}{\partial x} + \frac{rv}{h} + g(S_f - S_o) = 0 \qquad (6.89)$$

For turbulent, subcritical flow that arises as the result of rainfall on an initially-dry plane [$h(x, 0) = 0$ for all x; $v(x, 0) = 0$ for all x], with an upstream boundary condition of no-flow [$v(0, t) = 0$], and a downstream boundary condition of critical overfall [$v(L, t) = \sqrt{gh}$, from (6.39)]; the boundary value problem can be solved using the Lax–Wendroff scheme outlined in the previous section with suitable minor adaptations. However, there are certain numerical problems that arise in the simulation of overland flow that were not discussed in the section on streamflow generation. They result from the starting problems connected with the dry channel, and from the more likely occurrence of laminar flow and supercritical flow. It is beyond the scope of this chapter to enter this netherland and the reader is referred to Woolhiser and Liggett (1967) for rescue.

6.4.2   The kinematic approximation

Many of the numerical problems can be removed if we make use of a simplified analysis based on a kinematic flow model. Kinematic flow occurs on a plane whenever a balance between gravitational and frictional forces is achieved. Under such circumstances the first four terms of (6.89) become negligible with respect to the fifth, and the momentum Equation (6.89) reduces to

$$S_o = S_f \qquad (6.90)$$

The equations of flow thus become (6.88) and (6.90) with S_f defined by a stage-discharge relationship such as the Manning equation. For sheetflow, (6.53) becomes

$$S_f = \frac{v^2 n^2}{N h^{\frac{4}{3}}} \tag{6.91}$$

or, using (6.90) and a consistent set of units so that $N = 1 \cdot 0$

$$v = \frac{S_o^{\frac{1}{2}} h^{\frac{2}{3}}}{[n]} \tag{6.92}$$

This can be re-written as

$$v = \alpha h^{M-1} \tag{6.93}$$

where $\alpha = \sqrt{S_o}/n$ and $M = 5/3$. Multiplying both sides of (6.93) by h yields

$$Q = vh = \alpha h^M \tag{6.94}$$

which makes the stage-discharge nature of the model clearer. If we put (6.88) into the form:

$$\frac{\partial h}{\partial t} - r + \frac{\partial}{\partial x}(vh) = 0 \tag{6.95}$$

and insert (6.94), we obtain:

$$\frac{\partial h}{\partial t} - r + \frac{\partial}{\partial x}(\alpha h^M) = 0 \tag{6.96}$$

Knowing S_o, n, and r (6.96) can be solved for $h(x,t)$ using a numerical method similar to the Lax–Wendroff scheme, and $v(x,t)$ can be obtained directly from (6.93).

Woolhiser and Liggett (1967) have shown that the kinematic approximation is valid whenever a factor they denote by k is greater than 10. The dimensionless form of k is given by:

$$k = \frac{S_0 L_0}{F_0^2 H_0} \tag{6.97}$$

where

S_0 = slope of channel
L_0 = length of channel
H_0 = normal depth at $x = L_0$ the maximum outflow rate Q_{max}
$Q_{max} = r_{max} L_0$
r_{max} = maximum rainfall rate
F_0 = Froude number for normal flow at $x = L_0$ and $Q = Q_{max}$
    $= V_0/\sqrt{gH_0}$
$V_0 = Q_{max}/H_0$

Analysis of this parameter shows that the kinematic approximation is best for rough, steep slopes with low rates of lateral inflow. It is valid on almost all overland flow planes.

Woolhiser and Liggett (1967) also show that only with the kinematic approach can a unique overland flow hydrograph be generated. When the full equations are needed, for slopes with $k < 10$, the complications introduced by the alternative possibilities of subcritical and supercritical flow make the rising hydrograph non-unique. Woolhiser and Liggett (1967) present their results for such cases as a function of k and F_0.

6.4.2 The kinematic cascade

In order to bring overland flow simulations closer to reality it is necessary to attempt simulations on planes that are not restricted to a single slope S_o. The most advanced approach of this type is that of a *kinematic cascade*, which is defined as a sequence of a n discrete overland flow planes in which the kinematic wave equations are used to describe the unsteady flow. Kibler and Woolhiser (1970) provide the most complete analysis of such a system. They give credit to Brakensiek (1967) for introducing the concept.

In the kinematic-cascade approach, the equations of flow (6.96) and (6.93) are solved independently for each plane at each time step. S_o and n must be specified for each plane. r is usually taken as a step function common to all the planes. At $t = 0$ all planes in the cascade are considered dry; at later times the inflow hydrograph from the upstream plane establishes flow at the upper boundary of any plane. No flow enters the upstream boundary of plane 1. Kibler and Woolhiser (1970) describe a version of the Lax–Wendroff scheme for the solution of $h(x, t)$ on a kinematic cascade.

An undesirable feature of the kinematic-cascade model of overland flow is the development of shock-wave phenomena in the solutions. Kibler and Woolhiser (1970) have examined this aspect of the methodology rather carefully and they conclude that the shocks are a property of the mathematical approach rather than an observable feature of the hydrodynamic process.

Sample results of overland flow simulations will be deferred until the following section where results are presented for Smith and Woolhiser's (1971a, and 1971b) coupled model of a kinematic cascade with infiltration.

6.5 COUPLED MODELS

6.5.1 Overland flow on an infiltrating surface

Smith and Woolhiser (1971a and 1971b) couple an overland flow model in the form of a kinematic cascade, to a subsurface flow model, in the sense that they determine infiltration from the plane at any point with a one-dimensional, vertical, saturated–unsaturated flow calculation. Figure 6.14 is a schematic representation of their overland flow model. Each plane in the cascade may have a different slope S_i, length L_i and roughness n_i. The soil may be layered, and the one-dimensional subsurface model can be solved at as many points along the surface as are necessary to define the horizontal variations in soil properties.

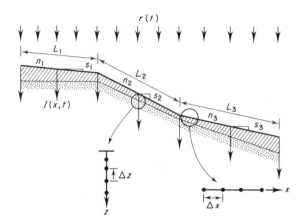

Figure 6.14 Schematic representation of overland flow model with saturated–unsaturated infiltration (after Smith and Woolhiser, 1971b)

The mathematical model consists of the simultaneous solution of the kinematic equations of overland flow (6.96) and (6.93) and the one-dimensional form of the subsurface flow Equation (6.13). The lateral inflow to the overland flow plane, designated by r in (6.96) becomes $q(x, t)$, where

$$q(x, t) = r(t) - I(x, t) \qquad (6.98)$$

and $r(t)$ is the rainfall rate. $I(x, t)$ is the rate of infiltration provided by the solution of the subsurface flow model. Smith and Woolhiser used the Lax–Wendroff method to solve the surface flow equations, and the Crank–Nicolson scheme with Jacobi iteration to solve the subsurface flow equation. The latter method is an implicit–iterative scheme, similar in concept to the LSOR technique presented in Section 6.2.6.

Smith and Woolhiser tested their model both in the laboratory and in the field. Figure 6.15 shows two numerically-simulated hydrographs compared with data taken from a 40-ft laboratory soil flume modified to create a prototype infiltrating slope. The match is quite impressive, and soil-moisture profiles from model and flume were also reported to show reasonable agreement. The only discrepancy in Figure 6.15 is a rapid jump in runoff near the end of the rainfall event in the moist case. The jump is probably caused by air entrappment, a phenomena not accounted for in the numerical model.

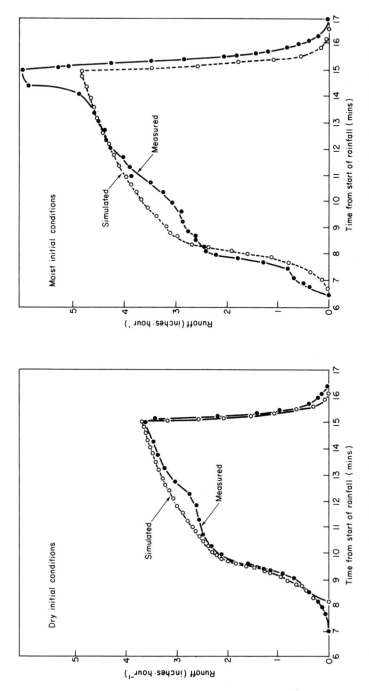

Figure 6.15   Measured and simulated overland flow hydrographs for laboratory prototype infiltrating slope (after Smith and Woolhiser, 1971b)

The model was also tested against published rainfall-runoff records from field plots in an experimental watershed in Nebraska, and a sample result is shown in Figure 6.16. The hydraulic properties of the soils were obtained from available records but the roughness parameter and initial subsurface conditions had to be fitted by trial. In light of these data shortcomings, the authors note that attempts at simulating watershed runoff with this approach are best considered as an exercise in fitting physical parameters into a theoretical framework.

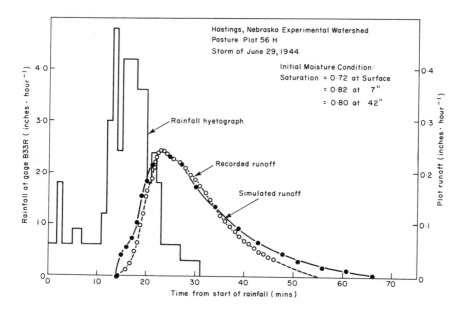

Figure 6.16   Measured and simulated overland flow hydrographs for a field plot in an experimental watershed (after Smith and Woolhiser, 1971b)

6.5.2   Overland flow feeding a stream channel

Several workers in the field of surface-water hydrology have created watershed models for upstream catchment areas that couple models of overland flow to models of channel flow. With this approach the shallow water equations, or their kinematic approximation, are first applied to the overland flow phase with rainfall as the lateral inflow, and then to the channel flow phase with overland flow as the lateral inflow. Wooding (1965a, 1965b, and 1966) used a single-slope, kinematic representation of overland flow to feed a single-slope, kinematic representation of channel flow. Harbaugh and Chow (1967) and Chen and Chow (1968) used the full equations for each component. Kibler and Woolhiser (1970) fed a kinematic cascade into a kinematic cascade. In all these watershed models, coupling exists only between the surface flow components. Subsurface flow is either ignored, or, in the case of Wooding's (1965a) study and that of Chen and Chow (1968), specified as a simple external function representing loss by infiltration.

Harbaugh and Chow (1967) refer to their model as a coneptual watershed and, recognizing the impossibility of measuring the roughness n on all hillslopes liable to overland flow throughout the watershed, they refer to it as the conceptual watershed roughness. In this light, these watershed models, which are usually thought of as physically based, become almost parametric, with n as the fitting parameter.

The original papers include many simulations for both hypothetical and real watersheds and the interested reader is directed to them for a sampling of representative results.

6.5.3   Subsurface flow feeding a stream channel

Freeze (1972a) has coupled the saturated–unsaturated subsurface-flow model outlined in the second section of this chapter with the channel-flow model described in the third section. The coupled model was applied to an analysis of (i) baseflow contributions to large perennial streams (Freeze, 1972a), and (ii) subsurface hillslope hydrologic processes in upstream source areas (Freeze, 1972b). We will present, as sample results, some of the output from the second part of the study.

The simulations were carried out on hypothetical hillslope-channel combinations for the express purpose of examining the role of subsurface stormflow as a runoff-generating mechanism. Early simulations revealed that necessary conditions for the dominance of the subsurface-flow mechanism included a deeply incised channel, a convex hillslope, and surface soils with very high hydraulic conductivity. The inset to Figure 6.17 shows such a configuration. The main portion of the figure displays the streamflow hydrographs at the source-area outlet for four possible values of saturated hydraulic conductivity of the hillside surface soil layer. The unsaturated characteristic curves were taken as being of the same shape as Figure 6.4 in each case, but with the differing K_o values. The rainfall pulse giving rise to the runoff events was a step-function of moderate intensity and 5-hr duration. In all four cases, the rainfall intensity was less than the saturated hydraulic conductivity of the surface soil layer, so that all rainfall infiltrated, and there was no overland flow. Each hydrograph was generated solely by subsurface stormflow. The four curves delivered 100 %, 35 %, 7 % and 1 % of the precipitation input, respectively, to the outlet as runoff.

Figure 6.18 shows the subsurface flow systems in more detail for Cases B and D of Figure 6.17. The hydraulic head patterns reveal that Case B is a true subsurface-stormflow case with water delivered through the entire profile to the stream channel via the unsaturated flow system. In Case D, on the other hand, moisture movement from the land surface is still feeding a residual soil moisture sink along the base of the soil layer. The source of subsurface outflow to the stream is restricted to water entering the soil profile very near the channel.

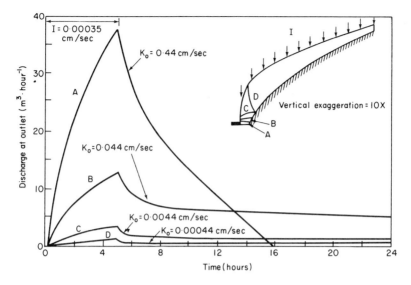

Figure 6.17   Simulated stream hydrographs at the outlet of a hypothetical upstream source area. The four cases of runoff are generated by subsurface stormflow from the same rainfall event on the same convex hillslope but with four different saturated hydraulic conductivity values for the surface soil layer. The inset shows the hillside cross-section and the position of the water table for each case at t = 5 hours (after Freeze, 1972b)

The purpose of producing these results here is simply to provide examples of output from this type of coupled mathematical model. For a more complete analysis of the hydrological interpretations of this study the reader is referred to the original papers and to Chapter 8 in this book.

The coupled model of Freeze (1972a) has also been applied to the simulation of a field event. Stephenson and Freeze (1974) report on the use of this model to complement a field study of snowmelt runoff in a small upstream source area in the

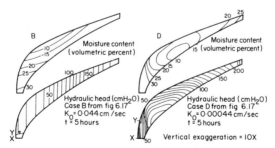

Figure 6.18   Moisture-content and hydraulic-head fields in the surface soil layer at t = 5 hours for two cases from Figure 6.17. Stippled zones are saturated (after Freeze, 1972b)

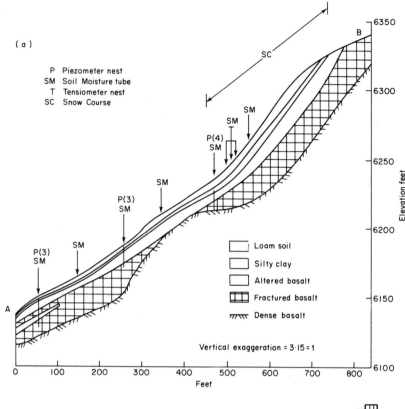

(a)

P   Piezometer nest
SM   Soil Moisture tube
T   Tensiometer nest
SC   Snow Course

Loam soil
Silty clay
Altered basalt
Fractured basalt
Dense basalt

Vertical exaggeration = 3·15 = 1

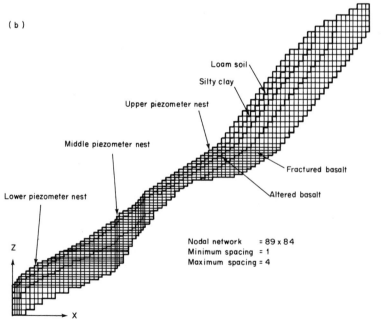

(b)

Loam soil
Silty clay
Upper piezometer nest
Middle piezometer nest
Fractured basalt
Lower piezometer nest
Altered basalt

Nodal network   = 89 × 84
Minimum spacing = 1
Maximum spacing = 4

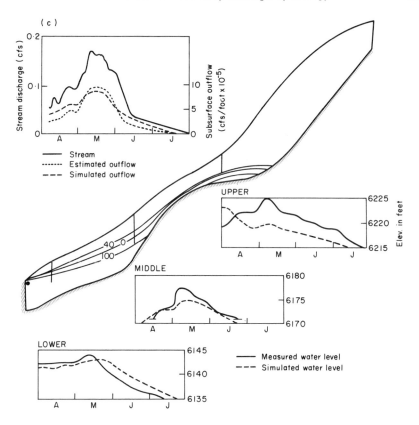

Figure 6.19 Simulation of subsurface flow contributions to snowmelt runoff from an instrumented hillslope in the Reynolds' Creek Experimental Watershed, Idaho. (a), Topography, geology and instrumentation; (b), nodal network for mathematical model; (c), calibrated transient flow system for the period April 5–July 13, 1971 (after Stephenson and Freeze, 1974)

Reynolds' Creek Experimental Watershed near Boise, Idaho. Field measurements from an instrumented hillslope on this small watershed show that the mechanism of streamflow generation is subsurface delivery of meltwater over limited distances through shallow, high-permeability, low-porosity formations of altered and fractured basalt. Figure 6.19(a) shows the hillslope topography, geology and instrumentation; Figure 6.19(b) shows the nodal network used in the numerical simulation; and Figure 6.19(c) shows the comparison between the measured and simulated system. The comparative checks between model and reality that are shown in Figure 6.19(c) are the water levels in three piezometers and the sub-surface outflow to the stream. The matches are neither particularly good nor particularly bad. They are probably representative of the type of results that can be expected from the application of complex theoretical models to complex field situations under the constraints created by data limitations and the resulting uncertainties in the calibration procedures.

6.6 LIMITATIONS OF MATHEMATICAL MODELLING OF HILLSLOPE HYDROLOGIC PROCESSES

Physically-based mathematical models of hillslope hydrologic processes of the type that have been discussed in this chapter have many limitations. For the purposes of discussion one might group them in the following way:

(a) Limitations due to the assumptions of the theoretical developments.
(b) Limitations due to a fundamental lack of correspondence between the theoretical models and reality.
(c) Limitations caused by the scarcity of data.
(d) Limitations caused by the inadequacy of computer capacity.
(e) Limitations of the calibration procedures.

To assess the importance of limitations of type (a) one must list the assumptions underlying the development of the flow equations. For the subsurface-flow model described in this chapter, it is assumed that flow is laminar and Darcian; that inertial forces, velocity heads, temperature gradients, osmotic gradients and chemical concentration gradients are negligible; that the soils are non-swelling; and that the air phase in the unsaturated zone is continuous and always in connection with constant external pressure. For the surface-flow equations, it is assumed that the fluid is incompressible and under isothermal conditions; that the requirements of one-dimensionality are satisfied; that the channel properties or those of the overland flow plane are gradually-varied; that the depth of water and the amplitude of the surface waves are much less than the wavelength; that the natural slope is small; and that the channel or plane is fixed with no sediment transport. In general, these types of assumptions do not lead to a large loss in generality of the models. Either the influence of the assumed conditions is relatively minor or the counter-occurrences are relatively rare. Perhaps the most serious discrepancies result at the infiltration end of the cycle from the lack of consideration of swelling soils and air entrappment in the unsaturated subsurface model.

Limitations of type (b) result, not from the failure of the model to account for some specific feature of the system (such as temperature gradients in extreme climatic conditions, which would result in a limitation of type (a)), but rather from its failure to represent the actual mechanisms at a fundamental level. The theoretical models for subsurface flow and for channel flow appear to be relatively accurate physical representations of reality, but the sheetflow representation of overland flow is open to serious question. It undoubtedly provides a satisfactory analysis of flow from a smooth homogeneous surface; it is less clear that it describes the complexity of quasi-channelized flow on natural grassy, forested or cultivated slopes. The overland-flow models are probably best viewed as parametric prediction models. I feel that they provide somewhat less insight into the mechanisms of flow than do the subsurface-flow or channel-flow models.

Once a model is deemed theoretically sound, the greatest practical limitations are those of types (c), (d) and (e). On the data front, the physically-based approach requires complete spatial specification of the saturated and unsaturated hydrogeological parameters and the surface roughness characteristics. Even when the distribution of these parameters is relatively even through space, there is seldom enough data available at field sites to provide the necessary input to hillslope hydrologic models. When the hydrologic or hydrogeologic patterns are highly heterogeneous, as is usually the case, the data problem is intensified even further. The usual approach is to use representative average values in simplified, idealized configurations of the actual conditions. The development of methods for arriving at representative values from the available statistical data, and the analysis of the uncertainties introduced into the solutions by these approximations, are fertile fields of research which bear greater emphasis than they have received in the past.

The computer limitations are severe for the subsurface-flow models, but not for the surface-flow models. In the coupled simulations of Freeze (1972a) 98 % of the computer time was spent on subsurface computations. In coupled models with large subsurface components, applications for some time to come will be restricted to individual hydrologic events on individual slopes feeding single streams. In the longer term, the anticipated growth in computer capacity should lead us to be optimistic about future applications on a larger scale.

Stevenson and Freeze (1974) discuss calibration problems of the subsurface model in some detail.

In summary, limitations of type (c) due to data availability and the associated calibration limitations of type (e), constitute the most serious long-term threat to the viability of the physically-based modelling approach. In the 1970s it is likely that this type of modelling will find its greatest value in the examination of flow mechanisms on hypothetical hillslopes or on small instrumented research watersheds rather than as a large-scale engineering tool for hydrologic prediction.

REFERENCES

Amein, M. and Fang, C. S., 1969, 'Streamflow routing (with applications to North Carolina rivers)', *Water Res. Res. Inst., Univ. North Carolina* Rept. 17, 106 pp.

Amorocho, J. and Hart, W. E., 1964, 'A critique of current methods in hydrologic systems investigation', *Trans. Am. Geophys. Unions*, **45**, 307–321.

Black, C. A. (Ed.), 1965, *Methods of Soil Analysis*, Part 1, American Society of Agronomy, Madison, Wisconsin, 770 pp.

Brakensiek, D. L., 1966, 'Hydrodynamics of overland flow and nonprismatic channels', *Trans. Am. Soc. Agric. Eng.*, **9**(1), 20–26.

Brakensiek, D. L., 1967, 'A simulated watershed flow system for hydrograph prediction: A kinematic application', *Proc. Intern. Hydrology Symposium*, Fort Collins, Colorado.

Carslaw, H. S. and Jaeger, J. C., 1959, *Conduction of Heat in Solids*, 2nd ed., Oxford Univ. Press, London.

Chen, C. L. and Chow, V. T., 1968, 'Hydrodynamics of mathematically-simulated surface runoff', *Hydraul., Eng.*, Ser. 18, Univ. Illinois, Urbana, 132 pp.

Chow, V. T., 1959, *Open-channel Hydraulics*, McGraw-Hill, New York.

Clarke, R. T., 1973, 'A review of some mathematical models used in hydrology, with observations on their calibration and use', *J. Hydrology*, **19**, 1–20.

Foster, G. R., Huggins, L. F. and Meyer, L. D., 1968, 'Simulation of overland flow on short field plots', *Water Res. Res.*, **4**(6), 1179–1187.

Freeze, R. A., 1971a, 'Three-dimensional, transient, saturated–unsaturated flow in a groundwater basin', *Water Res. Res.*, **7**(2), 347–366.

Freeze, R. A., 1971b, 'Influence of the unsaturated flow domain on seepage through earth dams', *Water Res. Res.*, **7**(4), 929–941.

Freeze, R. A., 1972a, 'Role of subsurface flow in generating surface runoff. 1. Base flow contributions to channel flow', *Water Res. Res.*, **8**(3), 609–623.

Freeze, R. A., 1972b, 'Role of subsurface flow in generating surface runoff, 2. Upstream source areas', *Water Res. Res.*, **8**(5), 1272–1283.

Harbaugh, T. E. and Chow, V. T., 1967, 'A study of roughness of conceptual river systems or watersheds', *Twelfth Congress of the International Association for Hydraulic Research*, Fort Collins, Colorado, **I**, 9–17.

Henderson, F. M. and Wooding, R. A., 1964, 'Overland flow and groundwater flow from a steady rainfall of finite duration', *J. Geophys. Res.*, **69**, 1531–1540.

Hornberger, G. M., Remson, I and Fungaroli, A. A., 1969, 'Numeric studies of a composite soil moisture ground-water system', *Water Res. Res.*, **5**(4), 779–802.

Kibler, D. F. and Woolhiser, D. A., 1970, 'The kinematic cascade as a hydrologic model', *Colorado State University Hydrology Paper*, No. **39**, 27 pp.

Kruger, W. E. and Bassett, D. L., 1965, 'Unsteady flow of water over a porous bed having constant infiltration', *Trans. Am. Soc. Agric. Engr.*, **8**(1), 60–62.

Liakopoulos, A. C., 1965, 'Retention and distribution of moisture in soils after infiltration has ceased', *Bull. Internat. Assoc. Sci. Hydrology*, **10**, 58–69.

Liggett, J. A. and Woolhiser, D. A., 1967, 'Difference solutions of the shallow-water equation', *J. Engr. Mechs. Div. Am. Soc. Civil Engr.*, **93**(EM2), 39–71.

Liggett, J. A. and Woolhiser, D. A., 1969, 'Closure to Liggett and Woolhiser (1967)', *J. Engr. Mechs. Div. Am. Soc. Civil Engr.*, **95**(EM1), 303–311.

McCracken, D. D. and Dorn, W. S., 1964, '*Numerical Methods and Fortran Programming*', John Wiley.

Morgali, J. R. and Linsley, R. K., 1965, 'Computer analysis of overland flow', *Proc. Am. Soc. Civil Engr., J. Hydraul. Div.*, **91**(HY3), 81–100.

Philip, J. R., 1957, 'The theory of infiltration: I. 'The infiltration equation and its solution', *Soil Sci.*, **83**, 345–357.

Ragan, R., 1966, 'Laboratory evaluation of a numerical flood routing technique for channels subject to later inflows', *Water Res. Res.*, **2**(1), 111–121.

Remson, I., Hornberger, G. M. and Molz, F. J., 1971, *Numerical Methods in Subsurface Hydrology*, Wiley–Interscience.

Rubin, J., 1968, 'Theoretical analysis of two-dimensional, transient flow of water in unsaturated and partly unsaturated soils', *Proc. Soil Sci. Soc. Am.*, **32**, 607–615.

Rubin, J. and Steinhardt, R., 1963, 'Soil water relations during rain infiltration, 1. Theory', *Proc. Soil Sci. Soc. Am.*, **27**, 246–251.

Smith, R. E. and Woolhiser, D. A., 1971a, 'Mathematical simulation of infiltrating watersheds', *Colorado State University Hydrology Paper* No. **47**, 44 pp.

Smith, R. E. and Woolhiser, D. A., 1971b, 'Overland flow on an infiltrating surface', *Water Res. Res.*, **7**(4), 899–913.

Stephenson, G. R. and Freeze, R. A., 1974, 'Mathematical simulation of subsurface flow contributions to snowmelt runoff, Reynolds' Creek Watershed, Idaho', *Water Res. Res.*, **10**(2), 284–298.

Strelkoff, T., 1969, 'One-dimensional equations of open-channel flow', *J. Hydraul. Div. Am. Soc. Civil Eng.*, **94**(HY3), 861–876.

Strelkoff, T., 1970, 'Numerical solution of Saint Venant equations', *J. Hydraul. Div. Am. Soc. Civil Engr.*, **95**(HY1), 223–251.

Verma, R. D. and Brutsaert, W., 1970, 'Unconfined aquifer seepage by capillary flow theory', *J. Hydraul. Div. Am. Soc. Civil Engr.*, **96**(HY6), 1331–1344.

Wooding, R. A., 1965a, 'A hydraulic model for the catchment-stream problem. I Kinematic-wave theory', *J. Hydrology*, **3**, 254–267.

Wooding, R. A., 1965b, 'A hydraulic model for the catchment-stream problem. II Numerical solutions', *J. Hydrology*, **3**, 268–282.

Wooding, R. A., 1966, 'A hydraulic model for the catchment-stream problem. III Comparison with runoff observations', *J. Hydrology*, **4**, 21–37.

Woolhiser, D. A. and Liggett, J. A., 1967, 'Unsteady, one-dimensional flow over a plane–The rising hydrograph', *Water Res. Res.*, **3**(3), 753–771.

CHAPTER 7

Field studies of hillslope flow processes

T. Dunne

Dept. of Geological Sciences and Quaternary Research Center, University of Washington, Seattle, Wash. USA

7.1 INTRODUCTION

When rain and meltwater reach the surface of the ground, they encounter a filter that is of great importance in determining the path by which hillslope runoff will reach a stream channel. The paths taken by the water (*see* Figure 7.1) determine many of the characteristics of the landscape, the uses to which land can be put, and the strategies required for wise land management.

If the rate of rainfall or melting is greater than the capacity of the soil to absorb water, the unabsorbed excess becomes overland flow, referred to later in this chapter as *Horton overland flow* (Path No. 1 in Figure 7.1).

If the precipitation is first absorbed by the soil, it may be stored there, or may move toward stream channels by a variety of routes. If the soil or rock is deep and of uniform permeability, the subsurface water moves vertically to the zone of saturation, and thence follows a curving path to the nearest stream channel (Path No. 2 in Figure 7.1). Inhomogeneities of geological structure may disrupt this simple pattern of flow in the groundwater zone, as illustrated throughout the text by Davis and DeWiest (1966). Because rates of groundwater flow are generally slow and the underground paths are long, most of the water following them contribute to baseflow between rainstorms. Some of this groundwater discharge contributes to the stormflow hydrograph, and the antecedent baseflow to which stormflow is added from other sources is an important factor in determining the size of flood peaks. In very permeable rock formations such as limestones and basalts with large joint systems, the rate of subterranean water movement may be so rapid that considerable amounts of stormflow originate from the groundwater. Generally, however, water taking the long path below ground dominates the baseflow of streams rather than their stormflow.

It at some shallow depth in the soil or rock, percolating water encounters an impeding horizon, a portion of the water will be diverted horizontally and will reach the stream channel by a much shorter route (Path No. 3 in Figure 7.1). Because of the shorter route, the high permeability of topsoil and weathered rocks relative to unweathered rock, and the generally greater potential gradients in these upper, sloping horizons, water following this path reaches the stream channel much more quickly than the groundwater flow described above. Some of this water arrives at the channel quickly enough to contribute to the storm hydrograph and is classified as *subsurface stormflow*.

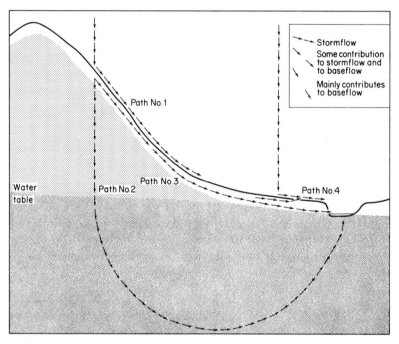

Figure 7.1   Possible flow paths of water moving downhill

In some parts of a hillslope, vertical and horizontal percolation may cause the soil to become saturated throughout its depth. When this happens some of the water moving by the shallow subsurface path emerges from the soil surface and reaches the stream channel as overland flow. Such water can be referred to as *return flow*. This was defined by Musgrave and Holtan (1964) as subsurface flow which returns to the ground surface and leaves the hillside as overland flow. Rainfall onto the saturated area cannot infiltrate, but runs over the surface. This contribution, termed *direct precipitation onto saturated areas* is difficult to separate from return flow. Storm runoff from these two sources are classified together as *saturation overland flow* (Path No. 4 in Figure 7.1). Its movement over the surface allows it to attain sufficient velocity to reach stream channels during rainstorms, and so runoff from this source contributes to storm hydrographs.

This chapter deals with the experimental evidence for each of the processes introduced above. Each process has a different response to rainfall or snowmelt, and the volumes, peak rates, and timing of contributions from each source will be considered. It will be stressed that the relative importance of each process in a particular region (and more precisely on each hillslope) is affected by climate, geology, physiography, soil characteristics, vegetation, and land use. Consequently, one should expect differences between the results of studies carried out in regions of differing geography, and no one researcher has documented the full range of runoff-producing mechanisms.

Some of the confusion associated with the study of runoff production appears to be due to varying conceptions of stormflow and baseflow. Separation of runoff into these categories has always been arbitrary and unfortunate. Whether runoff from a hillslope is contributing to stormflow or baseflow depends upon the character of storm hydrographs in the channels of that particular region. If the storm hydrographs in channels of small drainage basins usually extend over several days after a rainstorm, water flowing from hillslopes two days after rain contributes stormflow. But if the channel storm hydrograph of such a basin lasts for only a few hours after rain, only the drainage from hillslopes during those few hours is stormflow. Drainage basin area is important in this discussion, since the lag times of storm hydrographs and their duration increase with the size of the catchment. Therefore, drainage from hillslopes two days after a rainstorm may contribute stormflow to the channel near the exit of a 1000-km^2 river basin, while similar drainage in the small headwater catchments of the same basin is contributing to baseflow. Most attempts to model runoff physically and to understand the basic runoff processes, however, are confined to small catchments.

For some purposes and in some regions, the important questions about hillslope hydrology do not concern the production of stormflow, but only the moisture conditions on the hillslopes themselves. For the purposes of plant ecology, soil-moisture distribution during dry seasons may be critical, while farmers may be concerned with the spatial and temporal occurrence of saturated soils. Engineers may concentrate their attention upon the build-up of hydrostatic pressures in soils or rock. The study of these characteristics also depends upon an understanding of hillslope flow processes, and should not be forgotten in the discussion over the production of storm runoff.

7.2 THE HORTON MODEL OF STORM RUNOFF PRODUCTION

7.2.1 Infiltration and overland flow

The work of Horton (1933) and others on the process of infiltration has been introduced in Chapter 1 and is more fully described in Chapter 2. Representative variations of infiltration capacity over time are shown in Figure 2.9 and Figure 7.2. Many factors influence the shape of the infiltration-capacity curve. The most important controls are: rainfall characteristics (intensity, duration, drop size); soil characteristics (texture, structure, depth, initial moisture content, clay mineralogy, condition of the soil surface before rainfall); vegetation and land-use. Some of these controls are obvious from the curves of Figure 7.2. Table 7.1 gives a sample of some final constant infiltration rates in order to indicate the range of this parameter.

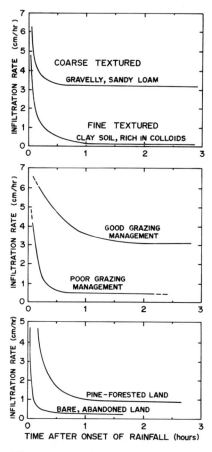

Figure 7.2   Infiltration capacity curves
for various soils (after Strahler, 1969)

If rainfall intensity at any time during the storm exceeds the infiltration capacity, water will accumulate on the soil surface and will run downslope as Horton overland flow (Chapters 5 and 6). Velocities of overland flow lie in the range 10–500 m/hr, so that flow from the top of a 100–m long hillside could reach the stream channel within 0·2–10 hr, and the discharge rate from a hillside plot would attain a constant value of

$$q_t = (i - f_t)$$

Table 7.1 Some final constant infiltration rates measured with a sprinkling infiltrometer or under rainfall

Region	Soil type	Cover	Final Infiltration rate (cm/hr)	Source
Forest Soils				
Vermont	Sandy loam and silt loam	Pasture, previously pine woodland	>8·0	Dunne (1969a)
Vermont	Sandy loam with concrete frost	Pasture, previously pine woodland	0·02	Dunne (1969a)
Ohio	Sandy loam and silt loam	Hardwood forest	>7·6	Whipkey (1969)
Sudbury, Ontario	Eroded sandy silt	Bare	0·2	Pearce (1973)
Agricultural Soils				
Midwestern US	Silt loams	Bluegrass	0·4–1·55	Musgrave and Holtan (1964)
Midwestern US	Silt loams	Corn	0·2–0·46	Musgrave and Holtan (1964)
Midwestern US	Silt loams	Old, permanent pasture	6·1	Musgrave and Holtan (1964)
Midwestern US	Silt loams	4–8 year-old pasture	4·6	Musgrave and Holtan (1964)
Midwestern US	Silt loams	3–4 year-old pasture	3·05	Musgrave and Holtan (1964)
Midwestern US	Silt loams	Permanent pasture, heavily grazed	1·4	Musgrave and Holtan (1964)
Midwestern US	Silt loams	Permanent pasture, moderately glazed	2·0	Musgrave and Holtan (1964)
Midwestern US	Silt loams	Hay	1·5	Musgrave and Holtan (1964)
Midwestern US	Silt loams	Weeds, grain	1·0	Musgrave and Holtan (1964)

Table 7.1 (*Continued*).

Region	Soil type	Cover	Final Infiltration rate (cm/hr)	Source
Midwestern US	Silt loams	Bare, crusted	0·68	Musgrave and Holtan (1964)
Midwestern US	Heavy, plastic clays	Bare	0·0–0·12	Musgrave and Holtan (1964)
Midwestern US	Clay loams, shallow sandy loams, soils low in organic matter and high in clay	Bare	0·12–0·38	Musgrave and Holtan (1964)
Agricultural Soils				
Midwestern US	Shallow loess and sandy loams	Bare	0·38–0·76	Musgrave and Holtan (1964)
Midwestern US	Deep sand, deep loess, aggregated silts	Bare	0·76–1·14	Musgrave and Holtan (1964)
Midwestern US	Silt loam	Corn	0·3–1·32	Bertoni *et al.* (1958)
Rangeland Soils				
W. Colorado	Soils developed on shales	Grazed range	Dry 1·6–2·2 Wet 1·4–2·1	Lusby *et al.* (1963)
W. Colorado	Soils developed on shales	Ungrazed range	Dry 1·6–2·0 Wet 1·4–1·9	Lusby *et al.* (1963)
W. Colorado	Soils developed on sandstones	Grazed range	Dry 4·7–5·2 Wet 3·1–3·3	Lusby *et al.* (1963)
W. Colorado	Soils developed on sandstones	Ungrazed range	Dry 6·4–8·4 Wet 4·2–5·5	Lusby *et al.* (1963)
Arizona/New Mexico	Gravelly sands and gravelly loams	Brush and grass	1·3–4·3	Kincaid *et al.* (1966)

where

q_t = discharge rate (cm/hr)
i = rainfall intensity (cm/hr)
f_t = infiltration capacity (cm/hr)

The relationship of infiltration and overland flow are shown in Figure 7.3 for an artificial rainfall of constant intensity.

In natural storms with rapid fluctuations of rainfall intensity, steady-state conditions are not attained, and more complex hydrographs are produced by

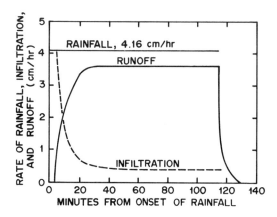

Figure 7.3   Relationship of Horton overland flow to infiltration for rainfall of constant intensity (after Sherman and Musgrave, 1942)

overland flow, as shown in Figure 7.4. At the beginning of the storm the infiltration capacity exceeds rainfall intensity, and there is no accumulation of water on the soil surface. When rainfall intensity later exceeds the infiltration capacity, the first portion of the excess fills surface depressions. When the depressions are filled, the excess precipitation spills downslope and coalesces as an irregular sheet of overland flow. Runoff then rises rapidly to a sharp peak, followed by a rapid decline that begins within a minute of the end of a burst of rainfall. When the rainfall intensity is suddenly reduced in this way, previously-accumulated surface water drains away to provide the runoff represented by the steep recession limb of the hydrograph. In the succeeding burst of rainfall, the process is repeated.

In areas where this process is the dominant producer of storm runoff, overland flow is general over large areas of hillside with a sufficiently low infiltration capacity. The infiltration capacity is not necessarily constant over a hillside or drainage basin as shown by the field measurements displayed in Figure 7.5. Betson (1964) used the concept of the spatial variability of infiltration capacity to improve his predictions of the volumes of storm runoff from small drainage basins. Using a

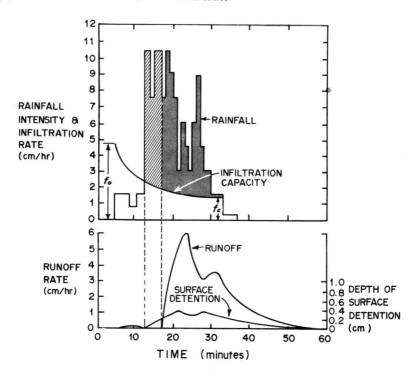

Figure 7.4   Relationship of Horton overland flow and surface detention to rainfall intensity and infiltration capacity for a storm of varying intensity. The shaded portion of the rainfall bar-graph represents water that does not infiltrate. The lightly-shaded portion represents depression storage, which must be exhausted before runoff is generated (after Horton, 1940)

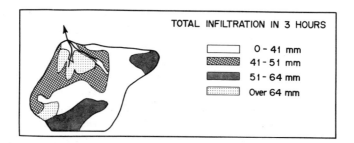

Figure 7.5   Spatial variation of infiltration characteristics, surveyed with an F-type infiltrometer for a 10·9-hectare catchment in claypan soils under grass (after Musgrave and Holtan, 1964)

Table 7.2   Runoff coefficients (runoff volume/rainfall volume) for plots and small drainage basins where storm runoff is generated by Horton overland flow

Location	Drainage area (km²)	Vegetation cover	Soil type	Runoff coefficient	Source
Sudbury, Ontario	0.000002	Bare	Sandy silt	0.84–0.88 (large storms)	Pearce (1973)
S. Negev, Israel	0.00004	Bare	Gravel	0.002–0.15	Yair and Klein (1973)
S. Negev, Israel	0.11	Grass and shrubs	Gravel and sand	0–0.88	Schick (1970)
S. Negev, Israel	0.58	Grass and shrubs	Gravel and sand	0–0.25	Schick (1970)
N. Negev, Israel	0.00008	Bare		0.14–0.27	Yair and Klein (1973)
Mpwapwa, Tanzania	0.00005	Bare	Sandy loam	0.2–0.8 (depending on storm size)	Temple (1972)
Lyamungu, Tanzania	0.00014	Young coffee bushes with various tillage practices	Clay loam	0–1.0 (depending on storm size and tillage practice)	Temple (1972)
Stillwater, Oklahoma	0.064	Grass		0.33–0.90; mean = 0.68 (for large storms)	US Dept. Agric. (1963–71)
Stillwater, Oklahoma	0.37	Grass		0.20–0.98; mean = 0.61 (for large storms)	US Dept. Agric. (1963–71)
Stillwater, Oklahoma	0.82	Grass		0.16–0.90; mean = 0.61 (for large storms)	US Dept. Agric. (1963–71)
Tombstone, Arizona	2.25	Grass and shrubs		0.13–0.80; mean = 0.37 (for large storms)	US Dept. Agric. (1963–71)
Tombstone, Arizona	8.85	Grass and shrubs		0.12–0.28; mean = 0.20 (for large storms)	US Dept. Agric. (1963–71)
Tombstone, Arizona	22.0	Grass and shrubs		0.05–0.42; mean = 0.2 (for large storms)	US Dept. Agric. (1963–71)
Tombstone, Arizona	112	Grass and shrubs		0.022–0.134; mean = 0.06 (for large storms)	US Dept. Agric. (1963–71)
Tombstone, Arizona	148	Grass and shrubs		0.03–0.25; mean = 0.14	US Dept. Agric. (1963–71)
Tombstone, Arizona	148	Grass and shrubs		0.04 (mean for all summer storms)	Osborn and Renard (1970)

modified form of Horton's infiltration equation, Betson predicted volumes of storm runoff in drainage basins of $0.015-85\,\text{km}^2$. In order for the predicted volumes to coincide with field measurements, a scaling factor had to be introduced to reduce the predicted volumes. This scaling factor varied between 0.046 and 0.858 and Betson interpreted it as indicating that the infiltration capacity was exceeded (i.e. overland flow was being produced) on only $4.6-85.8\%$ of the drainage basin. In a particular basin, the contributing area remained fairly constant throughout a storm and between storms, but the author suggested that it might enlarge during extreme storms. Betson used the term *partial-area concept* for his view of the storm runoff-generating process.

7.2.2 Characteristics of Horton overland-flow hydrographs

Volumes of Horton overland flow vary with storm size and intensity, and with the factors that affect infiltration. Some values of the runoff coefficient (volume of runoff/volume of rainfall) for Hortonian conditions are listed in Table 7.2. The lower bound on such coefficients is zero for all conditions, and the upper bound is almost 1.0 for impervious areas such as parking lots. Although Table 7.2 includes only a cursory sample of published data, it indicates that there are great differences in yields of runoff by Horton overland flow, but that on catchments of less than one

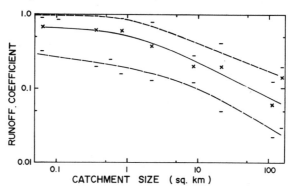

Figure 7.6   Variation of maximum, minimum and mean runoff coefficients for hydrographs of Horton overland flow during large rainstorms in the southwestern United States

square kilometre, storms commonly yield well over 25% of the rainfall as storm runoff. Maximum, minimum and mean values of the runoff coefficient for basins in the southwestern United States generally decrease with increasing drainage area (Figure 7.6) reflecting the lower average rainfall volumes over large areas and seepage losses in sandly floodplain and channel alluvium. For most of the catchments listed in Table 7.2, there is a positive correlation between rainstorm size and the runoff coefficient.

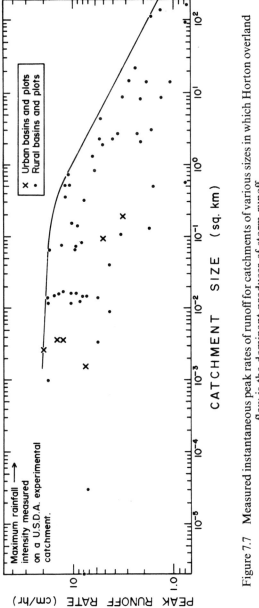

Figure 7.7 Measured instantaneous peak rates of runoff for catchments of various sizes in which Horton overland flow is the dominant producer of storm runoff

Table 7.3 Measured maximum peak rates of Horton overland flow for hillslopes and drainage basins of various sizes

Catchment	Area (km^2)	Peak runoff rate (cm/hr)	Source
Golan Heights, N. Israel			Inbar (1969)
Upper Mesashim	15	1·93	
Nahal Mesashim	160	0·76	
Nahal Yael, S. Israel			Schick (1970
Watershed 01	0·58	0·78	
Watershed 02	0·50	1·62	
Watershed 03	0·13	1·80	
Watershed 04	0·11	3.43	
Watershed 05	0·05	4.49	
Copper Cliff, Sudbury, Ontario	0·000002	8·00	Pearce (1973)
Presqu Isle, Maine	0·000031	7·10	Viessman (1968)
	0·00344	5·95	Knisel (1973)
Tombstone, Arizona			
Watershed W-1	148	1·37	US Dept. Agric.
Watershed W-2	113	1·70	(1958; 1963;
Watershed W-3	8.9	3·25	1968; 1970;
Watershed W-4	2·25	5·70	1971; 1972a;
Watershed W-5	22·2	2·41	1972b)
Watershed W-6	94	0·79	
Watershed W-8	15·35	2·82	
Watershed W-11	8·18	2·11	
Safford, Arizona			
Watershed W-1	2·08	2·11	US Dept. Agric.
Watershed W-2	2·75	3·68	(1958; 1963;
Watershed W-4	3·05	1·67	1968; 1970;
Watershed W-5	2·80	2·34	1971; 1972a;
			1972b)
Stillwater, Oklahoma			
Watershed W-1	0·067	17·78	US Dept. Agric.
Watershed W-3	0·370	12·00	(1958; 1963;
Watershed W-4	0·825	6·05	1968; 1970;
			1971; 1972a;
			1972b)
Riesel, Texas			
Watershed C	2·35	4·01	US Dept. Agric.
Watershed D	4·45	5·35	(1958; 1963;
Watershed W-1	0·705	11·45	1968; 1970;

Table 7.3 (*Continued*).

Catchment	Area (km²)	Peak runoff rate (cm/hr)	Source
Riesel, Texas cont.			
Watershed W-2	0·52	12·21	1971; 1972a;
Watershed W-6	0·169	10·15	1972b)
Watershed W-10	0·079	12·70	
Watershed Y	1·23	6·46	
Watershed Y-2	0·53	10·35	
Watershed Y-4	0·32	7·90	
Watershed Y-6	0·066	9·60	
Watershed Y-7	0·159	9·10	
Watershed Y-8	0·083	8·35	
Watershed Y-10	0·074	9·47	
Watershed SW-12	0·012	10·15	
Watershed SW-17	0·012	17·90	
Watershed P-1	0·00098	18·20	
Watershed P-3	0·00098	19·40	
Hastings, Nebraska			
Watershed W-3	1·95	5·08	US Dept. Agric.
Watershed W-8	8·45	1·32	(1958; 1963;
Watershed W-1	14·10	1·06	1968; 1970;
Watershed 1-H	0·0146	5·97	1971; 1972a;
Watershed 2-H	0·0138	8·80	1972b)
Watershed 3-H	0·0152	16·30	
Watershed 4-H	0·0148	19·40	
Watershed 5-H	0·0162	10·80	
Watershed 6-H	0·0162	14·40	
Watershed 7-H	0·0172	12·10	
Watershed 8-H	0·0160	9·30	
Watershed 18-H	0·0151	7·30	
Watershed 22-H	0·0154	8·08	
Watershed 25-H	0·0090	4·44	
Parking lots, Baltimore, Md.	0·0038	14·60	Viessman (1966)
Parking lots, Baltimore, Md.	0·0016	7·60	Viessman (1966)
Parking lots, Baltimore, Md.	0·0026	19·90	Viessman (1966)
Parking lots, Newark, Delaware	0·0038	12·65	Viessman (1968)
Suburban Baltimore	0·19	3·20	Miller and Viessman (1972)
Gray Haven residential area, Baltimore	0·092	5·08	Viessman *et al.* (1970)

Peak rates of runoff by this process vary with rainfall intensity, infiltration capacity and to a small extent with land gradient. In Figure 7.7, I have plotted maximum measured or published instantaneous peak rates of runoff for hillside plots and small drainage basins where Horton overland flow is known to be the dominant producer of storm runoff. The data do not constitute an exhaustive sample, and it is likely that higher peak rates have been measured in some experiments. The envelope can be revised upward as more data become available. However, Figure 7.7 provides an adequate first estimate of maximum instantaneous peak rates of runoff that are known to have been delivered by Horton overland flow. The upper envelope will be useful for later comparison with similar curves defined for other processes. Envelopes could be defined for particular regions, if the reader wishes to supplement the data listed in table 7.3 which are seen only as defining a provisional envelope for the whole range of conditions under which Horton overland flow is encountered. The envelope curve in Figure 7.7 has a negative gradient because of channel storage, seepage losses, and the lower average rainfall intensities over larger areas.

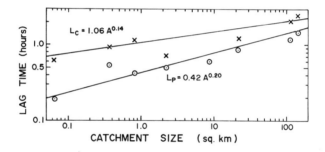

Figure 7.8   Relation of catchment size to the lag from a burst of intense rainfall to the hydrography peak (L_p) and to the lag between the centroids of rainfall and runoff (L_c) for hydrographs of Horton overland flow

Overland flow has sufficient velocity (0·3–15 cm/sec) to reach stream channels during and within a few minutes of the end of rainstorms, and contributes only to storm runoff. Stream hydrographs, therefore, respond quickly to the onset and fluctuations in the intensity of rainfall as Figure 7.4 demonstrates. The hydrograph peaks are generally sharp and their recession limbs steep. For a sample of Horton hydrographs taken from the US Department of Agriculture (1963–71) compilations of runoff data from small catchments, the lag times vary with drainage area as shown in Figure 7.8. If one fits an equation of the form

$$Q_t = Q_0 K^t$$

where

Q_t = discharge at time, t
Q_0 = peak discharge
K = a recession constant
t = time since the peak (hr)

to the recession limbs of Horton hydrographs in that publication, the values of K for eight drainage basins are as listed in Table 7.4. There is a general increase of these values with catchment size, but the correlation was not significant for the few data analysed.

Table 7.4   Values of the recession constant, K, for hydrographs of Horton overland flow in the southwestern United States

Location	Catchment area (km²)	K Range	Median	Mean
Stillwater, Oklahoma				
Watershed W-1	0·067	0·008–0·32	0·078	0·022
Watershed W-3	0·370	0·097–0·51	0·23	0·189
Watershed W-4	0·825	0·032–0·27	0·17	0·105
Tombstone, Arizona				
Watershed W-4	2·25	0·008–0·09	0·040	0·041
Watershed W-3	8·9	0·024–0·17	0·026	0·067
Watershed W-5	22·2	0·015–0·42	0·16	0·072
Watershed W-2	113	0·017–0·44	0·26	0·16
Watershed W-1	148	0·21–0·50	0·49	0·34

7.2.3   The range of validity of the Horton model

It is not known entirely to what geographic areas, types of climate, soil, and land use the Horton overland-flow picture of storm runoff applies. But there are large areas in humid landscapes where overland flow seldom, if ever, occurs over most of the landscape because infiltration rates are so high that few rainstorms have intensities that exceed them. The problem may best be visualized by reviewing one's own field experience and asking under what conditions a thin sheet of water flowing as unconcentrated overland flow has been observed.

I have seen overland flow on lawns, roads, tracks, around water-holes, and on many other areas where soils have been compacted by the passage of animals, people, or vehicles. Even in forested areas with deep, permeable soils, one frequently sees overland flow on haul roads, skid trails and landings. One cultivated lands of the American mid-west, I have also seen overland flow. In this region the process has been extensively studied in plot experiments by the US Department of Agriculture, and good descriptions of such work are provided in the literature of agronomy and agricultural engineering.

I

I have observed sheetflow on grazing lands of semi-arid western North America where vegetation density is low. Measurements of the process under such conditions have been made by Kincaid, Osborn and Gardner (1966). On ranges supporting only wild game populations feeding on grass and shrubs, I have seen unconcentrated sheetwash over very large areas of country in the Tsavo National Park of eastern Kenya and the Serengeti National Park of northern Tanzania. Around the edges of these parks, where heavy grazing by domestic livestock occurs, and where the grass is burnt annually, overland flow is particularly obvious in many rainstorms.

In humid regions, overland flow is much less common. I have observed it occurring over most of the landscape on mine spoil heaps devoid of vegetation in several humid regions of eastern North America and England. At Sudbury, Ontario, where acidic smelter fumes have caused complete deforestation of large tracts of land, I have observed sheetwash on the steep side slopes of gullies that scar the land surface and on unrilled slopes with gradients of 0·05–0·10. Measurements of runoff under these conditions have been made by Pearce (1973), and Schumm (1956) has provided detailed descriptions of the process on bare, steep slopes of gullies in a fine-textured fill in New Jersey. In humid regions that have not been disturbed by mining, urbanization, or logging, my experience of overland flow is much more limited. I have seen it occurring over frozen soils in pasture areas of northern Vermont, and in the tundra and boreal forest of the Labrador subarctic. During rainstorms, my sightings in both of these areas have been limited to poorly drained soils of valley bottoms, hillside hollows or seeps localized by geologic structure. I have not seen Horton overland flow over large areas of the landscape in a humid region.

7.3   STORM RUNOFF IN AREAS OF HIGH INFILTRATION RATE

7.3.1   Infiltration and flow in the Vadose and Phreatic zones

In most humid regions infiltration capacities are high because vegetation protects the soil from rain packing and dispersal, and because the supply of humus and the activity of microfauna create an open soil structure. Under such conditions, rainfall intensities generally do not exceed infiltration capacities and Horton overland flow does not occur on large areas of the landscape.

When the infiltration capacity of a soil exceeds the rainfall intensity, all rain enters the soil, and the situation has been described by Rubin (1966) for conditions where the soil surface is not drastically altered by rain packing. The infiltrating water raises the moisture content of the surface soil and, therefore, its hydraulic conductivity. Eventually, the hydraulic conductivity becomes equal to the rainfall intensity and the surface soil can pass the rainwater at the rate at which it is being received. In other words, rainwater is stored until it raises the ability of the soil to transmit water at a steady state under the given rainfall intensity. When this condition is reached, the soil-moisture content of the surface becomes constant and

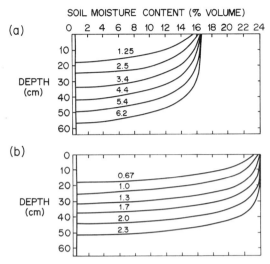

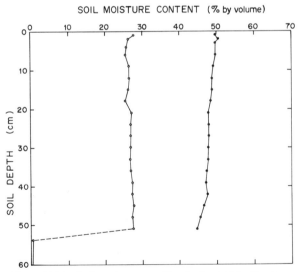

Figure 7.9   Changes of soil moisture with depth during infiltration in initially air-dry sand. Saturation moisture content for this soil is 38%. Figures on the curves indicate the duration of infiltration, in hours.
(a), Ratio of rainfall intensity to the saturated hydraulic conductivity of the soil = 0·026; (b), Ratio of rainfall intensity to the saturated conductivity of the soil = 0·098
(after Rubin, 1966)

Figure 7.10   Soil-moisture profiles during infiltration in a sand (open circles) and an aggregated clay (closed circles) during a 150-mm/hr rainfall (after Rubin *et al.*, 1964)

a wave of moisture percolates downward wetting successively deeper layers to a moisture content at which their hydraulic conductivity becomes equal to the rainfall intensity.

In Figure 7.9, the calculated situation is shown for infiltration into an initially air-dry sand under two rainfall intensities. Figure 7.10 presents laboratory data to show that the moisture content required to raise the hydraulic conductivity to the rainfall intensity is much lower in the case of a sand than in a clay. If rainfall intensity suddenly increases, water is again stored in the soil until the hydraulic conductivity is sufficient to transmit water at a rate equal to the new rainfall intensity, and a larger pulse of water moves down through the soil profile. If the rainfall intensity exceeds the saturated hydraulic conductivity of the soil, ponding and overland flow occur. When rainfall becomes less intense or stops, the soil begins to drain (see Figure 7.11), and the rate of downward moisture migration slows, eventually becoming very small as the soil-moisture content approaches field capacity.

As water percolates through soil to the water table it displaces water previously retained by the soil. At low rainfall intensities most of this movement takes place

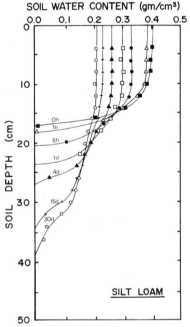

Figure 7.11   Soil-water profiles during drainage of a silt loam. Numbers on the curves indicate the number of hr or days since the cessation of infiltration which supplied 6·1 cm of water to the soil. Zero water content corresponds to an initial value of 0·02 g/cm^3 before infiltration (after Biswas *et al.*, 1966)

through the finer soil pores. Only when the soil is saturated or is close to saturation do considerable amounts of water move through large pores, such as wormholes and rootholes. Whipkey (1969) has reported observing flow through rootholes while soil was unsaturated. In laboratory experiments, Horton and Hawkins (1965) showed that even under abnormally-heavy rainfall intensities with the soil initially at field capacity, most of the water that infiltrates larger pores is drawn by capillary forces into the surrounding small pores before it can penetrate far along the larger opening. They passed water at rates simulating intense rainfalls into a column of coarse sand surrounded by another column of sandy clay. The water flowed from the large pores of the sand into the finer pores of the surrounding soil by the time the water had moved 30–60 cm down the column. Flow from the fine

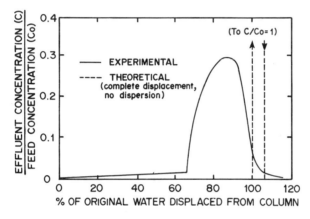

Figure 7.12　Displacement of tritiated water from a laboratory soil column (after Horton and Hawkins, 1965)

pores into the sand occurred at the base of their laboratory column where an artificial water table was created.

　In another experiment Horton and Hawkins saturated a column of soil and allowed it to drain to field capacity. A volume of water equivalent to one inch of rain, and containing radioactive tritium was then added to the top of the column. One inch of rain was added to the top of the column on each succeeding day of the experiment and the effluent from the bottom of the column was analysed daily for tritium. Figure 7.12 shows the results of the daily tritium measurements expressed as a function of the percentage of the original water displaced from the bottom of the column. If the preferred flow path for the labelled water had been through the large pores which contain air at field capacity, tritium would have been detected in the effluent shortly after its addition. Instead, 67 % of the water present in the soil was displaced before the effluent concentration reached 2 % of the field concentration, and 87 % of the original water was displaced before the peak tritium concentration was reached. There was a slight flattening and spreading of the tritium curve, indicating that some of the labelled water had moved more quickly and some more slowly than the bulk of the tracer. This was probably due to a velocity distribution within and between pores. All the tritium was not recovered after a single flushing of the column because some water was held at points of contact between soil particles.

In spite of storage and spreading of the tritium through time, the most striking result of this experiment was the indication that water entering the column in a single, simulated storm moves down the column as a coherent wave, displacing water below it, and pushing the latter down towards the water table. The experiment of Smith (1967) yielded photographs of this process occurring in the laboratory columns. Zimmerman *et al.* (1966), using a radioactive tracer, showed that the same process occurs in undisturbed profiles in the field. They concluded that, 'a single rainfall, labelled with isotope tracer forms a tagged layer of water that, although blurred by diffusion effects, moves downward as a distinguishable water mass, between the older rainwater below and the younger rainwater above.'

After rainfall has ceased, water continues to drain from the soil pores. Water in the larger pores is not held by capillary forces as strong as those in fine pores, and so the larger openings drain first, leaving only wedges of water held in the necks of pores. Some of the finer pores drain more slowly and eventually the continuity of their water films is also broken. In some of the finest pores, the water films may remain continuous, but the capillary and other forces acting to hold water in such pores causes drainage to be exceedingly slow. Robins *et al.* (1954), Ogata and Richards (1957), Nielsen, Kirkham and Van Wijk (1959), and Hewlett (1961), however, demonstrated that even after soil-moisture contents have declined below field capacity, slow vertical percolation provides an important supply of groundwater recharge to support baseflow during long dry periods. Observations by Hewlett and Hibbert (1963) showed that rates of recharge sufficient to account for observed minimum stream discharges in the mountains of North Carolina could be obtained from a trough of soil 13·7 m long and 0·9 m deep for at least 150 days after the cessation of rainfall. The experimental trough, however, was covered to prevent evaporative losses. In an undisturbed field situation, withdrawals of water by plants causes more rapid reduction of the soil-moisture content, the breaking of more water films around soil particles, and a more rapid decrease in the rate of drainage to the water table.

Freeze (1969) presented a numerical solution for a mathematical model of one-dimensional, unsaturated flow to a water table that fluctuates in response to the re-charge. For homogeneous, isotropic soils, he demonstrated how the water table begins to rise as fluid pressures are increased by percolating water in the unsaturated zone. His results show that the water table is more likely to rise in low-intensity rains of long duration than in short, intense storms; that shallow water tables are more likely to respond than deeper ones; and that responses will be greater for wet antecedent conditions and for soils with high hydraulic conductivity.

7.3.2 Field Studies of subsurface stormflow

The theoretical and experimental findings of soil physics can now be applied to a field situation, to show how these processes control the production of runoff in a humid region. Figure 7.13(a) shows the cross-section of a valley with straight hillslopes and no floodplain. In an ideal situation with uniform soils, the water table before a rainstorm has an approximately parabolic form, and soil-moisture content decreases with elevation above the water table. The exact form of the soil-moisture profile would vary with the texture of the soil, specifically with the amount of water that can be held by capillary forces at any elevation above the water table. Additions of water by recent storms and removal of water from the root zone by evapotranspiration would cause other complications, but the simple picture is adequate for the present purpose.

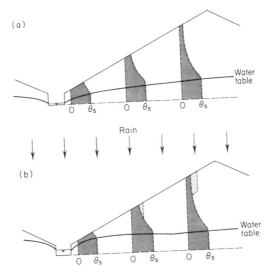

Figure 7.13 (a), Valley cross-section showing the water table and profiles of soil-moisture content at three locations on a hillside before rainfall. (b), Response of soil-moisture profiles and the water table to rainfall. Light shading shows the increase in soil moisture produced by infiltration. θ_s is the moisture content of the soil at saturation

At any depth below the soil surface the moisture content of the soil profile will increase with distance fom the hilltop. The hydraulic conductivity (a function of moisture content) at any depth will also increase downslope. Using the argument of Rubin, introduced previously, less storage of infiltrating water will be necessary in the topsoil near the base of the slope to raise the conductivity of the soil to a level sufficient to transmit rainwater at the applied rate. Thus, during rainfall, the rate of

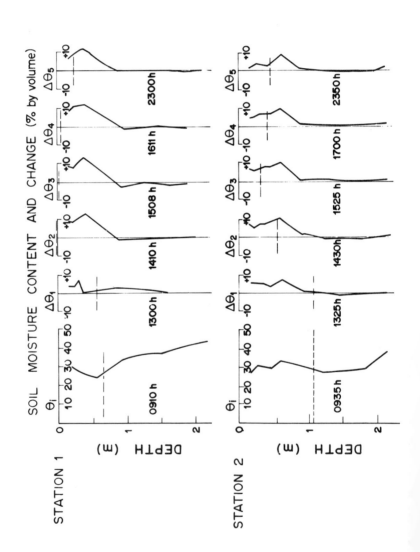

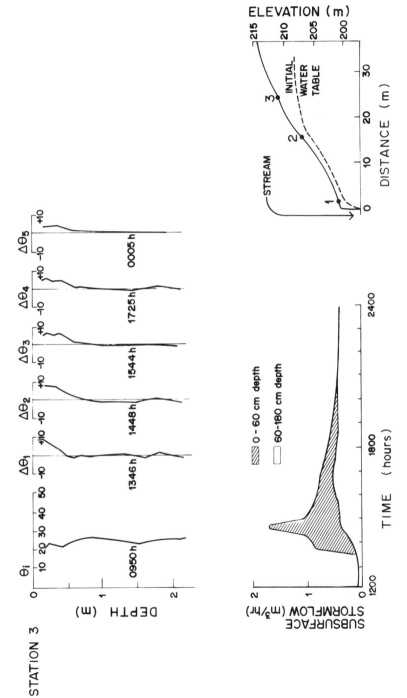

Figure 7.14 Profiles of initial soil-moisture content (θ_i), changes in soil moisture from the initial profile ($\Delta\theta$), and subsurface stormflow during a 44-mm rainstorm beginning at 12·25 hr and lasting for 2 hr on the hillside shown in the inset. The dashed lines represent the position of the water table

transmission of water to the phreatic zone increases downslope. This fact, and the smaller depth to the water table near the stream combine to ensure that vertical percolation will cause the water table near the base of the slope to rise early in a storm. Further upslope, the surface soil is generally drier, its conductivity is less, the depth of the water table is greater, and more water is stored during infiltration and percolation. Displacement of moisture into the saturated zone occurs more slowly than at the bottom of the slope. If the water table is deep enough, all the infiltrating water may go into storage in the vadose zone. In a rainstorm, therefore, one might expect a situation like that portrayed schematically in Figure 7.13(b). The steepening of the water table close to the stream would cause an increase at the rate of subsurface flow. The magnitude of this contribution to runoff (subsurface stormflow) depends upon the steepness of the water table, and the depth and conductivity of the soil.

This process of groundwater recharge in a field situation can be seen from Figure 7.14, which shows initial soil-moisture profiles, changes of soil moisture, and water-table elevations before, during, and after a rainstorm at three stations on a hillside. Locations of the measurement sites are shown in the inset. The changes were produced by an artificial storm occurring between 12·25 hr and 14·35 hr, and delivering 44 mm of rain to the slope. The three stations shown in Figure 7.14 were located in the middle of a hillside swale, where the water table at any distance upslope was closer to the ground surface than it would be at the same distance up a steeper slope, or one with a straight profile. On an adjacent straight slope, the same pattern of variation with distance downslope occurred, but the response of the water table and subsurface runoff were smaller because of the initially deeper water table (Dunne, 1969a).

If the topography of the valley cross-section is more complicated than the examples previously used, more variable patterns of groundwater recharge and subsurface stormflow are possible. Figure 7.15A shows the cross-sectional profile of a valley trenched into a deposit of uniform sand. The pre-storm soil-moisture profile is shown for two locations where the depths of the water table differ greatly. Figure 7.15B shows the response of the water table in this valley to a 28 mm rainstorm lasting 4 hr. Eighteen metres from the channel, where the depth to the phreatic zone was about one meter, the water table rose 0·06 m during the 26 hr after the beginning of rainfall. On the flat land near the channel, however, the phreatic zone was at shallow depth, recharge was rapid, and the water table responded more rapidly than under the hillslope further away from the stream. The result was the formation of a mound of groundwater, which initially at least must have caused some subsurface flow away from the stream, as well as towards it (Ragan, 1968).

These findings indicate that slope form and steepness, which control the depth to the phreatic zone and the vertical distribution of moisture content in the soil at the beginning of the storm, have an important effect upon the timing and magnitude of groundwater recharge and the production of subsurface stormflow. The vertical profile of the hillslope, and also its plan form (contours concave, convex, or

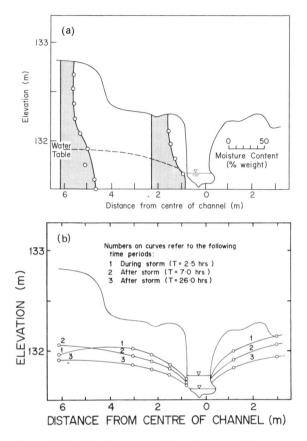

Figure 7.15 (a), Pre-storm soil-moisture profiles at two locations with differing water-table depths (after Gunn, 1967); (b), fluctuations of the water table during and after a 28-mm rainstorm lasting 4·5 hr (after Ragan, 1968)

straight) will also affect the depth of the water table below the soil surface and therefore the production of subsurface stormflow.

A slightly more complicated but common situation may also be visualized, where the upper soil horizon is underlain by a zone of lower hydraulic conductivity. This impeding layer might be an illuvial hardpan, a zone of partly-weathered bedrock, or a layer of unweathered rock. Water percolating through the topsoil in the manner described by Rubin would be stored in the zone immediately above the impeding layer, and if the rain persists for long enough a saturated zone with a 'perched' water table will form and water migrates laterally to the stream as subsurface stormflow. If the impeding layer is not completely impervious, some water may penetrate to a lower water table, but in most cases the important contributor of subsurface storm runoff is the perched saturated zone.

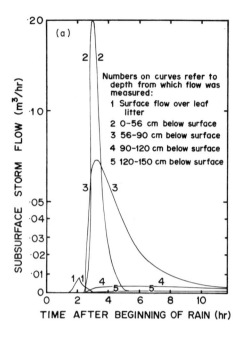

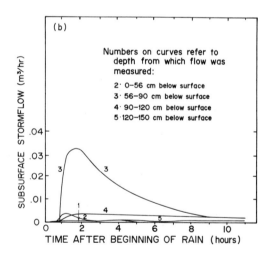

Figure 7.16   (a), Hydrographs of subsurface stormflow
from a hillside plot 17-m long and 2·4-m wide during and
after a 10·2-cm rainstorm in two hours on 'dry' antecedent
conditions. (b), Hydrographs of subsurface stormflow from
the plot during and after a 3·2-cm rainstorm in 1·67 hr on
'wet' antecedent conditions. Curve 1 represents surface
flow over leaf litter (after Whipkey, 1965)

Even where there is no sharp boundary between the topsoil and a less permeable layer, the impedance of percolation by a denser subsoil often causes a zone of saturation to be 'backed up' into the highly permeable topsoil. When this occurs, most of the subsurface stormflow occurs near the soil surface. The experiments of Whipkey (1965) demonstrates this process particularly well (Figure 7.16(a) and (b)). Similar hydrographs have been reported by Dunne (1969a), Hewlett and Nutter (1970) and Weyman (1970). A more detailed discussion of subsurface flow is given in Chapter 4.

7.3.3 Characteristics of subsurface stormflow hydrographs

Volumes of subsurface stormflow are more difficult to define than those of Horton overland flow, because there is no generally accepted definition of stormflow. Whipkey reported total seepage for the 24-hr period after the onset of rain, but by far the greater portion of the stormflow occurred within 8 hr after rainfall began. The yields varied with size of rainstorm, antecedent conditions and soil type (see Figure 7.17(a) and Table 7.5). I reported comparable volumes of subsurface stormflow within 8 hr of the beginning of rainfall for a sandy loam with an impeding layer at a depth of 60 cm and with a shallow water table. On an adjacent sandy loam hillside without these favourable conditions, much smaller volumes were measured. Published data from these and other studies are listed in Table 7.5, which shows that the yield of stormflow, expressed as a percentage of rainfall volume is generally much smaller for subsurface stormflow hydrographs than for those of Horton overland flow, listed in Table 7.2. The results listed in the tables are generally from large storms.

Peak rates of subsurface stormflow reported by Whipkey range up to 0·54 cm/hr from a 90 cm deep sandy loam, and up to 0·99 cm/hr for a 44 cm deep silt loam. These maxima varied with rainfall intensity and antecedent conditions as indicated by Figure 7.17(b). Peak subsurface stormflow rates from other plot studies are listed in Table 7.6. This table also contains published values of peak runoff rates (per unit area and per unit width of hillside) for regions where sub-surface stormflow is the dominant contributor of stormflow (as claimed by the writers of each source paper referred to in the table). Data from this table are used to construct Figure 7.18, which again defines a tentative envelope which will undoubtedly be raised as more studies become available. The envelope in this case lies one to two orders of magnitude below that for Horton overland flow from catchments of similar size. Part of the difference lies in the fact that the overland-flow envelope was constructed from relatively long records, while that for subsurface stormflow is defined from only a few published hydrographs and some experimental studies. But the published hydrographs of subsurface stormflow were virtually all recorded during storms with high recurrence intervals. It seems clear, therefore, that there are important differences between peak rates of runoff generated by the two processes. Soil thickness and hydraulic conductivity set constraints upon the rate of generation of subsurface stormflow.

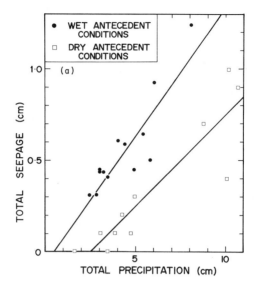

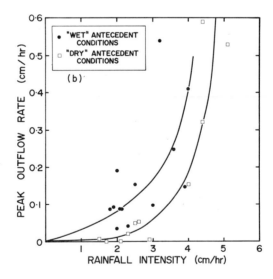

Figure 7.17 (a), Relation between rainfall volume and the volume of subsurface stormflow during a 24-hr period after the onset of simulated rainstorms. Conditions were classified as 'dry' if more than 4 days had passed since the last rain. Conditions were 'wet' if less than 4 days had passed since the last rain (b), Relation of peak runoff rate to rainfall intensity for subsurface stormflow from the upper 90 cm of a soil profile (after Whipkey, 1965)

Table 7.5   Runoff coefficients (runoff volume/rainfall volume) for subsurface stormflow from hillsides and small drainage basins

Location	Soil type	Calculation period (hours after onset of rain)	Runoff coefficients (range depending on storm size)	Source
Ohio	1·5 m deep, sandy loam	24	0·085–0·16 (for large storms on wet antecedent conditions)	Whipkey (1965)
Ohio	1·5 m deep, sandy loam	24	0–0·085 (for large storms on dry antecedent conditions)	Whipkey (1965)
Ohio	1·8 m deep, sandy loam	24	0	Whipkey (1969)
Ohio	Loam	24	0·18–0·53 (for large storms)	Whipkey (1969)
Ohio	0·44 m deep, silt loam	24	0·15–0·62 (for large storms)	Whipkey (1969)
Vermont	2·5 m deep, sandy loam and silt loam impeding horizon	8	0·008–0·107 (for large storms) 0 (for most natural storms)	Dunne (1969a)
Vermont	2·1 m deep, sandy loam	8	0·002–0·066 (for large storms) 0 (for most natural storms)	Dunne (1969a)
Vermont	0·3 m deep, silt loam	8	0–0·05	Dunne (1969a)
South Carolina	1·2 m deep, sandy loam	up to 19	0–0·02 (median = 0)	Wilson and Ligon (1973)
North Carolina	2·1 m deep, sandy loam	30–40	0·053 (for a 100-mm storm) 0·21 (for a 143-mm storm)	Hewlett and Nutter (1970)
North Carolina	Deep, sandy loam on a 0·152-km² catchment	30–40	0·02 (for a 100-mm storm) 0·06 (for a 143-mm storm) 0·20 (for a 500-mm storm)	Hewlett and Nutter (1970)
North Carolina	Deep, sandy loam on a 0·435-km² catchment	30	0·10 (for a 32-mm storm)	Hewlett and Nutter (1970)

Table 7.6   Measured instantaneous peak rates of subsurface stormflow for hillslopes and drainage basins of various sizes

Catchment	Area (km²)	Peak runoff rate[a] (m³/hr/m)	Peak runoff rate over catchment (cm/hr)	Source
Ohio, 44 cm deep silt loam	0.0025	1.82	0.99	Whipkey (1969)
Ohio, 1.5 m deep sandy loam	0.000042	0.089	0.54	Whipkey (1965)
Vermont, 2.5 m deep sandy loam	0.0012	0.110	0.14	Dunne (1969a)
N. Carolina, deep sandy loam	0.152	—	0.095	Hewlett and Nutter (1970)
N. Carolina, 2.1 m deep sandy loam	0.00074	0.044	0.072	Hewlett and Nutter (1970)
Wales, peaty podzol	0.0002	0.088	0.044	Knapp (1973a)
N. Carolina, deep sandy loam	0.435	—	0.0395	Hewlett and Nutter (1970)
Vermont, 2.1 m deep sandy loam	0.00065	0.015	0.0375	Dunne (1969a)
S. Carolina, 1.2 m deep sandy loam	0.00085	0.060	0.0355	Wilson and Ligon (1973)
Georgia, loamy sand, 0.6–2-m deep	0.0034	—	0.033	Knisel (1973)
Vermont, 0.3 m deep silt loam	0.0024	—	0.028	Dunne (1969a)
Kenya, deep volcanic ash	0.52	—	0.027	Pereira et al. (1962)
Wales, peaty podzol	0.003	0.060	0.020	Knapp (1973b)
N. Carolina, clay loam	0.0002	0.0004	0.0155	Tennessee Valley Authority (1966)
Georgia Piedmont, deep soils	0.24	—	0.0119	Hewlett and Nutter (1970)
Alaska, muskeg	1.8	—	0.0114	Dingman (1966)
Wales, peaty podzol	0.0005	0.052	0.0105	Knapp (1973b)
Wagon Wheel Gap, Colorado	0.81	—	0.0071	Hewlett and Nutter (1970)
Vermont, deep sand	0.46	0.049	0.004	Ragan (1968)
S. W. England, sandy loam	0.0003	0.0084	0.0028	Weyman (1970)

[a] Discharge per metre width of hillside.

Table 7.7 Saturated hydraulic conductivity of soils in which subsurface stormflow has been measured

Soil type	Saturated hydraulic conductivity (cm/hr)	Source
Upper 7·5 cm of a sandy loam (A₀ horizon)	118 (highest of a series of measurements)	Laboratory measurement, Dunne (1969a)
Sandy loam topsoil	34·2–37·2	Field measurements, Dunne (1969a)
Sandy loam	30·5	Field measurements, Hewlett and Hibbert (1963)
Sandy loam (56–90-cm depth)	28·6	Field measurement, Whipkey (1965)
Sandy loam (7·5–60-cm depth)	Mean 24·3 Range 17·2–46·0	Laboratory measurements, Dunne (1969a)
Silt loams and loams	Median 8·4–10·4 Range 0·15–16·5	Field measurements, Rawitz *et al.* (1970)
Varved sandy silt subsoil	Mean 8·9 Range 1·3–18·5	Laboratory measurements, Dunne (1969a)
Varved sandy silt subsoil	Mean 4·8	Field measurements, Dunne (1969a)
Clay loam topsoil	2·5–7·5	Field measurements, Betson *et al.* (1968)
Loam subsoil (90–120-cm depth)	1·7	Field measurement, Whipkey (1965)
Clay loam subsoil	0·75	Field measurements, Betson, *et al.* (1968)
Clay loam subsoil (120–150-cm depth)	0·2	Field measurement, Whipkey (1965)

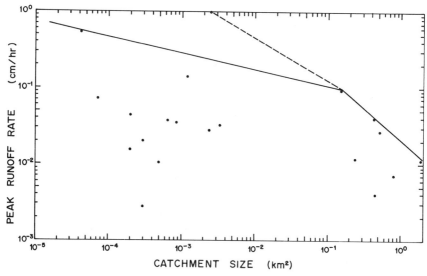

Figure 7.18   Measured instantaneous peak rates of subsurface stormflow for hillslopes
and drainage basins of various sizes

Table 7.8   Some published values of lag times between the onset of rainfall and the onset of
subsurface stormflow, and of lag times between the centre of the main burst of rainfall and the
the peak of the subsurface stormflow hydrograph

Catchment size (km^2)	Soil type	Lag to onset (hr)	Lag to peak (hr)	Source
0·000042	1·5 m deep, sandy loam	0·75–2·5	0·8–2·0	Whipkey (1965)
0·0001	Peaty podzol	1·85	0·9	Knapp (1973a)
0·0002	Peaty podzol	3·3	5·2	Knapp (1973a)
0·0003	0·75 m deep, sandy loam	—	26	Weyman (1970)
0·0005	Peaty podzol	—	22·5	Knapp (1973b)
0·00065	2·1 m deep, sandy loam	1–6·5	8[a]	Dunne (1969a)
0·00085	1·2 m deep, sandy loam	3–3·8	0·3–>2	Wilson and Ligon (1973)
0·0008–0·0033	0·90 m deep, silt loam	1–3	3–9	Monke et al. (1967)
0·0011	Sandy loam and silt loam	0·3–1·0	1·1	Whipkey (1969)
0·0012	2·5 m deep, sandy loam	0·5–2·0	5[a]	Dunne (1969a)
0·0025	0·44 m deep, silt loam	0·6–1·2	1·7	Whipkey (1969)
0·23	Deep sandy loam	—	12	Hewlett and Nutter (1970)
0·435	Deep sandy loam	—	8·3	Hewlett and Nutter (1970)
1·8	Muskeg	—	21	Dingman (1966)

[a] Median value.

Maximum rates have been recorded during large rainstorms on the deep forest soils of Ohio, North Carolina and Vermont.

Average velocities of subsurface stormflow through a permeable forest topsoil range from approximately 30 cm/hr down to much smaller values. It is possible that in the topmost few centimetres, velocities are much higher; I have calculated velocities of 45 cm/hr for the top 7·5 cm of a sandly loam. Table 7.7 lists saturated hydraulic conductivities of soils in which subsurface stormflow has been measured. Because of the damping effect of storage and percolation in soils, the response of subsurface stormflow to rainfall is much slower than the response of Horton overland flow. In Table 7.8 I have listed the lag times between the centre of a burst of rainfall and the peak of subsurface stormflow for various field studies. These lag times tend to increase with catchment size, as shown in Figure 7.19. Comparison of this diagram with Figure 7.8 shows that subsurface stormflow hydrographs have lags approximately 40 times longer than those of Horton overland flow for catchments in the range 0·1–1·0 km^2. Lags between the onset of rainfall and of runoff are also much longer than those of Horton overland flow.

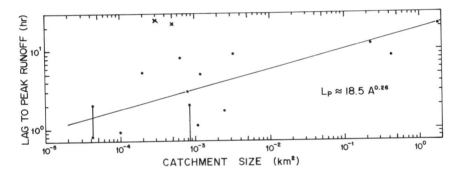

Figure 7.19 Relation of catchment size to the lag from a burst of intense rainfall to the hydrograph peak of subsurface stormflow. The two crosses, from British studies, were ignored in fitting the line

The recession limbs of subsurface stormflow hydrographs also reflect the slow response of this form of runoff. Very few published data are available from which recession constants can be calculated, but I have compiled some data in Table 7.9. The recession constants for subsurface stormflow on very small plots and drainage basins are generally greater than 0·80, indicating much slower recessions than the Horton hydrographs, which have recession constants less than 0·35 even for catchments of more than 100 km^2 (Table 7.4).

There is ample evidence, therefore, that subsurface stormflow occurs. It is particularly common in permeable forested soils. Yields and peak rates of stormflow from this process are much lower than those of Horton overland flow, and the vast majority of the runoff reaches the stream as baseflow. The response of the storm hydrograph (as expressed by time to onset of runoff, lag time, or recession rate) is much slower than that of Horton hydrographs for drainage basins of similar size.

7.3.4 Subsurface stormflow, return flow and direct precipitation onto saturated areas

In most humid areas, storm runoff appears in channels almost as soon as rainfall occurs, and frequently the stream hydrograph from small drainage basins declines rapidly within minutes, or at most several hours, after the end of the storm. Moreover, the published examples of subsurface stormflow hydrographs are mostly from storms that would have a rather high recurrence interval of, say, 5 to more than 100 years in their respective regions (Whipkey, 1965; Dunne, 1969a; Hewlett and Nutter, 1970). Some exceptions are the subsurface stormflow hydrographs published by Ragan (1968), and Dunne (1969a) for shallow, poorly-drained soils, and by Weyman (1970). Thus, although subsurface stormflow has been convincingly demonstrated, it is not the only mechanism by which storm runoff is produced in humid regions. One must look for another process that will supply runoff more quickly.

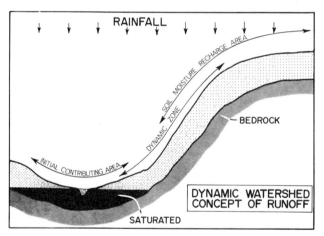

Figure 7.20 Illustration of the dynamic watershed concept of the Tennessee Valley Authority. An almost identical diagram was used by the United States Forest Service (1961) to illustrate their variable source concept (after Tennessee Valley Authority, 1964)

Hydrologists working in the southeastern United States have proposed a solution to this problem. A conceptual model of storm runoff was developed more or less simultaneously by the United States Forest Service (1961) and the Tennessee Valley Authority (1964). The model is illustrated in Figure 7.20. Hewlett and

Hibbert (1967) explained that streamflow in small upland watersheds is generated in a small saturated area in valley bottoms and hollows: 'The yielding proportion of the watershed shrinks and expands depending on rainfall amount and antecedent wetness of the soil. When subsurface flow of water from upslope exceeds the capacity of the soil profile to transmit it, the water will come to the surface and the channel length will grow.' I will refer to this conceptual model as the *variable source concept*, which is approximately the term proposed by Hewlett and Hibbert. The terms 'dynamic watershed concept' (TVA, 1964) and 'partial area concept' (Ragan, 1968; Dunne and Black, 1970b) have also been used to refer to the same idea, but it is better to reserve the latter term for the Horton runoff conditions to which Betson (1964) first applied it.

This conceptual model suggested the need for field investigation of the mechanisms by which water reaches a stream, the relative contribution of each process, and the timing of this contribution relative to the hyetograph and the channel hydrograph. It was also necessary to delimit the range of conditions of soil, topography and climate under which each of the processes that contribute stormflow becomes important. Questions arose about the location, extent and variation of these runoff-producing areas, their predictability, and their relation to soil, topography and antecedent wetness.

The pioneer field study was carried out near Burlington, Vermont by Ragan (1968), who measured inflows to a 190-m reach of channel, draining a 0.46 km^2 forested catchment in a deep uniform sand overlying a dense silt. His analysis was based upon the following form of the continuity equation for short time periods

$$Q_U + Q_B + Q_P + Q_W + Q_L - Q_D = \frac{\Delta S}{\Delta t}$$

where

Q_U = flow entering reach from upstream
Q_B = baseflow entering along reach before storm
Q_P = precipitation falling directly into main channel
Q_W = flow from a series of eight small wet areas along the channel (seeps)
Q_L = distributed lateral inflow, consisting of flow through the forest litter, subsurface stormflow, and flow from small unmeasured seeps along the channel
Q_D = flow leaving the reach at its downstream end
ΔS = change in channel storage
Δt = time period for computations

All terms in the equation were measured directly, except Q_L, which was calculated from the equation.

In a series of 18 storms with low recurrence intervals, the zone producing runoff was found to vary between 1.2% and 3% of the watershed area. Water falling directly into the main channel contributed 2–5% of total stormflow; seep flow (return flow and direct precipitation onto the saturated area) supplied 55–62%, and distributed lateral inflow (litter flow, subsurface stormflow, and flow from small unmeasured seeps) contributed 36–43%. The timing of the various contributions is shown for one storm in Figure 7.21. Although some of the distributed lateral inflow occurred as overland flow from small unmeasured seeps, the majority of the Q_L component originated as subsurface flow from the valley floor material. In this zone, the water table was close to the ground surface at the

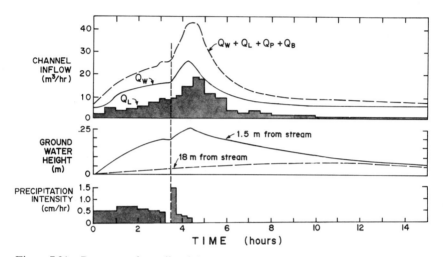

Figure 7.21   Response of runoff and the water table to a 28-mm rainstorm lasting 4·5 hr along a 190-m reach of channel in a 46-hectare catchment in Vermont. Soil-moisture and water-table conditions are also illustrated in Figure 7.15 (after Ragan, 1968)

beginning of the storm, and a small quantity of water infiltrating the soil converted the capillary fringe into a zone of positive pressure and produced a rapid, localized response of the water table as shown in Figure 7.15. Computed lateral inflows were closely related to variations in water-table elevation recorded in a well 1·5 m from the stream. The volume of ungauged lateral inflow could be accounted for by the rain falling on a strip of land 9·5–12·8-m wide along the channel.

I instrumented three plots on a steep well-drained hillside, one plot in a shallow moist swale, and four reaches of the wet valley floor in a 3·9 hectare catchment at the Sleepers River Experimental Watershed of the Agricultural Research Service in northeastern Vermont. On the well-drained hillslope, which had an average gradient of 63%, and an area of 0·24 hectares the three instrumented plots had contours that were respectively convex, concave, and straight (plots 1, 2, and 3 in Figure 7.22). At the base of the slope, a trench was dug to a depth of 2·4–3 m, and drains and a channel were installed to collect water flowing through the subsoil, through the topsoil and over the ground surface (Figure 7.23). Over a period of 2 years, runoff from the three levels in each plot was measured continuously, and

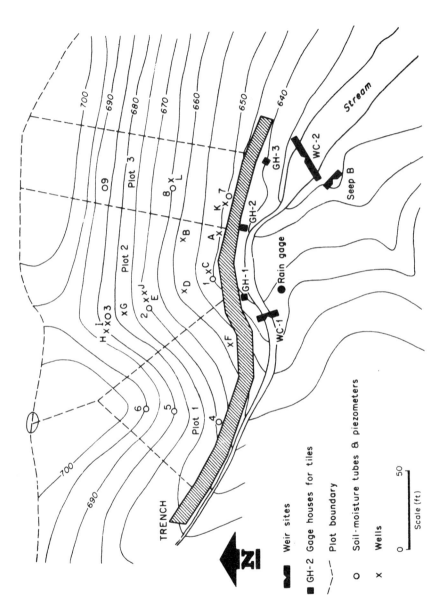

Figure 7.22   The three instrumented plots on the well-drained hillside

periodic measurements were made of soil moisture, water-table elevation, piezometric head, and meteorological phenomena. These variables were also measured intensively before, during and after storms. Measurements of surface runoff, piezometric head and water-table elevation were also made in the shallow moist swale, and in the wet valley bottom. The study provided data on the processes of water movement under conditions of snowmelt and during natural and artificial rainstorms with recurrence intervals varying from less than two years' to several hundred years. A more detailed discussion of this work was published by Dunne and Black (1970a; 1970b; 1971).

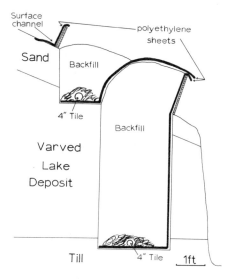

Figure 7.23 The trench and channel system collecting runoff from the well-drained hillside

During most storms in the summer months, rain was stored in the upper metre of soil on the steep, well-drained slopes which cover most of the catchment. The water table averaged more than 1·3 m below the surface of the plots, and the unfilled storage capacity of the soil above the saturated zone was so great that only rarely was water displaced to the phreatic zone at rates sufficient to produce even a minor increase of subsurface flow after the storm. Small amounts of runoff were measured from the plot in the shallow moist seep where the water table was close to the ground surface. In these wetter soils, a small amount of rain was required to bring the phreatic surface to ground level, whereupon subsurface water returned to the surface, was augmented by direct precipitation and contributed to channel runoff as overland flow. During these summer months, the magnitude of peak flows and the general form of the channel hydrograph were controlled by rain falling directly onto the wide, marshy depression in the wet valley floor. This contributing area occupied approximately 1·5–2 % of the catchment area.

During autumn, when soil conditions become more moist, the area contributing storm runoff expanded to 2–5 % of the catchment area. Subsurface contributions increased somewhat during this season but were still too small, too late, and too insensitive to fluctuations of rainfall intensity to control the form of the channel hydrograph. They contributed significantly to the recession limbs of hydrographs, but storm runoff was mainly produced on saturated areas of the valley floor and the moist swale.

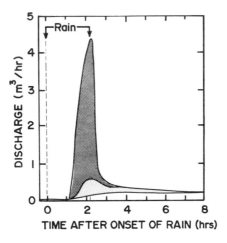

Figure 7.24   Runoff from a 44-mm, 2-hr artificial storm on the concave plot (No. 2 in Figure 7.22) of the well-drained hillside. The unshaded area represents subsurface stormflow into the lower drain shown in Figure 7.23; the light shading represents subsurface stormflow into the upper drain; and the dark shading represents return flow and direct precipitation onto saturated areas, measured in the surface channel

Artificial rainstorms provided additional data from events of constant intensity and high recurrence interval. It was possible to collect more data on changes of soil moisture, water-table elevation and piezometric head at various locations in each plot during these storms. One experiment illustrates well most of the principles of the variable source concept. Figure 7.24 shows the runoff at various levels in the soil on the steep, well-drained hillside resulting from a 2-hr long, artificial rainstorm which supplied 44 mm of rain to the concave plot (No. 2 in Figure 7.22) and 49 mm to the straight slope (No. 3). Antecedent moisture conditions were those only attained at the height of the snowmelt period which is the wettest time of the year in Vermont.

Table 7.9   Values of the recession constant, K, for subsurface stormflow hydrographs

Location	Catchment size (km²)	Approximate values of K	Source
N. Carolina, sandy loam 2·1-m deep	0·000074	0·87–0·99	Hewlett and Nutter (1970)
N. Carolina, drainage basin	0·152	0·83–0·88	Hewlett and Nutter (1970)
N. Carolina, drainage basin	0·435	0·97	Hewlett and Nutter (1970)
Georgia Piedmont	0·23	0·98	Hewlett and Nutter (1970)
Kimakia, Kenya	0·52	0·985	Pereira *et al.* (1962)
Southern England	0·0003	0·97[a]	Weyman (1970)
Ohio, sandy loam, 1·5-m deep	0·000042	0·50–0·75	Whipkey (1965)
Vermont, sandy loam topsoil, 0·6-m deep	0·0012	0·27–0·82	
Vermont, silty sand subsoil, 0·6–2·5-m deep	0·0012	0·89–0·96	
Vermont, sandy loam topsoil, 0·6-m deep, during snowmelt	0·0012	0·65–0·82	
Vermont, silty sand subsoil, 0·6–2·5 m, during snowmelt	0·0012	0·94–0·96	
Vermont, sandy loam topsoil 0·6-m deep, during snowmelt	0·00054	0·42–0·65	

[a] Very approximate value.

Although the recurrence interval of the storm on the straight slope was 50 years, no surface runoff was measured on this plot. The infiltration capacity of the soils of both plots was known to be greater than 80 mm/hr. The water table in this and other storms lay so far (1·2–2·1 m) below the ground surface of the straight hillside that it did not rise appreciably during the course of the storm. Consequently the only runoff yielded by the straight plot was slow subsurface stormflow beginning 1·4 hr after the onset of rain and reaching a peak rate of 0·048 cm/hr 9 hr after the end of rainfall.

In this and other artificial storms, and in artificial storms augmented by natural rains, large amounts of storm runoff were produced on the concave plot (No. 2), as shown in Figure 7.24. Water that remained beneath the soil surface on its way down hillsides was measured in the buried title drains. The total amount, peak runoff rates and lag times of this subsurface stormflow were comparable with those reported in other studies of this process (see Figure 7.18 and Tables 7.5, 7.6, 7.8, and 7.9.

The study confirmed, therefore, that subsurface stormflow occurs in the deep well-drained soils of Vermont; that it produces hydrographs of the commonly assumed form; and that from some parts of Vermont drainage basins with permeable soils and deep water tables, subsurface stormflow is the *only* runoff mechanism.

Relative to other runoff processes, however, water that remained beneath the soil surface on its way down hillsides was a minor contributor to the storm hydrograph, even during large rainstorms on wet soils. Despite the high conductivity of the soil (see Table 7.7), storage and transmission within the unsaturated and saturated zones heavily dampened the response of subsurface stormflow to rainfall or snowmelt. Figure 7.24 shows the relatively minor response of subsurface flow in one storm which completely saturated the soil of the concave plot.

Over most of the plot in the early part of each storm, rainfall percolated vertically, displacing water in the soil to the water table. Because of the resistance of even this permeable soil to subsurface stormflow, virtually all the infiltrating rainwater was stored in the soil, raising the level of the water table. Eventually the water table intersected the ground surface in the central hollow at the foot of the

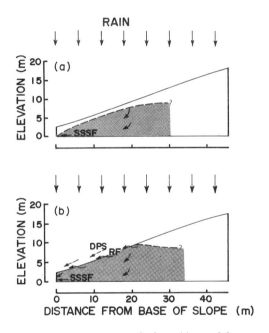

Figure 7.25   Runoff processes during a 44-mm, 2-hr storm on the well-drained, concave plot (No. 2 in Figure 7.22). SSSF, subsurface stormflow; DPS, direct precipitation onto saturated area; RF, return flow. (a), Early in the storm. Rainfall has just begun; the water table is approximately at its pre-storm position; and subsurface stormflow has just begun. Flow lines based on piezometric measurements, are shown in the saturated zone. The dashed line represents the water table. (b), Late in the storm. The water table has risen to intersect the ground surface over the lower portion of the hillside. Return flow and direct precipitation onto the saturated area produce overland flow. Subsurface stormflow has increased

Table 7.10 Amounts of stormflow (cm) produced on the concave and straight plots of the well-drained hillside by natural and artificial storms

Plot	Oct. 10–11, 1967	Oct. 5, 1967	Oct. 16, 1967	Oct. 25, 1967	Oct. 26, 1967	July 17, 1968	July 17, 1968
Plot No. 2 (Concave)							
Rainfall	3.43	2.60	2.95	4.36	2.24	6.10	4.65
Overland flow	0.018	0	0.074	0.708	0.148	0.056	0.800
Flow from root (0–0.6 m) zone	0.036	0	0.033	0.180	0.106	*a*	*a*
Flow from 0.6–2.5-m zone	0.030	0.025	0.106	0.285	0.104	0.005	0.091
Total runoff	0.084	0.025	0.213	1.173	0.358	0.061	0.891
Plot No. 3 (Straight)							
Rainfall	3.43	2.24	2.41	4.92	2.24	5.02	4.65
Overland flow	0	0	0	0	0	0	0
Flow from root (0–0.6 m) zone	0	0	0	0	0.028	0	0.193
Flow from 0.6–2.1 m zone	0.008	0.005	0.028	0.094	0.014	0	0.112
Total runoff	0.008	0.005	0.028	0.094	0.042	0	0.305

a Values not known.

concave plot, and outcropped over an expanding area of the hollow as the storm proceeded. The extent of this saturated zone was mapped a few minutes before the end of each storm. Intersection of the ground surface by the water table allowed the return of infiltrated water to the soil surface where its velocity increased by a factor of 1000–5000 (to 3–15 cm/sec depending upon hillslope gradient and the depth of flow). Without the impedance of many meters of subsurface flow, water could move quickly to the channel in larger amounts than could be contributed by the subsurface system. In addition to this return flow, direct precipitation onto the saturated area of the hillside (which constituted from 5–10 % of the total irrigated area) contributed significant amounts of runoff, at rates varying from 24–58 % of the peak rate of overland flow. The processes are indicated to scale in Figure 7.25.

Table 7.10 shows the yields of stormflow from the two irrigated plots on the well-drained hillside for the previously described storm and for several others. Rise of the water table to the ground surface and the duration of this condition determined the importance of the hillside as a contributor of stormflow. Areas of the watershed where the phreatic surface did not intersect the ground surface (e.g. plot No. 3) were not important contributors of storm runoff. Comparison of the volumes and timing of hillslope and channel hydrographs from four stream gauging stations in the small catchment confirmed that return flow and direct precipitation onto saturated areas dominated the storm hydrograph.

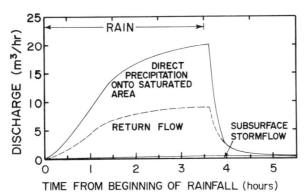

Figure 7.26   Runoff from the shallow, moist swale during
an artificial rainstorm of 54 mm in 3·5 hr

Most rainstorms did not produce any runoff from the steep, well-drained hillside, and the runoff that did result from large storms on this slope lagged the onset of rainfall by one to several hours, during which time the storage capacity of the soil was exhausted over small areas. Other parts of the drainage basin produced large amounts of overland flow without such delay. Artificial storms on the gently-sloping, poorly-drained, shallow soil of the moist swale yielded hydrographs such as that shown in Figure 7.26. Before the storm, water-table depths ranged from one

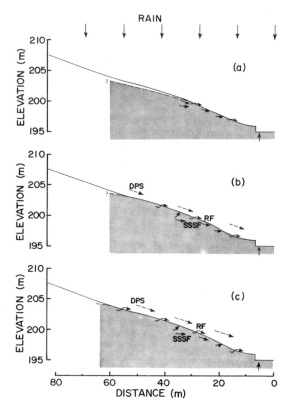

Figure 7.27 Runoff processes during a 54-mm, 3·5-hr
rainstorm on the shallow, moist swale. (a), Beginning of the
storm; (b), after 1 hr of rainfall; (c), after 3·5 hr of rainfall
The saturated area is shaded, and the water table is
represented by the dashed line. Abbreviations as in
Figure 7.25

foot in the centre of the swale to more than three feet below the steeper slopes at its
edge. The magnitude and rate of rise of the water table varied with topographic
position in the way indicated in Figure 7.15. The processes of runoff generation are
shown to scale in Figure 7.27, for the 3·5-hr long storm which produced the
hydrograph shown in Figure 7.26. The amount of direct precipitation onto the
saturated area was obtained by mapping the spread of saturated conditions.

   In the marshy channel depression and its surrounding seeps, the water table was
at the ground surface throughout the year. This area yielded storm runoff from
direct precipitation in every rainstorm, and it expanded seasonally and during
storms. The contribution of stormflow from this source was measured by frequent
mapping of the saturated area. Artificial storms generated only along the channel
depression produced hydrographs that were very similar in peak runoff rate,

volume and timing to those produced by natural rainstorms over the whole catchment. Hydrographs of stormflow from channel areas can be found in the reports by Dunne (1969a) and Dunne and Black (1970b).

The conclusions drawn from these experimental studies which appear to apply throughout most of Vermont, and to many areas of similar soils, physiography and climate that I have visited since, are as follows:

(1) In most rainstorms, Horton overland flow did not occur except on roads and in similar, disturbed locations.

(2) The three major processes generating stormflow were subsurface stormflow, return flow, and direct precipitation onto saturated areas, the two last processes forming saturation overland flow.

(3) These three processes generated stormflow on the poorly-drained soils of the stream depression, the moderate-to-poorly-drained soils of the shallow moist swale, the well-drained soils of the steep convex, concave, and straight hillsides. The same processes were measured on poorly-drained and moderately well-drained soils in another catchment of the Sleepers River Watershed.

(4) Only the frequency, timing and relative importance of the three processes varied, depending upon soil and topography. Because of differences in topographic position, water-table depth, or depth to an impeding layer, the unfilled storage capacity of the soils varied from zero in the channel depression to small in the shallow moist swale, and to large on the steep, well-drained hillside.

(5) In short intense rainstorms, practically all of the stormflow was contributed by direct precipitation onto the saturated areas around the channels, its neighbouring seeps, and the lower part of the shallow moist swale. If the rainstorm was small, but of long duration, contributions of subsurface stormflow occurred from the areas close to the channel. This response occurred by vertical percolation of water infiltrating a wet soil, as shown in Figures 7.13, and 7.15. Subsurface stormflow was important in controlling the recession limb of the hydrograph, and in some small storms of long duration dominated the volume of stormflow produced.

(6) In large rainstorms, the contribution of subsurface stormflow increased, and was measured even in the very deep, well-drained soils on steep slopes. It contributed to the peak of some hydrographs (Figure 7.24), and was important in controlling the later recession of the hydrograph. It was the only producer of stormflow in the deepest, well-drained soils.

(7) In such large storms, however, the contributions of return flow and direct precipitation onto saturated areas increased much more rapidly than those of subsurface stormflow. The saturated areas expanded rapidly from poorly-drained soils into areas of initially better drained soils and steeper topography. Hollows were preferred avenues for such expansion, partly because of their role in concentrating downslope subsurface flow, but to a much greater extent, because their concave profiles caused the water table to

be closer to the surface at the beginning of the storm, than it was under a hillside of similar overall gradient but straight profile.

(8) Because of velocities of saturation overland flow were 1000–5000 times those of subsurface stormflow, water that emerged from the ground surface could contribute storm runoff to the channel at high rates and dominate the channel hydrograph. Contributions that remained below the soil surface were of relatively minor significance in controlling the shape of the hydrograph, except during the recession phase. Channel hydrographs supplied by saturation overland flow rose steeply during the first few minutes of rain. They responded sharply to changes of rainfall intensity, and declined rapidly. Most of the hydrographs produced by natural and artificial storms on the experimental plots and the 3·9-hectare catchment were reminiscent of hydrographs of Horton overland flow.

(9) The relative importance of overland flow from small saturated areas of the basin increased rapidly with storm size, and during large storms on poorly drained shallow soils, subsurface stormflow became a negligible contributor of stormflow. This was confirmed by field experiments on another catchment with extensive areas of poorly-drained soils on gentle hillslopes.

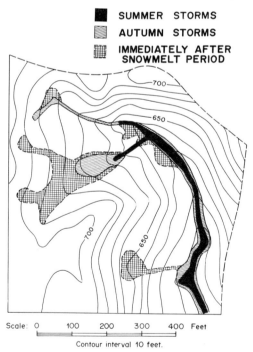

Figure 7.28 Seasonal variation of the pre-storm saturated area for a 3·9-hectare catchment (WC-4) with steep, well-drained hillsides and a confined valley bottom

(10) The area over which the water table reached the surface was dynamic in the sense that it varied seasonally and throughout a storm (see Figures 7.28, 7.30, and 7.32).

(11) The location and extent of saturated areas that produced overland flow were determined by the dynamics of the subsurface flow system. This included not only the subsurface stormflow occurring during a storm, but the slower, larger-scale changes in the phreatic zone during periods of dry weather. Consequently, a knowledge of the subsurface flow system is necessary for understanding the production of storm runoff.

Table 7.11   Approximate reccurrence intervals of rainstorms that produce significant storm runoff from various portions of the 3·9-hectare basin shown in Figure 7.28

Part of drainage basin	Approximate recurrence interval of runoff producing events (years)
Lower part of valley floor	10^{-2}
Upper part of valley floor	10^{-1}
Lower part of shallow, moist swale	10^{-1}
Upper part of shallow, moist swale	10^{0}
Lower, concave portion of well-drained hillside	$10^{1}-10^{2}$
Straight well-drained hillside	$10^{2}-10^{3}$

A major feature of the variable-source concept is that the area over which return flow and direct precipitation are generated vary seasonally and throughout a storm. Figure 7.28 shows the seasonal variation of the pre-storm saturated area for the 3·9-hectare WC-4 basin, referrred to earlier. The fluctuations of this contributing area can be related to topography, soils, antecedent moisture, and

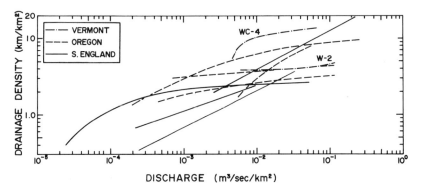

Figure 7.29   Seasonal variation of drainage density with stream discharge for small drainage basins in Vermont, Oregon, and South England (after Gregory and Walling, 1968; Weyman, 1970; Roberts and Klingeman, 1972; Blyth and Rodda, 1973; Dunne)

K

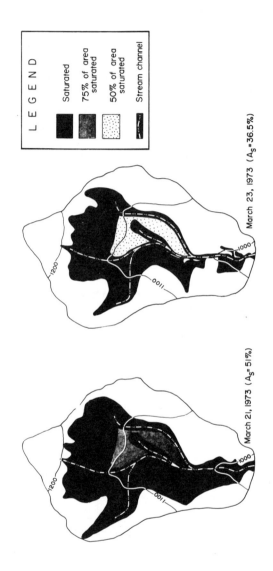

March 23, 1973 (A$_s$=36.5%)

March 21, 1973 (A$_s$=51%)

LEGEND

Saturated

75% of area saturated

50% of area saturated

Stream channel

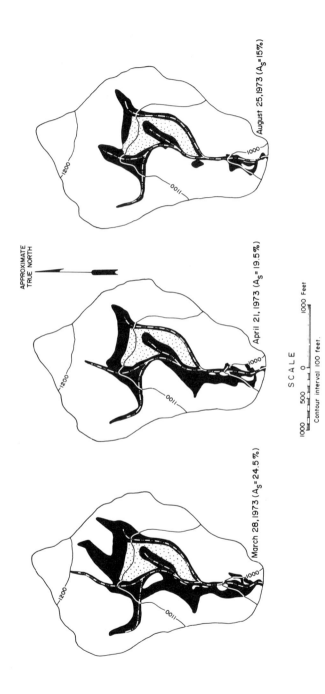

Figure 7.30   Seasonal variation of the pre-storm saturated area for a 59-hectare catchment (W-2) with gentle hillslopes, a broad valley bottom, and extensive areas of moderate-to-poorly drained soils

rainfall characteristics. Rather than having distinctly different reactions to rainfall, the various soils of the watershed form a continuum. The valley floor, the shallow moist swale, and various parts of the well-drained hillside differ in the magnitude of the storm required to bring the water table to the surface, the delay with which this is accomplished, and the total amount of runoff produced by saturation overland flow. Different soils of a watershed in humid regions contribute most of their storm runoff by the same process, but with differing frequencies. Field observations during natural and artificial storms allowed the characterization of various portions of the 3·9 hectare drainage basin as to the approximate recurrence interval of rainstorms which would cause each to contribute significant storm runoff (see Table 7.11).

The small catchment illustrated in Figure 7.28 has generally steep hillsides, covered with deep, well-drained soils. Large amounts of rainfall or snowmelt are needed to bring the water table to the surface in these soils, and so the expansion of the saturated area is confined to the narrow valley bottom and the small area of shallow, poorly-drained soils on hillslopes of low gradient. The fluctuation is mainly lineal.

The length of the saturated valley floor before a storm depends upon the height of the water table which is also related to the stream discharge. Figure 7.29 shows the variation of the saturated length (expressed as drainage density) with baseflow for catchment WC-4, for a second small drainage basin (W-2) in the Sleepers River Watershed, and for several basins in Oregon (Roberts and Klingeman, 1972) and southern England (Gregory and Walling, 1968; Weyman, 1970; Blyth and Rodda, 1973). There is no general mathematical expression that fits all the relationships because of the differences in physiography, drainage area, and relative amounts of change between the catchments, but most of the curves show a rough parallelism.

In catchments that do not have steep well-drained hillsides, and a confined valley bottom, the fluctuation of the length of saturated areas is not a particularly good

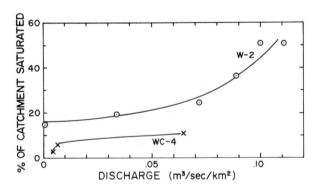

Figure 7.31   Seasonal variation of the extent of the saturated area with baseflow for a steep catchment with well-drained soils (WC-4) and a catchment with gentler hillsides and moderate-to-poorly drained soils (W-2)

measure of antecedent wetness. For in such basins saturated conditions can spread not only along the valley floor and hillside swales but for considerable distances up the gentle hillsides as well. This can be illustrated by mapping the extent of the saturated area at several times throughout the year. Figure 7.30 portrays a series of such maps for the W-2 catchment of the Sleepers River Experimental Watershed. The W-2 catchment has average hillside gradients of 10–13 %, and long gentle footslopes, covered with moderate-to-poorly drained soils. The upslope extension of saturated conditions is important not only for predicting storm runoff, but for understanding water quality (Dunne, 1969b; Kunkle, 1970). The relationship of the extent of the saturated area to pre-storm baseflow is shown in Figure 7.31 for both WC-4 and W-2. The more rapid spread of saturated conditions in W-2 as

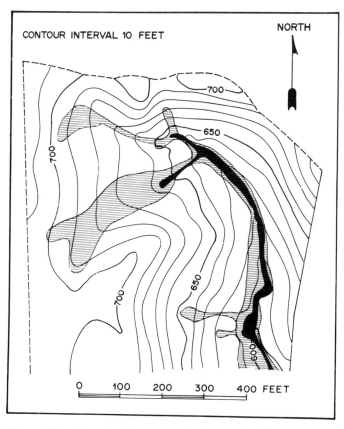

Figure 7.32   Growth of the area producing overland flow as return flow and direct precipitation onto saturated areas during a 46-mm rainstorm in a basin with steep well-drained hillsides and a narrow valley bottom. The solid area produced stormflow at the beginning of the storm, but by the end of the storm, the water table had reached the ground surface over the shaded area

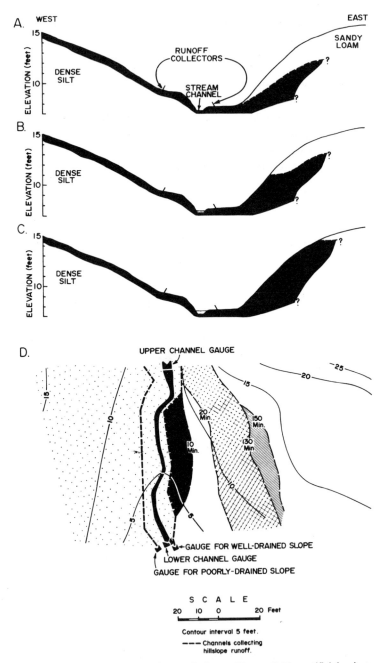

Figure 7.33 Growth of the saturated area during a 62-mm, 2·5-hr artificial rainstorm on strip of land 35 m along the channel by 32 m on either side of the channel in the W-2 catchment. (A), The beginning of the storm; (B), after 1 hr of rain; (C), after 2·5 hr of rain; (D), shows the growth of the saturated area after 10, 20, 130, and 150 min of rainfall. The dotted area on the west side of the channel and in the stream depression was saturated from the beginning of the storm

baseflow increases reflects differences in the topography and soils of the two catchments.

The contributing area also varies during a storm, as shown in Figure 7.32 for a 46-mm rainstorm on the 3·9 hectare WC-4 catchment. Again, the variation was lineal, because the deep well-drained soils on the steep hillslopes did not become saturated, and even in a 50-year storm, the zone contributing stormflow covered only 7 % of the basin. On the gentler hillslopes and wetter soils of catchment W-2, the contributing area grew rapidly upslope as shown in Figure 7.33 for a 62 mm artificial storm generated over a strip of land which extended 35 m along the channel and 32 m on either side of the channel. The footslope on the west side of the channel had a shallow, poorly-drained soil that was saturated when rainfall began; on the east side the slope was steeper, and the soil was deeper and better drained. During the artificial storm, runoff from the channel area, and saturation overland flow from the hillsides were gauged separately at locations shown in Figure 7.33D, and water-table elevations and the extent of the saturated area were measured throughout the storm. The growth of the saturated area in plan and vertical view are documented in the diagram. Runoff was generated by the same processes as on the plots described previously.

During a storm the rate of runoff depends upon the properties of the soil, the pattern of subsurface fluid pressures, the area over which water is emerging from the ground and the intensity of rainfall. Consequently, for any single storm there is probably a hysteretic relationship between runoff rate and the extent of the saturated area (or its length). Within a few minutes of the end of rainfall, runoff from direct precipitation onto the saturated area has ceased, return flow is also declining rapidly, but the extent of the saturated area diminishes slowly. I have observed this after many storms and snowmelt periods, but have not developed any statistical generalizations of the process. Another complication is that the relation of saturated area to time or to accumulated rainfall can be a step function under some topographic conditions.

Much work remains to be done in the field recognition and prediction of these runoff-producing areas. They are usually visible to an observer in the field and on false-colour infra-red film, and are related to the distribution of certain plants and pedological characteristics. Each of these techniques can be used for mapping saturated zones. It is possible that their extent can be predicted by means of an antecedent precipitation index. Blyth and Rodda (1973) have used multiple regression to predict the length of flowing channels from the effective precipitation of the preceeding 6 days and the length of flowing channels on the seventh day prior to the prediction.

The variable source concept has been given a more coherent structure by the mathematical model of Freeze (1972a; 1972b), which describes the generation of runoff from hillsides and small, upland catchments. The model yields numerical solutions to the equations describing saturated–unsaturated subsurface flow, return flow, direct precipitation onto saturated areas and flow in small channels.

Freeze investigated runoff production under a range of rainfall duration and intensity, soil conductivity, hillslope shape and soil thickness. Subsurface stormflow occurred, as recognized in the field studies, but because of the impedence of even highly-permeable soils, most of the infiltrating water was stored within the soil, raising the water table to the land surface over an expanding area (see Figure 7.34). Return flow occurred over the saturated area, as the lower part of the hillside became a seepage face. Direct precipitation onto the saturated zone augmented this overland flow. The rise of the water table was mainly fed by vertical percolation, rather than by horizontal seepage and the production of overland flow depended upon developments within the subsurface flow system.

A sample of Freeze's results is shown in Figure 7.34, and Table 7.12 is a summary of the simulated hydrographs from a 120 m × 33·5 m hillside. Variations in rainfall intensity and duration, or in soil thickness and slope did not have major effects on

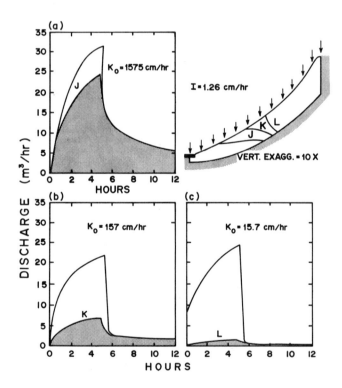

Figure 7.34   Calculated hydrographs of subsurface storm flow and return flow (shaded), and direct precipitation onto saturated areas for a 5-hr rainstorm with an intensity of 1·26 cm/hr, on soils of various hydraulic conductivities. The hillslope and the response of the water table at the end of the storm is shown for each case in the inset (after Freeze, 1972b)

Table 7.12 Summary of calculated peak runoff rates and 12-hr runoff yields (after Freeze, 1972b)

Rainfall intensity (cm/hr)	Rainfall duration (hr)	Hillslope shape	Gradient (%)	Soil thickness (m)	Saturated hydraulic conductivity of soil (cm/hr)	Subsurface stormflow and return flow		Direct precipitation onto saturated area		Percentage of hillside that is saturated
						Peak rate (cm/hr)	Yield (% of rain)	Peak rate (cm/hr)	Yield (% of rain)	
1·27	5	Convex	7·5	0·20 at top of slope; 1·0 at base	157	0·327	22·5	0·019	0	0
1·27	5	Convex	7·5	0·20–1·0	15·7	0·075	7	0·031	0	0
1·27	5	Convex	7·5	0·20–1·0	1·6	0·025	1	0·094	1·45	6·5
1·27	5	Convex	15	0·20–1·0	157	0·325	24	0	0	0
1·27	5	Convex	15	0·20–2·0	157	0·335	29	0	0	0
2·54	2·5	Convex	7·5	0·20–1·0	157	0·467	24·5	0	0	0
1·27	5	Concave	7·5	1·0 at top of slope; 0·2 at base	157	0·175	10·2	0·378	17·1	45
1·27	5	Concave	7·5	1·0–0·2	15·7	0·036	2·1	0·575	24·1	59

runoff rates. Saturated hydraulic conductivity of the soil was the most important control of the volume and peak rate of runoff. Values of saturated hydraulic conductivity are given in Table 7.7 for a number of field studies of subsurface stormflow, and the conductivities at which Freeze found subsurface stormflow to be important were at the very high end of the scale for field soils. It should be noted that Freeze's definition of subsurface stormflow includes runoff that was separately classified in the earlier discussion as subsurface stormflow (which remains below the soil surface on its way down hillsides), and return flow (which originates within the soil but emerges from the ground surface and reaches the stream as overland flow). I made the distinction earlier because the two processes can be measured separately in the field, and because the fact that return flow occurs over the soil surface has important consequences for geomorphology (Kirkby and Chorley, 1967) and water quality (Dunne, 1969b; Kunkle, 1970b) as well as for hydrology.

Freeze concluded that there are stringent limitations on the occurrence of subsurface stormflow as a quantitatively significant storm runoff component. He showed that only on hillslopes which feed deeply-incised channels—and then only when saturated soil conductivities are very large—does subsurface stormflow contribute storm runoff at rates and volumes sufficient to dominate the hydrograph. In soils with lower conductivities, and especially on concave slopes, hydrographs are dominated by overland flow from direct precipitation onto saturated areas.

7.3.5   Characteristics of variable-source hydrographs

In areas generating runoff as subsurface stormflow and saturation overland flow, it is difficult to decide upon a definition of storm runoff, and various workers have used different definitions. Table 7.13 lists some values of the runoff coefficient for catchments in which return flow and direct precipitation dominate the storm hydrograph volumetrically though subsurface stormflow adds importantly to the recession limb. On the whole, runoff coefficients are slightly higher than for hydrographs produced only by subsurface stormflow, but not greatly so. Table 7.14 and Figure 7.35 indicate, however, that the envelope of the peak rates of runoff is 5–40 times greater than that for subsurface stormflow from catchments of similar size. Even in those catchments where storm runoff is dominated volumetrically by subsurface stormflow, peak rates of runoff generated by overland flow for short periods at the beginning of the storm can be at least 50% higher than the later subsurface stormflow peaks. But even these early overland flow peaks are smaller than peaks of similar origin in other humid regions, because subsurface stormflow dominates the hydrograph only where shallow poorly-drained soils and valley bottoms (and therefore overland flow contributions) are of limited extent. The envelope for saturation overland flow peaks in humid regions is 4–10 times smaller than that for Horton overland flow (Figure 7.7) because in the latter case, overland flow occurs more generally throughout the catchment. Some of this difference, however, may be due to the small sample presented in Figure 7.35. I have included

Table 7.13   Runoff coefficients (runoff volume/rainfall volume) for catchments subject to variable-source stormflow contributions

Location	Condition of the drainage basin	Catchment size (km²)	Runoff coefficient	Source
S. Ontario	Glacial till; rainstorms	18	Mean 0·065 Median 0·05 Range 0·01–0·25	Dickinson and Whiteley (1970)
S. Ontario	Glacial till; snowmelt	18	Up to 0·50	Dickinson and Whiteley (1970)
S. Ontario	Clays; rainstorms	42	Mean 0·20 Median 0·10 Range 0–0·59	Dickinson and Whiteley (1970)
Pennsylvania	Pasture and agricultural land; summer storms	0·162	0·02–0·10	Rawitz *et al.* (1970)
Burlington, Vermont	Deep sands	0·46	0·012–0·03	Ragan (1968)
Sleepers River, Vermont	Sandy loams; steep slopes	0·0012	0–0·27	Dunne (1969a)
Sleepers River, Vermont	Silt loams; shallow slopes	0·0024	0–0·27	Dunne (1969a)
Sleepers River, Vermont	Sandy loams and silt loams; large storms	42·5	Median 0·30 Range 0·02–0·42	US Dept. Agric. (1963–1971)
Sleepers River, Vermont	Sandy loams and silt loams; large storms	0·59	Median 0·07 Range 0·05–0·19	US Dept. Agric. (1963–1971)
Sleepers River, Vermont	Sandy loams and silt loams; large storms	8·30	Median 0·13 Range 0·05–0·39	US Dept. Agric. (1963–1971)
Sleepers River, Vermont	Sandy loams and silt loams; large storms	110	Median 0·13 Range 0·02–0·44	US Dept. Agric. (1963–1971)

Table 7.14 Measured instantaneous peak rates of runoff from hillslopes and drainage basins which generate variable-source stormflow

Catchment	Area (km)	Peak runoff rate (cm/hr)	Source
Sleepers River, Vermont	0·0012	2·05	
	0·0024	0·81	
	0·0056	1·50	
	0·0078	5·00	
	0·009	1·27	
	0·027	0·68	
	0·039	0·54	
	0·47	0·43	US Dept. Agric. station records
	0·59	0·54	US Dept. Agric. station records
	0·68	0·28	US Dept. Agric. station records
	1·02	0·24	US Dept. Agric. station records
	2·00	0·27	US Dept. Agric. station records
	2·25	0·18	US Dept. Agric. station records
	8·3	0·19	US Dept. Agric. station records
	15·5	0·30	US Dept. Agric. station records
	16·2	0·60	US Dept. Agric. station records
	21·4	0·31	US Dept. Agric. station records
	42·5	0·30	US Dept. Agric. station records
	43·0	0·30	US Dept. Agric. station records
	110·0	0·38	US Dept. Agric. station records
Eaton River, S. Quebec	87	0·12	J. Chyurlia, personal communication
	3·38	0·11	J. Chyurlia, personal communication
S. England, Upper East Twin Brook	0·10	0·12	Weyman (1970)

Return flow and direct precipitation in areas where storm runoff is dominated volumetrically by subsurface flow.

S. England, Lower East Twin Brook	0·11	0·041	Weyman (1970)
Georgia Piedmont	0·24	0·011	Hewlett and Nutter (1970)
N. Carolina	0·435	0·054	Hewlett and Nutter (1970)
Burlington, Vermont	0·46	>0·0054	Ragan (1968)
Kimakia, Kenya	0·52	0·042	Pereira et al. (1962)
Wagon Wheel Gap, Colorado	0·81	0·011	Hewlett and Nutter (1970)

only storms in drainage basins in which there is general agreement in the literature about the runoff processes. Hack and Goodlett (1960) list peak runoff rates of 7·3 and 5·4 cm/hr from two Virginia Appalachian basins of 6·2 km² and 65 km² respectively. Their descriptions strongly suggest that return flow and direct precipitation onto saturated areas were the dominant runoff processes, but because direct field observations were not made, these points are not included in Figure 7.35. The envelope in this diagram, refers mainly to regions with low rainfall intensities, and should be viewed only as an indication of relatively low peak runoff rates. It will be increased several-fold as more data become available from detailed runoff studies in other areas.

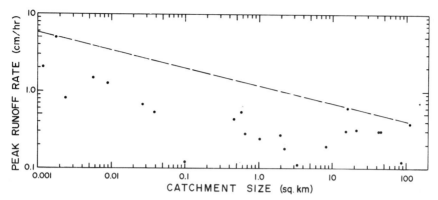

Figure 7.35   Measured instantaneous peak rates of runoff for catchments of various sizes which generate variable source stormflow

Because large amounts of storm runoff from the variable sources reach the channel as overland flow, the response of runoff to rainfall is rapid. The hydraulics of this overland flow is not understood at present. The runoff rate does not necessarily increase linearly with distance downslope, as described by the Horton model. The mat of vegetation in runoff-producing areas is so thick that the application of Manning's equation to the overland flow is questionable. On a hillslope with a gradient of 0·40, I have measured velocities of overland flow ranging from 3–15 cm/sec. On hillslopes with lower gradients and thick vegetation covers, however, velocities of less than 0·1 cm/sec are common. In these rough, marshy areas, I have measured more than 2 cm of depression storage and up to 1 cm of detention storage and velocities of 0·1 cm/sec (by the salt injection technique described by Calkins and Dunne, 1970).

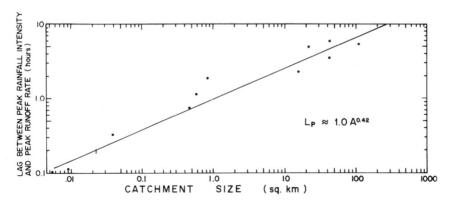

Figure 7.36   Variation of lag between peak rainfall and peak runoff for catchments of various sizes in the Sleepers River experimental watershed, Danville, Vermont

The hydrographs of return flow and direct precipitation presented earlier show extremely rapid responses to rainfall with steep recessions, but even several hours after the end of rainfall, water can be seen draining slowly over the surface of flat marshy contributing areas and supplying the recession limb of the hydrograph. Figure 7.36 shows a relationship between catchment size and the lag from a burst of intense rainfall to the peak of the hydrograph for drainage basins in the Sleepers River Watershed.

Table 7.15   Values of the recession constant, K, for variable-source hydro-graphs for the Sleepers River Watershed, Danville, Vermount

Catchment	Drainage area (km^2)	K
Concave, permeable hillside	0·0012	0·0065–0·014
Shallow moist swale	0·0024	<0·01
Watershed WC-1	0·0056	>0·001–0·76
Watershed WC-2	0·009	0·006–0·75
Watershed W-2	0·59	0·42–0·60 (large storms)
		0·81 (all storms)
Watershed W-3	8·30	0·77–0·89 (large storms)
		0·90 (all storms)
Watershed W-1	42·5	0·84–0·88 (large storms)
		0·90 (all storms)
Watershed W-5	110	0·84–0·94 (large storms)
		0·90 (all storms)

Comparison of Figures 7.36 and 7.19 shows that variable source hydrographs have lag times approximately 20–40 times less than those of subsurface stormflow hydrographs for catchments of the same size. Drainage basins whose hydrographs are dominated volumetrically by subsurface stormflow often have an early peak from overland flow onto saturated areas. The lag of these early peaks is much less than the lag to peak of the subsurface stormflow. The rates of recession of hydrographs composed of both overland flow and subsurface stormflow are indicated by the recession constants in Table 7.15. While overland flow is the dominant producer of runoff on small plots the recession constant is as low as those for Horton hydrographs, but during the later stages of the hydrograph the contribution of subsurface stormflow (and in larger catchments, the effect of channel storage) increase the recession constant which then approaches the values shown in Table 7.9 for subsurface stormflow.

7.3.6   The range of stormflow processes in humid regions

Where soils are well drained, deep and permeable and steep hillslopes border a narrow valley floor, subsurface stormflow dominates the hydrograph volumetrically. Contributions of direct precipitation onto the saturated valley floor and small amounts of return flow may produce sharp peaks, but the

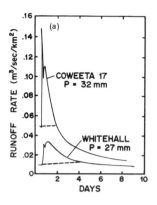
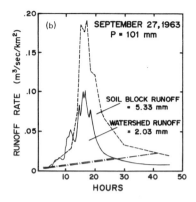

Figure 7.37 Hydrographs from regions where subsurface stormflow dominates the hydrograph (a), Hydrographs from a 43·5-hectare catchment at Coweeta in the Southern Appalachians of Northern Carolina, and from a 24-hectare catchment at Whitehall in the Georgia Piedmont; (b), comparison of runoff from a 15·2-hectare catchment and a 0·0074-hectare experimental plot after a rainstorm at Coweeta.
The dashed lines indicate a separation of runoff into 'quickflow' and 'delayed flow' proposed by Hewlett and Hibbert (1967) in basins where subsurface stormflow dominates the hydrograph (after Hewlett and Nutter, 1970)

narrowness of the valley floor and the fact that saturated conditions spread slowly up steep hollows and sideslopes limit the spread of saturation overland flow. Later, larger contributions of subsurface stormflow have low peak rates. Several examples of such hydrographs are given by Hewlett and Nutter (1970) for basins in the mountains of North Carolina and the Georgia Piedmont (see Figure 7.37). A more extreme example is shown in Figure 7.38 for a catchment in the Aberdare Range of Kenya. The basin is developed in permeable volcanic ash which is hundreds of feet thick, but which contains impeding layers at several depths. Figure

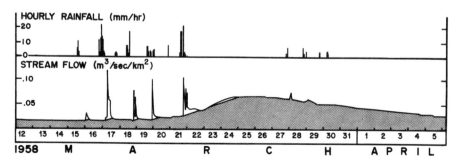

Figure 7.38 Stormflow resulting from several large rainstorms at the end of the long dry season in the 52-hectare Kimakia catchment, Aberdare Highlands, Kenya. The hydrographs are separated into quickflow and delayed flow (shaded) by the method of Hewlett and Hibbert (after: Pereira, *et al.*, (1962))

7.38 shows the response of runoff to several days of heavy rain at the end of a long dry season. Sharp hydrographs were produced by each rainstorm, but they yielded only a few percent of the rain falling onto the catchment. One day after the seventh rainstorm, the stream began a slow rise sustained by a subsurface flow which peaked several days later.

Patric and Swanston (1968) irrigated weathered till soils on steep (26° and 38°) forested hillsides in southeastern Alaska with more than 1000 cm of rain in 30 days. Return flow occurred on only small areas below the sprinkled plots; most of the runoff occurred as subsurface stormflow. We have no information on the relative runoff contributions of the poorly-drained soils of this region in similar storms, but the data of Patric and Swanston confirm that subsurface stormflow can occur at high rates under extreme rainfall conditions.

Subsurface stormflow, in agreement with the Freeze predictions, is of greatest significance to the hydrograph (though not necessarily any greater in absolute terms relative to its magnitude in other regions) in those areas where conditions are unfavourable for other stormflow-generating mechanisms. In particular, the occurrence of heavy forest vegetation on steep slopes with thick soils and deeply-incised streams ensure that infiltration and percolation rates are relatively high, yet

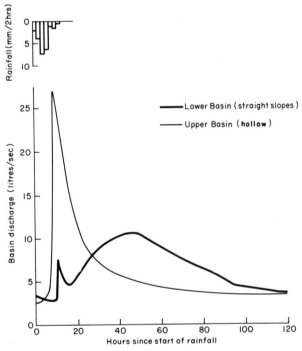

Figure 7.39   Storm-runoff hydrographs from two areas of the East Twin Basin, Somerset, each of 0·1 km² (after Calver, Kirkby and Weyman, 1972, Figure 7.5)

return-flow and direct-precipitation contributions are minimized by the absence of poorly-drained soils on gentle slopes.

Even in areas where subsurface stormflow dominates the volume of the storm hydrographs (e.g. Figures 7.37 and 7.38), the highest peaks, are contributed by precipitation which 'fell in the perennial channel and its immediate banks' (Hewlett and Nutter, 1970). Calver *et al.* (1972) have presented hydrographs that show the responses of two zones of a catchment. One of the zones produces a rapid response due to saturation overland flow; the second responds more slowly due to subsurface stormflow (see Figure 7.39). In other regions, such as Vermont, where the saturated and near-saturated valley bottoms are more extensive and where footslopes are gentler and soils thinner, this early contribution becomes larger in both peak rate and total volume. The mechanisms responsible for this contribution are return flow and direct precipitation onto the saturated area. Subsurface stormflow occurs in such regions but is less important relative to the mechanisms contributing overland flow. A range of conditions exists between those which tend to produce a preponderance of subsurface stormflow and those which favour the occurrence of return flow and direct precipitation contributions.

7.4 SUMMARY

Water can reach stream channels by several routes. The processes that deliver stormflow, and the volumes and timing of their contributions vary with

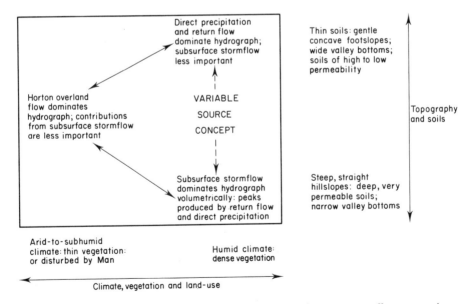

Figure 7.40   Schematic illustration of the occurrence of various runoff processes in relation to their major controls

topography, soil properties and rainfall characteristics, and indirectly with climate, vegetation and land use. Even within a particular basin, the dominant runoff process may vary with the characteristics of rainstorms. Even the highest infiltration capacities of forest soils will not accommodate the highest recorded rainfall intensities. The various models of storm runoff, therefore, are complementary rather than contradictory. Differences of emphasis between studies of runoff reflect the physical geography of the regions in which experiments were carried out.

Horton overland flow is most common in arid and semi-arid lands, or in humid areas where the original vegetation and soil structure have been destroyed. The spatial and temporal occurrence of other storm runoff processes is less well understood, but Figure 7.40 indicates schematically the relation of the various runoff processes to their major controls.

ACKNOWLEDGEMENTS

Much of my own work reported in this chapter was sponsored by the Agricultural Research Service of the US Department of Agriculture. I have also profited from discussions or correspondence with J. Chyurlia, R. A. Freeze, A. J. Pearce, B. J. Knapp, S. H. Kunkle, T. R. Moore, R. M. Ragan, D. R. Weyman, and R. Z. Whipkey. Several of these people assisted with fieldwork or provided me with unpublished data. I am very grateful to all of them and to the Agricultural Research Service.

REFERENCES

Bates, C. G. and Henry, A. J., 1928, 'Forest and streamflow experiment at Wagon Wheel Gap, Colorado', *Monthly Weather Review*, Supplement No. 30, 30–79.
Bertoni, J., Larson, W. E. and Shrader, W. D., 1958, 'Determination of infiltration rates on Marshall silt loam from runoff and rainfall records', *Proc. Soil Sci. Soc. Am.*, **22**, 571–74.
Betson, R. P., 1964, 'What is watershed runoff?', *J. Geophys. Res.*, **69**, 1541–52.
Betson, R. P., Marius, J. B. and Joyce, R. T., 1968, 'Detection of saturated interflow in soils with piezometers', *Proc. Soil Sci. Soc. Am.*, **32**, 602–4.
Biswas, T. D., Nielsen, D. R. and Biggar, J. W., 1966, 'Redistribution of soil water after infiltration', *Water Res. Res.*, **2**, 513–24.
Blyth, K. and Rodda, J. C., 1973, 'A stream length study', *Water Res. Res.*, **9**, 1454–61.
Calkins, D. J. and Dunne, T., 1970, 'A salt-tracing method for measuring channel velocities in small mountain streams', *J. Hydrology*, **11**, 379–92.
Calver, A., Kirkby, M. J., and Weyman, D. R., 1972, 'Modelling hillslope and channel flows', in *Spatial Analysis in Geomorphology*, Chorley, R. J. (Ed.), Methuen, pp. 197–220.
Davis, S. N. and DeWiest, R. J. M., 1966, *Hydrogeology*, John Wiley, New York, 463 pp.
Dickinson, W. T. and Whiteley, H., 1970, 'Watershed areas contributing to runoff', *Internat. Assoc. Sci. Hydrology, Proc. Wellington Symposium*, Publication 96, pp. 12–26.
Dingman, S. L., 1966, 'Characteristics of summer runoff from a small watershed in central Alaska', *Water Res. Res.*, **2**, 751–54.
Dunne, T., 1969a, *Runoff Production in a Humid Area*, Ph.D. dissertation, Department of Geography, Johns Hopkins Univ., Baltimore, Maryland.

Dunne, T., 1969b, 'The significance of "partial-area" contributions to storm runoff for the study of sources of agricultural pollutants', unpublished manuscript, *Agric. Res. Ser.*, Danville, Vermont.

Dunne, T. and Black, R. D., 1970a, 'An experimental investigation of runoff production in permeable soils', *Water Res. Res.*, **6**, 478–90.

Dunne, T. and Black, R. D., 1970b, 'Partial area contributions to storm runoff in a small New England watershed', *Water Res. Res.*, **6**, 1296–1311.

Dunne, T. and Black, R. D., 1971, 'Runoff processes during snowmelt', *Water Res. Res.*, **7**, 1160–72.

Emmett, W. W., 1970, 'The hydraulics of overland flow on hillslpoes', *US Geolog. Surv.*, Prof. Paper **662-A**.

Freeze, R. A., 1969, 'The mechanism of natural groundwater recharge and discharge: One-dimensional, vertical, unsteady, unsaturated flow above a recharging or discharging groundwater flow system', *Water Res. Res.*, **5**, 153–71.

Freeze, R. A., 1972a, 'Role of subsurface flow in generating surface runoff. 1. Base flow contributions to channel flow', *Water Res. Res.*, **8**, 609–23.

Freeze, R. A., 1972b, 'Role of subsurface flow in generating surface runoff. 2. Upstream source areas', *Water Res. Res.*, **8**, 1272–83.

Gregory, K. J. and Walling, D. E., 1968, 'The variation of drainage density within a catchment', *Bull. Internat. Assoc. Sci. Hydrology*, **12**(2), 61–62.

Hewlett, J. D., 1961, 'Soil moisture as a source of baseflow from steep mountain watersheds', (Station Paper 132.) *US Forest Service, Southeastern Forest Experiment Station*, Asheville, North Carolina.

Hewlett, J. D. and Hibbert, A. R., 1963, 'Moisture and energy conditions within a sloping soil mass during drainage', *J. Geophys. Res*, **68**, 1081–87.

Hewlett, J. D. and Hibbert, A. R., 1967, 'Factors affecting the response of small watersheds to precipitation in humid areas,' in *Forest Hydrology*, Sopper, W. E. and Lull, H. W., (Eds.) Pergamon, pp. 275–90.

Hewlett, J. D. and Nutter, W. L., 1970, 'The varying source area of streamflow from upland basins', *Proceedings of the Symposium on Interdisciplinary Aspects of Watershed Management*, Montana State University, Boseman, American Society of Civil Engineers, pp. 65–83.

Holtan, H. N., 1961, 'A concept for infiltration estimates in watershed engineering', *US Department of Agriculture, Agriculture Research Service, ARS 41–55*, Beltsville, Maryland, 25 pp.

Horton, J. H. and Hawkins, R. H., 1965, 'Flow path of rain from the soil surface to the water table', *Soil Sci.* **100**, 377–83.

Horton, R. E., 1933, 'The rôle of infiltration in the hydrologic cycle', *Trans. Am. Geophys. Union*, **14**, 446–60.

Horton, R. E., 1940, 'An approach towards a physical interpretation of infiltration capacity', *Proc. Soil Sci. Soc. Am. Proc.* **4**, 399–417.

Inbar, M., 1969, 'A geomorphic analysis of a catastrophic flood in a Mediterranean basaltic watershed', *Mimeographed report*.

Kincaid, D. R., Osborn, H. B. and Gardner, J. H., 1966, 'Use of unit-source watersheds for hydrologic investigations in the semi-arid Southwest', *Water Res. Res.*, **2**, 381–92.

Kirkby, M. J. and Chorley, R. J., 1967, 'Overland flow, throughflow and erosion', *Bull. Internat. Assoc. Sci. Hydrology*, **12**(2), 5–21.

Knapp, B. J., 1973a, 'A system for the field measurement of soil water movement', *Tech. Bull.*, No. 9, *Brit. Geomorphology. Res. Gp.*, 26 pp.

Knapp, B. J., 1973b, 'Throughflow and the problem of modelling', unpublished manuscript.

Knisel, W. G., 1973, Comments on 'Role of subsurface flow in generating surface runoff. 2. Upstream source areas' by Freeze, R. A., *Water Res. Res.*, **9**, 1107–10.

Kunkle, S. H., 1970, 'Sources and transport of bacterial indicators in rural streams',

Proceedings of the Symposium on Interdisciplinary Aspects of Watershed Management, Montana State University, Bozeman, American Society of Civil Engineers, pp. 105–32.

Lusby, G. C., Turner, G. F., Thompson, J. R. and Reid, V. H., 1963, 'Hydrologic and biotic characteristics of grazed and ungrazed watershed of the Badger Wash Basin in western Colorado, 1953–58', US Geolog. Surv. Water-supply, Paper 1533-B.

Miller, C. R. and Viessman, W., 1972, 'Runoff volumes from small, urban watersheds', Water Res. Res., 8, 429–34.

Monke, E. J., Huggins, L. F., Galloway, H. M. and Foster, G. R., 1967, 'Field study of subsurface drainage in a slowly permeable soil', Trans. Am. Assoc. Agric. Engrs., 9(6), 573–576.

Musgrave, G. W. and Holtan, H. N., 1964, 'Infiltration,' in Handbook of Applied Hydrology, Chow, V. T. (Ed.), McGraw-Hill, New York, Section 12.

Nielsen, D. R., Kirkham, D. and Van Wijk, W. R., 1959, 'Measuring water stored temporarily above the field moisture capacity', Proc. Soil Sci. Soc. Am. 23, 409–12.

Ogata, G. and Richards, L. A., 1957, 'Water content changes following irrigation of bare-field soil that is protected from evaporation', Proc. Soil Sci. Soc. Am., 21, 355–56.

Osborn, H. B. and Renard, K. G., 1970, 'Thunderstorm runoff on the Walnut Gulch Experimental Watershed, Arizona, USA', Internat. Assoc. Sci. Hydrology, Proc. of the Wellington Symposium, Publication 96, pp. 455–64.

Patric, J. H. and Swanston, D. N., 1968, 'Hydrology of a slide-prone glacial till soil in southeast Alaska', J. Forestry, 66, 62–66.

Pearce, A. J., 1973, Mass and Energy Flux in Physical Denudation, Defoliated Areas, Sudbury, Ph.D. dissertation, Department of Geological Sciences, McGill University, Montreal, 235 pp.

Pereira, H. C., Dagg, M. and Hosegood, P. H., 1962, 'The water-balance of bamboo thicket and of newly planted pines', East African Agricultural and Forestry Journal, 27, 95–103.

Philip, J. R., 'The theory of infiltration', Soil Sci., 83, 345–57.

Ragan, R. M., 1968, 'An experimental investigation of partial area contributions', Internat Assoc. Sci. Hydrology, Symposium of Bern, Publication 76, pp. 241–251.

Rawitz, E., Engman, E. T. and Cline, G. D., 1970, 'Use of the mass balance method for examining the role of soils in controlling watershed performance', Water Res. Res., 6, 115–23.

Roberts, M. C. and Klingeman, P. C., 1972, 'The relationship of drainage net fluctuation and discharge', Proc. Internat. Geographical Congress, pp. 181–91.

Robins, J. S. Pruitt, W. O., and Gardner, W. H., 1954, "Unsaturated flow of water in field soil and its effects on soil moisture investigations', Proc. Soil Sci. Soc. Am., 18, 344–47.

Rubin, J., 1966, "Theory of rainfall uptake by soils initially drier than field capacity and its applications', Water Res. Res., 2, 739–49.

Rubin, J., Steinhardt, R. and Reiniger, P., 1964, 'Soil water relations during rain infiltration. II. Moisture content profiles during rains of low intensities', Proc. Soil Sci. Soc. Am., 28, 1–5.

Schick, A. P., 1970, 'Desert floods', Internat. Assoc. Sci. Hydrology, Symposium of Wellington, Publication 96, pp. 479–93.

Schumm, S. A., 1956, 'Evolution of drainage systems and slopes in badlands at Perth Amboy, New Jersey', Bull. Geolog. Soc. Am., 67, 597–646.

Sherman, L. K. and Musgrave, G. W., 1942, "Infiltration', in Hydrology, Meinzer, O. E., (Ed.), McGraw-Hill, New York, Chapter 7, 712 pp.

Smith, W. O., 1967, 'Infiltration in sands and its relation to groundwater recharge', Water Res. Res., 3, 539–55.

Strahler, A. N., 1969, Physical Geography, 3rd ed., John Wiley, New York, p. 240.

Strahler, A. N. and Strahler, A. H., 1973, Environmental Geoscience, Hamilton Publishing Co., Santa Barbara, 583 pp.

Temple, P. H., 1972, 'Measurements of runoff and soil erosion at an erosion plot scale, with particular reference to Tanzania', *Geografiska Annaler.* **54A**. 203–20.

Tennessee Valley Authority, 1964, 'Bradshaw Creek—Elk River, a pilot study in area–stream factor correlation', Research Paper No. 4, *Office of Tributary Area Development, Tennessee Valley Authority*, 64 pp.

Tennessee Valley Authority, 1966, 'Cooperative research project in North Carolina: Annual report for water year 1964–65', *Division of Water Control Planning, Hydraulic Data Branch*, 31 pp.

United States Department of Agriculture, 1958, 'Annual maximum flows from small agricultural watersheds in the United States', *Agric. Res. Ser.*, Beltsville, Maryland.

United States Department of Agriculture, 'Hydrologic data for experimental agricultural watersheds in the United States, 1956–59, 1960–61, 1962, 1963, 1964, 1965, 1966', *US Dept. Agric. (miscellaneous publications)*, 945, 994, 1070, 1164, 1194, 1216, 1226, Beltsville, Maryland, 1963, 1965, 1968, 1970, 1971, 1972a, 1972b.

United States Forest Service, 1961, 'Some ideas about storm runoff and baseflow', *Annual Report, Southeastern Forest Experiment Station*, pp. 61–66.

Viessman, W., 1966, 'The hydrology of small impervious areas', *Water Res. Res.*, **2**, 405–12.

Viessman, W., 1968, 'Runoff estimation for very small drainage areas', *Water Res. Res.*, **4**, 87–93.

Viessman, W., Keating, W. R. and Srinivasa, K. N., 1970, 'Urban storm runoff relations', *Water Res. Res.*, **6**, 275–79.

Weyman, D. R., 1970, 'Throughflow on hillslopes and its relation to the stream hydrograph', *Bull. Internat. Assoc. Sci. Hydrology*, **15**(2), 25–33.

Whipkey, R. Z., 1966, 'Subsurface stormflow from forested slopes', *Bull. Internat. Assoc. Sci. Hydrology*, **10**(2), 74–85.

Whipkey, R. Z., 1969, 'Storm runoff from forested catchments by subsurface routes', *Internat. Assoc. Sci. Hydrology, Symposium of Leningrad*, Publication 85, pp. 773–779.

Wilson, T. V. and Ligon, J. T., 1973, 'The interflow process on sloping watershed areas', *Water Res. Res. Inst., Clemson Univ.*, Report No. 38, 58 pp.

Yair, A. and Klein, M., 1973, 'The influences of surface properties on flow and erosion processes on debris covered slopes in an arid area *Catena*, **1**, 1–18.

Zimmerman, U., Munnich, K. O., Roether, W., Kreutz, W., Schubach, K. and Siegel, O., 1966, 'Tracers determine movement of soil moisture and evapotranspiration', *Science*, **152**, 346–47.

CHAPTER 8

Implications for modelling surface-water hydrology

R. P. Betson and C. V. Ardis Jr.

Hydrologic Research and Analysis, Tennessee Valley Authority, Knoxville, Tenn. 37902, USA

The mathematical modelling of hydrologic processes is far from a straightforward process. The modelling of hillslope surface-water hydrology in particular brings many problems into focus. But there are also a host of problems associated with modelling in general that must be considered in evaluating and developing models. This chapter will review modelling problems and in turn present some hillslope and basin surface-water hydrology models. Only those mathematical models that identify hydrologic processes and consider the variable runoff source areas will be presented, although some examples of other types of models will be used for illustration purposes.

8.1 THE MODELLING PROBLEM

8.1.1 Why are there so many models?

'The hydrology of surface waters is characterized by the multiplicity of prediction formulas that have been devised to facilitate the prediction of runoff rates and volumes'. This observation by Merva *et al.* (1969) applies as well to most aspects of hydrologic modelling from the hillslope to the river basin. Despite the fact that each of the hydrologic processes is well understood and working formulations describing them were available by the 1950s (Betson, 1973), hydrologic modelling remains an art. There are, however, some very good reasons. These reasons will serve to place this chapter on modelling into perspective.

One of the basic reasons for the very large number of process formulations and models that have been developed involves advancing technology. Prior to the relatively recent development of electronic computers, the need for uncomplicated technology required so many simplifying assumptions that many diverse approaches could yield rational results. It was not until the early 1960s when the partial-area runoff process was quantified that the true complexity of watershed hydrology became generally appreciated. Not until the last few years have hydrologic models began to incorporate these concepts.

A more basic reason for the plethora of models that have been developed lies in the fact that models are, at best, nature-imitating. The typical heterogeneity in the soils on a hillslope, for example, introduces so much variability into the modelling problem that exact process modelling becomes an impossibility even were it possible to exactly model the vertical movement of water. 'Therefore in the solution of practical problems, some idealization of the physical system and of the input must be made' (Nash, 1967). These idealizations or simplifying assumptions mean that the model can only imitate nature. But in hydrologic modelling, nature is quite easy to imitate. For example, 'almost any model will reproduce the semblance of a proper hydrograph. The poorest models can be expected to cause an increase in streamflow in response to a large enough rainfall and to produce a recession of flow after rainfall ceases' (Dawdy and Thompson, 1967). Other hillslope hydrologic processes are also easy to imitate with various models that are capable of providing reasonable results. The ability with which a particular model can imitate hydrologic processes, as will be shown, is but one consideration in the development and evaluation of a model.

The development of a hydrologic model involves a compromise between those processes and detail the modeller would like to incorporate and those that he can obtain data for and reasonably expect to control in solving the model. 'The simpler the model is, the easier it is to use, and probably the cheaper it is to use' (Dawdy, 1969). As models grow in detail, data for individual processes become increasingly difficult to obtain and the utility of the model is often lessened. For example, a complete hydrologic-system model of a hillslope would require, among other things, a mass of soils data that would never be available in real-life model applications. Therefore, models aimed at practical applications are limited to reasonably available data. Further, as the detail in a model grows, it becomes increasingly difficult to manage the interrelationships among components and processes, as will be explained in more detail below. So, each model must be, at least today, a compromise between theory and reality.

Another reason for the number of models that have been developed stems from the fact most modelling efforts are uni-disciplinary. Models developed under the uni-disciplinary approach typically necessitate fairly gross assumptions to force nature imitation, and the many possible simplifying assumptions that can be made assure a continuing growth in the number of models. Yevjevich (1968), for example, observed that one reason the science of hydrology has been retarded lies in the fact that hydrology is considered an appendage to hydraulics and hydraulic engineering. Much of the research effort in hydrology through the 1960s centered around unit hydrographs and flow-routing techniques. It was primarily hydraulic research. Hydrologic research should involve the air–plant–soil complex and its effect upon water movement. This necessitates an interdisciplinary approach.

In summary then, there are many reasons why *the* hydrologic model does not exist. As has been described in earlier chapters, we are still learning about the hydrology of even simple hillslopes. Models are, at the present state-of-the-art, only nature-imitating and a host of assumptions can be made in developing models that will yield good imitations. And, in the final analysis, the ultimate application for the model and the data that will be available dictate how a model will be developed.

8.1.2 Approaches to modelling

There are conceptually, at least, three approaches to quantitative mathematical modelling. The three can be termed *stochastic, parametric*, and *deterministic*. The first of these, stochastic modelling, statistically relates model input to output. 'Mathematically speaking, a stochastic process is a family of random variables $X(t)$ which is a function of time (or other parameters) and whose variate X_t is running along in time t within a range T '(Chow, 1964). Since the classic stochastic model in hydrology does not identify subcomponents within the system, this approach does not fall within the framework of this chapter.

The second concept for modelling, the parametric approach, is an important one in hydrology. This approach 'is defined as the development and analysis of relationships among the physical parameters involved in hydrologic events and the use of these relationships to generate or synthesize hydrologic events. Historical hydrologic data and known physical data are utilized to develop these relationships' (Committee on Surface Water Hydrology, 1965). In other words, relationships among causative independent variables and the characteristics of elements within the system are defined using rational empirical parametric expressions. Values for the parameters are typically obtained by adjusting predicted system information to observed dependent-variable data using an optimization procedure. The parametric approach has value in complex systems where the relationships cannot be explicitly expressed, as is the case in most natural systems. However, since the physical laws operating upon the system are usually not well defined and are difficult to manage in a model, the form of parametric models will vary considerably among modellers. As an example of a simple parametric model, the triangle can be used as a unit hydrograph as shown below:

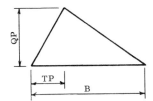

Figure 8.1 Parametric triangular
unit hydrograph

Since the area under the triangle must equal $1 \cdot 0$ basin inch (cm), one parameter must be used to satisfy this condition and therefore $B = 2 \cdot 0/QP$. The two measures, time to peak, TP, and unit graph peak, QP, resemble corresponding hydrograph attributes, yet they are parameters because the unit hydrograph is an empiricism.

The third approach, deterministic modelling, is also important in hydrology. The deterministic approach in hydrology 'is physically based in that models have a theoretical structure based primarily on the laws of conservation of mass, energy, or momentum; deterministic in the sense that when initial and boundary conditions and inputs are specified the output is known with certainty' (Woolhiser, 1971). The saturated–unsaturated subsurface-flow model described by Freeze (1971) is one example of a deterministic model and Grace and Eagleson's (1965) study of similitude for modelling runoff from small, impervious areas is another. To illustrate the process of developing a deterministic model, the conceptual model of the instantaneous unit hydrograph described by Chow (1964) is presented. A linear reservoir is defined in which the storage S is directly proportional to outflow Q

$$S = KQ \qquad (8.1)$$

where K is the storage coefficient. Since the difference between inflow I and outflow Q equals the rate of change in storage, the continuity equation becomes

$$I - Q = \frac{dS}{dt} \qquad (8.2)$$

With suitable substitutions the following equation for an instantaneous unit hydrograph of a single linear reservoir receiving an instantaneous unit input ($I = S = 1{\cdot}0$) can be derived as

$$u = \frac{1}{K}e^{-t/K} \qquad (8.3)$$

The resulting model is deterministic in the sense that it is physically based even though the linear reservoir is a fiction.

Unfortunately, at the present state-of-the-art in deterministic modelling, functional relationships and boundary values cannot be expressed with sufficient confidence. Hence the output is no longer 'known with certainty' and often may not even be in touch with reality. For example, even under conditions where the instantaneous unit hydrograph in Equation 8.3 could reasonably be applied, such as for small impervious areas, the storage coefficient would have to be estimated. There is, however, a form of this general approach that can be termed 'quasi-deterministic' (called by some, 'synthetic modelling') that is often used in hydrology. In this latter approach, model components are developed using rational relationships usually verified through external studies. The components are then combined and model output synthesized. This approach has value in applications where interactions among model components are to be studied and/or there are limited 'real-world' data with which to calibrate the model.

In reality, the differences between the parametric and quasi-deterministic models generally are not very great. As the knowledge of hydrologic processes increases, the parametric models tend to incorporate additional rationale in their parametric representations so that some of the parameters become deterministic coefficients. The builders of quasi-deterministic models on the other hand, frequently find that to keep model output in touch with reality, it is necessary to 'adjust' model coefficients that represent difficult-to-measure characteristics. This adjustment amounts to a form of optimization and the model therefore, in practice, is not much different from a parametric model.

8.1.3   The model verification problem

Hydrologic-model verification typically consists of evaluating the capability of the model to simulate one of the output components, usually streamflow. If the simulations imitate observed data in a reasonable manner, the model, its assumptions, its algorithms, and its parameter or coefficient values are all assumed to be verified. Inconsistencies between the simulated and the observed data may then be implicitly assigned to inadequate input data such as rainfall or evapotranspiration.

However, while a 'reasonable' agreement between simulated and observed data is a necessary condition, it is not a sufficient criterion for model verification. Using optimization techniques, for example, experience has shown that it is surprisingly easy to devise (parametric) models that explain 80–85 % of the initial variance (TVA, 1972 and O'Connell, Nash and Farrell, 1970).

In the case of quasi-deterministic models, the verification problem is much less obvious and therefore potentially more dangerous. Despite the greater use of process formulation and laws and the greater use of detail accounting for water movement in each of the processes, deterministic models still must incorporate simplifying assumptions and because data for some component algorithms typically are lacking, some 'tuning' of the model becomes necessary. In other words, to compensate for missing watershed descriptive data or the lack of adequate model input data, the coefficients are adjusted until a reasonable agreement between the predicted output and observed output is obtained. This coefficient adjustment process is akin to parameter optimization although the degree and the motives may differ. In either case, however, the final reasonable adjustment that may result does not necessarily verify the model.

A classic example of the failure of the goodness-of-fit criteria to verify a model was provided by Betson and Green (1968). Using a three-parameter gamma function to express the unit hydrograph, they used optimization to solve these parameters and the storm-runoff vector in the convolution equation

$$Q = \sum_{i=1}^{D} \sum_{j=1}^{N-i+1} ro_i q_j \qquad (8.4)$$

where Q is stormflow, ro are the incremental volumes of storm runoff or precipitation excess in duration D, each multiplied by an appropriately lagged unit hydrograph vector q. The summation continues for a given hydrograph duration of N. The goal of the approach was to determine the precipitation excess distribution analytically from observed hydrographs so that these values could in turn be used as dependent variables in another model aimed at analytically determining the source areas of runoff (TVA, 1965). Equation 8.4 was adjusted to storm hydrographs associated with as many as 10–15 precipitation excess parameters. The model was found capable of consistently explaining well over 99 % of the initial storm hydrograph variance. Using the goodness-of-fit criteria, the model certainly was verified, and yet the model was found to be unacceptable and later was abandoned (TVA, 1973a).

The reason for the unacceptability of the above hydrograph model despite the high degree of adjustment obtained was that similar high degrees of adjustment could be obtained with many different arrangements of the parameter values (Betson and Green, 1968). In other words, consistent parameter values could not be obtained each time the model was adjusted to a particular hydrograph. The reason for the inconsistent results stemmed from the fact that the model parameters were highly interrelated in their effect such that many combinations of parameter values could achieve similar results. Since the goal of the study was to relate parametrically-determined precipitation excess to source areas in a second step to quantify the partial-area runoff process, inconsistent parameter results left the approach intractable.

This problem with intercorrelations or interactions is not unique to parametric models, although it manifests itself differently in deterministic models. In hydrology most processes are interrelated by nature (TVA, 1973a). This means that as a deterministic model becomes more and more complete, the processes become increasingly interrelated. The consequence of high interrelations among processes in a deterministic model lies in the fact that a reasonable adjustment of the model could be achieved with many different process algorithmic developments and with a range of values for model coefficients. Regarding this problem. Nash (1967) observed that 'the postulation of too complex a model in the first instance may render difficult or impossible the extrapolation of results obtained from one basin to another, or the recognition of relationships between physical characteristics and the parameters of the model—relationships which must be recognized if the postulation of a suitable conceptual model for ungauged basins is ever to be achieved'.

Two goals, then, in the development of a rational model are simplicity and independence. The model should be no more complicated than is necessary and to this, end, possible interactions among processes and among parameters or coefficients should be minimized. Once a suitable degree of adjustment has been achieved with a model, the rationale of the component process contributions and of the parameter of coefficient values obtained becomes important criterion to consider in verifying a model.

One final problem in model verification involves the manner in which it will be adjusted to data. For the parametric model builder, while the problem is obvious, the solution is not. The objective function to be used and the optimization technique (for non-linear models) both can create model verification problems. Optimization involves adjusting model predictions to observed data in such a manner that, typically, some measure of the difference or error, termed the 'objective function', is minimized. The model parameters are not optimized *per se*; rather, they are assumed to be optimal values when a minimum objective function is achieved. There are, however, a variety of ways for measuring objective functions that can be used including: sums of squared errors, sums of absolute errors, sums of weighted errors squared, etc. Each technique for defining an objective function will result in different parameter values. Since parameter values must be considered in verifying a model, the selection of an objective function becomes a problem. 'The choice of the objective function itself should be optimal in some sense to the model user. There is no objective method for choosing the objective function, however' (Dawdy, 1969).

To illustrate how the choice of objective function can influence the value obtained for parameters, a simple example is presented. The triangular unit hydrograph shown in Figure 8.1 was adjusted to the complex storm hydrograph shown in Figure 8.2 using the convolution relationship in Equation 8.4. The

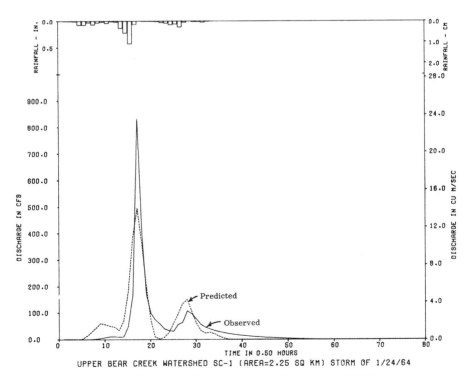

UPPER BEAR CREEK WATERSHED SC-1 (AREA=2.25 SQ KM) STORM OF 1/24/64

Figure 8.2 Adjustment of triangular unit hydrograph to 1/24/64 storm hydrograph on 2·25 km² Upper Bear Creek, Watershed, Alabama

precipitation excess vector was determined using the Soil Conservation Service (1966) method and therefore the model contains only the two parameters TP and QP. Table 8.1 below shows the optimized parameter values obtained using sums of squared error and two additional objective functions, the minimum sum of the absolute errors, and a sum of weighted errors squared. To force peak prediction, for the latter objective function, all errors associated with observed discharges below 10% of the peak were multiplied by 0.5. Optimization was performed using the Pattern Search method (Green, 1970). Figure 8.2 shows the results obtained using the sum of absolute errors as the objective function.

Table 8.1 Optimized parameter values obtained by three different methods

Parameter	Method		
	Sum of squared error	Minimum sum of the absolute errors	Sum of the square of weighted errors
QP (m³/sec)	0.618	0.404	0.800
TP (hr)	2.5	1.88	2.5
r^2	0.595	0.514	0.577

All optimization techniques do not necessarily yield similar results. Green (1970) in a study of the Pattern Search technique, for example, found some significant differences in the optimized parameters relative to those obtained by Betson (1964) using the method of differential corrections (Betson and Green, 1967). Each optimization technique presents its own problems and advantages, and the selection is usually based upon availability. The text by Wilde (1964) described a variety of approaches.

For the quasi-deterministic model builder, the adjustment problem is less obvious. Here, adjustment of the model typically consists of successively adjusting the value of each coefficient for which either field data are inadequate or unavailable, until a satisfactory fit is achieved. This process is in reality a form of subjective optimization. Admittedly, the range of values that will be accepted for the coefficients may be more limited than in the case for a parameter; nevertheless, in complex models with interacting processes, a host of values for the deterministic coefficients will all be capable of yielding reasonable results.

In summary, model verification must include a variety of considerations. The model must be capable of achieving a reasonable adjustment to data. In addition, the processes within the model must react rationally and the same parameter or coefficient values must be obtained each time the model is adjusted to a given data set. The parameter or coefficient values, to the extent possible, should agree with field data and be rational. The final and most critical test in verifying a model lies in its application. The results obtained must be applicable at other locations, at ungaged areas, or under other conditions such as different land-use, or the model is no more than an academic exercise.

8.2   HILLSLOPE MODELLING CONSIDERATIONS

8.2.1   Some early developments

The complexity of hillslope hydrology has long been recognized. Hursh, in a report (1944) of the American Geophysical Union's Subcommittee on Subsurface Flow, for example, discussed the significance of lateral movement of water in the soil as the water table approached the surface and observed that 'the hydrologic significance of different soil horizons as related to storage and movement of water...[constitutes]...some of the most critical of the practical problems of land-use hydrology, water control on the land and watershed management.' Until the process of partial-watershed-area contributions to runoff was generally recognized in the early 1960s, however, most models were based upon assumptions of homogeneity. Then, researchers cognizant of the partial-area process began to take a closer look at what goes on within a watershed.

One of the early promising methods for handling the partial-area runoff process in models involved the concept of unit-source areas. The unit-source area referred 'to a subdivision of a complex watershed. Ideally, this subdivision has a single cover, a single soil type, and is otherwise homogeneous' (Amerman, 1965). Unit-source areas would be instrumented and later the response from these areas modelled. It was assumed that biologic, physiographic, and climatic factors influencing runoff production could be isolated and evaluated precisely enough so that runoff could be accurately predicted on any unit-source area under a combination of these factors. By prediction of each unit-source area and summation, it would thus be possible to predict runoff on ungaged complex watersheds. Hickok and Osborn (1969) concluded that 'the unit-source watershed is a mechanism which "bridges the gap" between the rain site (experimental plot and infiltrometer data) and the hydrology of the complete surface water system with its multiple runoff sources and significant channel influences on the hydrograph'. Despite the optimism, the unit-source approach failed because, as has been pointed out earlier in this text, even single hillslopes are far from being simple, homogeneous systems. Ironically, the fate of the unit-source area approach was foretold about the same time that the procedure was being proposed in a paper by Minshall and Jamison (1965) in which they concluded that return-flow gains and/or transmission losses precluded using weighted-plot runoff (in the presence of claypan soils at least) to represent true watershed runoff. Engman *et al.* (1971) attributed the inadequacy of the unit-source area approach to the fact 'that small hydrologic systems are not linear components of large systems and that below a minimum size, scale operates in an unpredictable manner'.

The partial-area runoff process, of course, has been implicitly incorporated into many models even though the process may or may not have been recognized, *per se.*

For example, Hickok Keppel and Rafferty (1959), based upon a study of lag time, observed that the steeper portions of a watershed seemed to control the time to peak of the hydrograph even though 'rainfall excess occurs over the entire area'. They developed a concept of a controlling source area defined as the half of the watershed with the highest average land slope and developed equations to predict lag time that incorporated a measure of the average width and slope of the source area. Kharchenko and Roo (1963), as another example, used probability curves for the parameters of infiltration from which it became 'possible to determine the percentage of the area of the watershed from which runoff is possible with a given depth of precipitation'. Probably one of the better known techniques incorporating an implicit partial area runoff feature is the Stanford Watershed Model (Crawford and Linsley, 1966). In that model, infiltration is assumed to be variable across the watershed and for each storm that portion of the watershed for which the infiltration capacity is excceeded is determined, which in effect, allows for partial-area storm runoff contributions. The variable-area algorithm used in both of these latter techniques is shown conceptually in Figure 8.3.

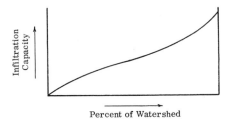

Figure 8.3   One type of variable-area
response algorithm

To what extent these, as well as many other earlier models, may have been modified had the partial-area runoff process been recognized earlier is conjectural. As we shall see, modelling even when the process is recognized is not easier, nor are superior results necessarily assured.

8.2.2   Hillslope processes and models

The complexity of hillslope hydrology should be considered in the development of watershed-hydrology models regardless of the size of catchments to be handled. Development of watershed models compatible with the hydrology of hillslopes should help minimize the number of unproductive approaches along with erroneous simplifying assumptions, resulting in more general and transferable models.

Perhaps the feature of hillslope hydrology that has the greatest impact on modelling is that of partial-area runoff, or dynamic variations in the areas within a watershed contributing to stormflow. This feature greatly complicates the modelling process because it is difficult to handle; source areas and their variation are quire variable from catchment to catchment.

There have been a series of watershed research studies in recent years, many of which are described in earlier chapters, that have each produced additional information about the partial-area runoff process under different conditions. A few of these will be reviewed from the viewpoint of hydrologic modelling.

Possibly one of the studies that brought into focus many of the problems of hillslope hydrology that have modelling implications was that conducted by the Tennessee Valley Authority and North Carolina State University (1970). The goal of the study was to relate agricultural land-use and management to streamflow and to find techniques for expanding the findings obtained from 4–6-acre watersheds to larger catchments. This study led to a detailed investigation into the hydrology of several steep mountainous areas in western North Carolina. Figure 8.4 is a map of one of the study watersheds showing soil A-horizon depth and the location of soil piezometers that were installed to study the source areas of storm runoff (Betson and Marius, 1969).

The picture of the runoff process that emerged from this study is interesting. During frequent light-to-moderate storms, storm runoff originates primarily from an artificial swampy area created above the measuring flume by the installation of sheet metal cut-off walls. This finding explained the results obtained earlier by Betson (1964) on these watersheds in an analytic study of an infiltration equation where he concluded that the effective runoff producing drainage area was much smaller than the total drainage area. However, storm runoff is not always confined to the swampy area. During larger storms or moderate storms under conditions of high soil moisture, because of relatively lower permeabilities in the B horizon, soils in those areas in the catchment with shallow A horizons become saturated. When this occurs, subsequent rainfall will initiate overland flow in these areas. However, whether or not this flow reaches the measuring flume depends upon the capacity of a downslope deeper A-horizon soil to re-infiltrate this moisture. As storms become progressively larger, additional areas become saturated and the area contributing to stormflow continues to grow, resulting in increasingly higher storm runoff yields. Some areas such as the deep A-horizon zone in the west-central portion of the watershed seldom contribute to runoff despite the presence of some shallow A-horizon areas on the upslope ridge.

How does this picture of the runoff process agree with other findings? Quite well according to some recent research relative to how the contributing area varies. Kirkby and Chorley (1967) concluded that overland flow resulting from saturation of the soil column most frequently occurs adjacent to flowing streams, where lines of greatest slope converge, where local concavities occur, or where the soil cover is locally thin or less permeable. Hickock and Osborn (1969) observed that 'we must develop in our watershed hydrology research the means for evaluating the interflows which commonly occur between surface and subsurface waters.' Hewlett and Nutter (1970) visualize the growth of the contributing areas as 'an

L

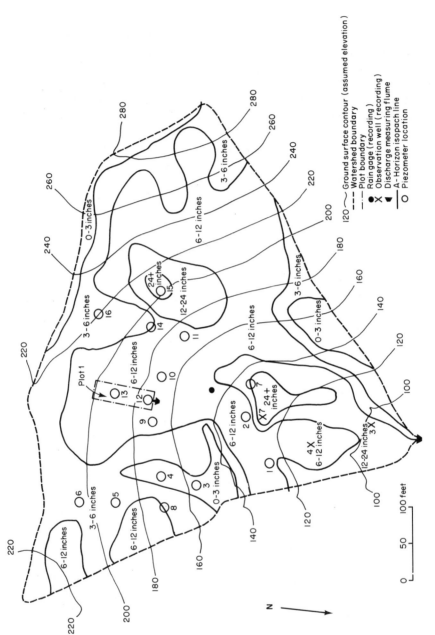

Figure 8.4   Soil A horizon depth map. Western North Carolina Watershed 1 (after TVA–NCSU, 1970)

expansion of the perennial channel system into zones of low storage capacity, and thus rapid subsurface seepage into small draws, swampy spots and intermittent channels.' Dunne and Black (1970) found in the upland watersheds of Vermont that storm runoff was produced as overland flow from small saturated areas close to streams which are supplied by water escaping from the ground surface to reach the channel as overland flow. Rawitz, Engman and Cline (1970) in Pennsylvania concluded that 'the contributing area, as defined by the storage volume in the soil, appears more important than the total watershed area'. Engman *et al.* (1971) describe the watershed as being 'delineated into runoff zones and recharge zones and the size of these zones depends on storm size, soil, geology, antecedent moisture, etc.' And Weyman (1973) in an area of England where no overland flow was observed concluded that the response of the hillslope is dominated by saturated throughflow within the mineral layers of the soil.

Although the mechanisms vary somewhat on different watersheds, all of these recent findings indicate that the response of a watershed is far from uniform. Runoff characteristically originates initially from certain areas with the size of this area growing in proportion to the amount of rainfall. The soil infiltration and percolation rates are important, but so are soil horizon depths and the relative location on slopes. Watershed hydrology must be considered in three dimensions. Whether or not it *must* be modelled in three dimensions is another matter. Figure 8.5 illustrates the possible dimensions that can be used in modelling a watershed.

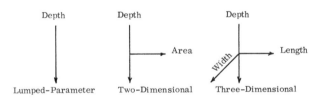

Figure 8.5   Conceptual watershed modelling dimensions

Hydrologic models conventionally have been one-dimensional. That is, the total watershed area is considered lumped together at a point since both the input and output are assumed to be uniformly distributed across the area. These models have been termed 'lumped-parameter' since the effect of all spatial variation is lumped into the parameters and only vertical variation is considered. Two-dimensional models allow runoff source areas to vary with time, but do not account for the spatial distribution within a watershed. Only in the three-dimensional model is there a capability to account for detailed spatial variations. Since the complexity of a model is proportional to the degree of spatial variability accounted for, the full three-dimensional approach cannot always be justified for model applications. Regardless of the number of watershed dimensions included in a model, however, spatial variability remains a consideration.

The three-dimensional nature of hydrology has many implications in modelling. For one, it explains why the use of constant unit hydrographs is avoided, particularly on small areas. Where the partial-runoff area is particularly dynamic, the principle of linearity is violated. It also explains why the yield to runoff is so non-linear and why rainfall–runoff relationships vary from storm to storm and watershed to watershed. Further, for some applications three-dimensional modelling becomes a necessity. For example, since land-use planning is concerned with the impact of change on the water resource, planning models must differentiate between primary-runoff source areas and primary recharge areas. The impact upon streamflow of paving over a thin-soil primary source area, for instance, would be far less than a similar amount of paving over a primary recharge area. To explain this difference quantitatively, the prediction model must incorporate a three-dimensional capability. And finally, the three-dimensional nature of a watershed also affects water-quality modelling. The yields of fertilizers, herbicides, or animal wastes, for example, can be highly dependent upon where the material is placed in relation to the runoff source areas. Kunkle (1970), for instance, concluded from a study in Vermont that 'because of the runoff processes involved, upland contributions of bacteria to streams were small compared to contributions from land surfaces near channels, the channel itself, or direct inputs'. If water-quality models are to account for specific management practices on individual watersheds, then they must consider the source of the constituent in relation to the hydrology of the watershed.

8.3   HILLSLOPE HYDROLOGY MODELS

There have been, and there are being developed, a great number of traditional models which could be applied to characterize the hydrology of hillslopes and small watersheds. They vary from the pure hydraulic routing type of model to the rational–empirical infiltration-based models. Chow's 'Handbook of Applied Hydrology' (1964) presents a good summary of these traditional approaches. This section will describe models that address the variable runoff contributing-area problem in either two or three dimensions. Emphasis will be placed upon the algorithms used to handle variable contributing areas.

It has not been until the last decade or so that much of the hydrologic variability intuitively present in larger watersheds was also found to be present on hillslopes. Hillslopes were considered to be homogeneous. For this reason and because attention has been focussed on modelling larger catchments, there are not a large number of hillslope hydrology models. A range of parametric and quasi-deterministic approaches, however, will be presented to indicate the present state-of-the-art.

8.3.1   Parametric models

One very simple two-dimensional model was developed TVA and NCSU (1970).

For this model, the area contributing to runoff was assumed to vary in proportion to the accumulated rainfall. A power function was arbitrarily used to express the relationship as

$$A_i = a + b \left(\sum RE_i \right)^c \tag{8.5}$$

where

A_i = the area contributing at time i
    a, b, c are parameters
$\sum RE_i$ = the accumulated rainfall in excess of B-horizon permeability (B) as given by

$$\sum RE_i = \sum_{i=1}^{D} (RE_i - B) \tag{8.6}$$

Use of the soil B-horizon parameter (B) allowed the contributing area to diminish during lulls in the rainfall, thus providing for an increase or a decrease in the contributing area depending upon storm intensity. Rainfall in excess of the B-horizon parameter was partitioned into surface runoff and saturated A-horizon interflow using an additional parameter. Runoff was assumed to occur only from the contributing area and was determined by multiplying rainfall for each interval by the contributing area. The non-contributing or moisture-storage areas were associated with deep permeable topsoils, depression storage areas, etc., and thus

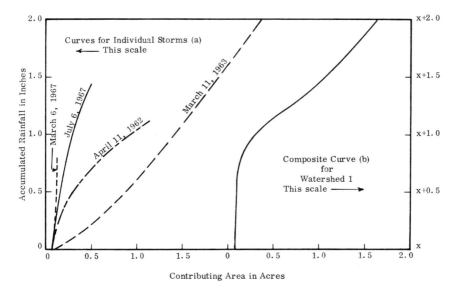

Figure 8.6   Apparent accumulated rainfall versus contributing area functions (after TVA–NCSU, 1970)

potential sources for future contributions. Two linear routing parameters, one for each type of flow, were used to route the runoff to the gage. The model thus had seven parameters—two for allocating rainfall, two used in routing, and three contributing area parameters. Figure 8.6A shows the accumulated rainfall versus contributing area relationships obtained analysing four storms on a 1·88-hectare (4·64 ac) mountainous watershed.

The three contributing area parameters in Equation 8.5 should be related to antecedent moisture conditions. Hence, each of the four relationships shown in Figure 8.6(a) probably is only a measured portion of the family of curves in the total relationship. Therefore, the four curves were composited into an apparent contributing area function as shown in Figure 8.6(b), where the parameter X was used as a measure of antecedent moisture. When this composite function was used on a particularly complex storm not used in the development of the function, excellent predictions were obtained.

This approach illustrates how a blend of rationale and pure parameters in a model can be used both to predict system behaviour and to obtain information about the system. Also, the apparent composite contributing area function illustrates the complex manner in which the contributing area may vary.

Hewlett and Nutter (1970) described the varying source area of streamflow with a conceptual model. The process was intended for application under conditions where infiltration is not usually a limiting condition. Discharge was conceived as coming from a 'two-dimensional sloping flow system in which two pathways are represented in a differential equation

$$\frac{dq}{dt} = \frac{da}{dt}\left[\frac{dp}{dt} + k\frac{dH}{dL}\right] \tag{8.7}$$

where

q = the rate of discharge from the watershed
t = time
a = the area occupied by the expanding or shrinking channel system
p = rainfall
k = the saturated permeability of the stream bank and bottom materials
dH/dL = the hydraulic gradient driving flow through the wetted perimeter into the channel.

The first term in the brackets (dp/dt) accommodates channel precipitation ("overwater flow") and the second term in the brackets accounts for subsurface flow'. Although the authors realized they could not solve the equation, 'as things stand' they presented it to illustrate the hypothesis upon which they were working.

8.3.2 Quasi-deterministic models

Perhaps the complexity that may be necessary to include in hillslope hydrology models can best be seen in a few of the quasi-deterministic models, some of which are still being developed. At this point, since there is neither a general appreciation in the literature that hillslope heterogeneity should be considered in models nor a full understanding of the mechanics of partial area runoff under markedly differing site conditions, there is not a great deal of new information to guide modellers.

One interesting approach that appears capable of identifying and quantifying the actual source areas of runoff on a hillslope is that described by Engman and Rogowski (1974). The storm-hydrograph model proposed 'utilizes a physically-based infiltration-capacity distribution for computation of rainfall excess and it incorporates two stages of kinematic routing. In the first stage, the rainfall excess is routed over a flow plane to become the lateral inflow hydrograph for the second or channel phase. The overland flow plane expands upslope as the infiltration capacity is exceeded, with the size of the contributing area and the length of the flow plane being calculated from infiltration curves'.

The infiltration capacity used in the technique is the Philips (1969) two-coefficient equation:

$$v_0 = 0 \cdot 5St^{-0 \cdot 5} + A \tag{8.8}$$

where v_0 is the infiltration capacity, S is the sorptivity computed using Parlange's (1972) approximation, t is the time, and A is the conductivity of the wetting front computed using modified models proposed by Rogowski (1971; 1972a; 1972d). Coefficients associated with the infiltration model are either abstracted from constructed curves or estimated from available data and from literature. Once the model coefficients are determined for each soil class in a study area, subsequent infiltration-capacity curves for individual storms are based upon the antecedent soil-water contents of different soil zones. In practice, the soil-water content of the most obvious source area is measured or estimated, and then assuming that the frequency associated with this value remains constant over the entire watershed, the probability-distribution functions (Rogowski, 1972b; 1972c) for the other soils can be used to estimate the initial water content for each at the same frequency as the most obvious source-area soil. Re-infiltration can occur in the model if the overland flow plane extends downslope into an area that has an infiltration capacity greater than the zone where it originated.

Three measures are needed for storm hydrograph simulations: the antecedent soil-water content and two Manning's n values. One Manning's n value was selected for the overland flow plane and another for the channel. Once determined, these two kinematic routing parameters were held constant and not changed for the watershed studied.

The model has been tested on a 58-hectare (144-acre) agricultural catchment located in Pennsylvania. Although the individual soil series configuration in the watershed is complex, they were grouped into the five zones on the basis of physical characteristics. For one test, a 16-hectare (40-acre) segment was selected for simulations. Figure 8.7 shows the lateral-inflow hydrographs calculated for four July 1971 storms, along with that portion of the watershed which contributed surface runoff to the channel. Figure 8.8 shows the computed infiltration-capacity

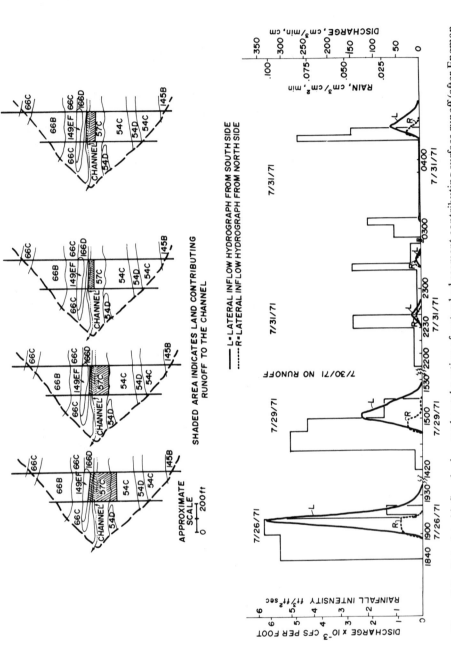

Figure 8.7   Calculated inflow hydrographs and portion of watershed segment contributing surface runoff (after Engman and Rogowski, 1974)

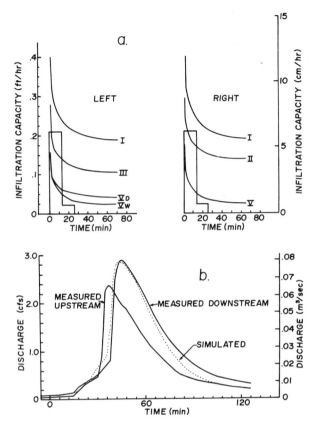

Figure 8.8   (a) Calculated infiltration capacity curves for each soil zone along with (b) simulated and observed surface runoff hydrographs. Storm of 8/1/69 (after Engman and Rogowski, 1974)

curves for each soil zone and the measured rainfall intensities along with the simulated and measured hydrographs for a storm that occurred on August 1, 1969. Hydrographs simulated for the storms were considered satisfactory by the authors (Engman and Rogowski, 1974) although some under-simulation during the recessions was attributed to the fact that interflow contributions have not been taken into account in the model.

Two studies have been completed that chose to assume that infiltration could be treated as a point-source problem. Smith and Woolhiser (1971) developed a mathematical model governing infiltration using the partial differential equation for vertical, one-phase, unsaturated flow of water in soils. The resultant excess of surface water, along with kinematic routings, was used as the basis for a mathematical surface runoff model. Woolhiser (1971) concluded that the technique seemed limited to experimental plots and very small watersheds because 'application to large watersheds with substantial spatial variations of rainfall or soil properties appears to require an excessive amount of input data and computational time'. Haan (1972) attempted to evaluate the diffusivity equation for describing infiltration and moisture movement under field conditions and found that 'the treatment of soil water flow on a watershed scale as a one-dimensional process appears to be a poor approximation to reality'.

A modelling effort, which was begun in the early 1970s, that deserves to be mentioned because of its scope and unparalleled gathering together of disciplines is that being undertaken under the International Biological Program (Burgess and Swank, 1973; Munro et al., 1976). Presently, sub-models of the environmental system are being developed which, when brought together, promise to provide a capability for simulating chemical, biotic, and hydrologic processes and their interactions. For the Eastern Deciduous Forest Biome, the Stanford Watershed Model (Crawford and Linsley, 1966) is being used as the basic hydrologic model to transport materials in ecosystems (Patterson et al., 1974). While the Stanford Watershed Model was not developed as a hillslope model, in this application it is, for example, being applied to segments within a small watershed that are relatively homogeneous in vegetative cover and soil type. Detailed movement of water in the plant-soil system is being described by a model developed by Goldstein and Mankin (1972). Plant-water movement is modelled using an electrical analogy with evapotranspiration based upon computed atmospheric potential measures and subject to soil-moisture limitations. Moisture accounting is performed in five soil layers, three of which are in the root zone. (Goldstein, Mankin and Luxmore, 1974).

And finally, although outside the strict framework of this chapter, the hydrology of small impervious areas must be acknowledged as a facet of hillslope hydrology. Many applied hydrology problems are associated with urbanization, and urban areas include both pervious and paved hillslope areas. Relative to the hydrology of pervious hillslopes, that of small impervious only areas is relatively simple as described, for example, by Viessman (1968). Using only a depression storage function for loss, he derived one-minute unit hydrographs using rationale similar to that shown in Equations 8.1–8.3. He studied four impervious areas of less than an acre and found peak flow rates computed using optimized values for lag-time, for 66 storms, resulted in an average absolute error of only 7·5 %. Future small-area urban research efforts, therefore, need to focus on combining an existing impervious-area model with a hillslope-hydrology model to account for the complex runoff processes from pervious areas.

8.4 WATERSHED MODELLING

Conceptually, large areas are composed of a series of hillslopes, and it should be possible to model basin hydrology by integrating a hillslope model over an entire watershed. In practice, this is a near-impossible task. One obvious reason, as was pointed out by Woolhiser (1971), is that accounting for the spatial variability in soil properties requires an excessive amount of input data and computational time. There are, however, additional reasons for developing basin hydrology models along different lines.

Basin hydrology models are designed to be applied under conditions where typically little input data exist. Applied hydrology models are often called upon to be used where only a single rain gauge, at best, may be located in the catchment; where the nearest source of hydro-meterological data is usually many miles outside of the catchment; and where oftentimes no soils and geologic data and sometimes no topographic data exist.

Under these circumstances there is no point in developing or using a model based upon the detail present in a hillslope model.

In addition, however, the output of large hydrologic systems is quite dampened relative to the input. On large catchments, because of variations in the physical characteristics within the watershed, large variations in input, unless they are extreme, will not cause unusual variations in the output. Also, in large basins, the effect of channel characteristics and channel storage is considerable and serves to dampen input variations and mask land-use effects. The success of lumped-parameter models such as the unit hydrograph for predicting storm hydrographs for large areas is due to this dampening effect. As a consequence of this dampening, there is little to be gained by including more detail than is necessary in a basin model. 'The optimal rule to follow in developing a model would be that of Occam's Razor, which states that if a simple model will suffice, none more complex is necessary' (Dawdy 1969).

Are there any guides as to the scale at which to go from the hillslope model to the basin model? Probably the best guide is the ultimate use for which the model is being designed. If requisite data are available, if detailed answers involving point source locations are needed, and if there are few computer limitations, then the hillslope-modelling approach conceivably could be expanded to watershed scale. Seldom is this the case. Data adequate for hillslope models become increasingly difficult and expensive to obtain as the size of a catchment increases until finally another approach becomes necessary. Engman *et al.* (1971) in a study of scale problems related to inter-disciplinary water-resources research concluded that it was possible to bring the disciplines of hydrology, geology, meteorology, and water quality together at a scale of about 100 acres or so. Although each of the disciplines has difficulty working at this scale, it does provide a scale 'toward which all can extend their present level of expertise'. Perhaps this size provides a 'rule-of-thumb' for the upper limit for detailed hillslope models.

While the detail of a hillslope model may be neither required nor desirable in a basin model, it is desirable to use empiricisms, algorithms, and assumptions that are compatible with the principles of hillslope hydrology. Incorporating hillslope-hydrology rationale into a basin model will facilitate extrapolation of the model to other conditions and extension to ungaged areas. The following section describes a sampling of basin hydrology models that include or were designed to incorporate some hillslope hydrology concepts.

8.4.1 Basin models

A two-dimensional parametric basin model with an explicit dynamic contributing area formulation was successfully developed by Lee and Delleur (1972) based upon adaptations of a hillslope model presented earlier (TVA and NCSU, 1970). The model expressing the dynamic area relationship was used in a study of about 200 storms from 14 Indiana watersheds, ranging in size from 5–780 km^2:

$$A_i = A_o \frac{\left[\left(D \sum_{k=0}^{i-1} (RF_k - B)\right) + (RF_i - B)\right]^N}{\sum_{k=0}^{T} (RF_k - B)} \qquad (8.9)$$

where

A_i = the contributing area
A_o = the total drainage area
RF = the rainfall intensity
B = the B-horizon permeability
D = the fraction of the antecedent rainfall contributing to the response area
N = a parameter
T = the total number of sampling points of runoff hydrographs
k = an index to count time of antecedent rainfall excess, $k < i$
i = an index indicating current time

Direct runoff then is given by the equation

$$Q_i = A_i(RF_i - B) \qquad (8.10)$$

where Q_i is the direct runoff. To preserve continuity the parameter N was determined so as to equate the summation of Equation 8.10 over time T to the volume of rainfall excess. The B-horizon permeability parameter was set to be zero based upon a study of sub-layer soil-permeability data for these Indiana watersheds. Results showed the values to be very low compared to rainfall intensities. The authors tested the antecedent rainfall weighting parameter D and concluded that it was a relatively insensitive parameter. A value of $D = 0.8$ was selected for the model. Once values for B and D were established, Equation 8.9 became deterministic.

Using Equation 8.9 the area contributing to runoff during a storm can be computed. It will begin at zero and increase more or less in proportion to the rainfall that has occurred. To spatially distribute runoff within the catchment, Lee and Delleur (1972) subdivided the watershed into stream reach intervals. Then they assumed that at any given time the ratio of the response area to the drainage area within various stream-reach intervals was equal to the ratio of the contributing area for the whole basin to total watershed area. Once the response areas are determined, the rainfall excesses from these are routed through the stream network to obtain direct runoff of the basin outlet.

Attentuation of rainfall excess for isochronal stream reaches in the Lee and Delleur (1972) model is done using a linearized complete solution derived by Harley (1967). An upstream inflow instantaneous unit hydrograph is determined that serves to attenuate and transport the rainfall excess computed for each stream reach. The downstream direct runoff is determined by convolution. The routing technique requires two parameters—a reference discharge and roughness. Optimization of these two model parameters was done using the uni-variate technique. The difference between the observed and calculated peak discharge and time-to-peak discharge was the preferred objective function.

Lee and Delleur (1972) then developed relationships and recommendations for predicting or estimating the model parameters based upon watershed measures and other data. Using these data and the relationships presented, some of the storm hydrographs used were then regenerated. It was found that the average absolute error for the peak flows was 20 % and that the time to peak 'had better regeneration performance'.

The effect of partial-area runoff on the storm hydrograph was incorporated in two ways in a model developed by Ardis (TVA, 1973b). He expressed the unit hydrograph as a double triangle composed of the early heavy runoff from riparian wet areas with other areas contributing later as soils become saturated as in Figure 8.9.

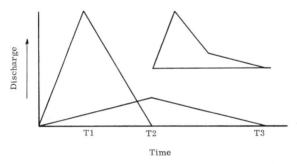

Figure 8.9   Partial-area runoff concept represented by an initial and a delayed response (after TVA, 1973b)

In use, the two triangles are combined to form a quadrilateral with four parameters: peak, time to peak, time to inflection point, and unit hydrograph duration. From the analyses of 140 storms on 11 watersheds, the model was found capable of explaining an average 96 % of the initial variance. The watersheds varied in size from 7 to 17 square miles.

To allow for variability resulting from partial-area runoff, the unit hydrograph was not assumed to be constant in the TVA model. Rather, each parameter was allowed to vary according to storm characteristics:

$$PAR = a \cdot SRO^b \cdot NPE^c \qquad (8.11)$$

where

PAR = a model parameter
SRO = storm runoff
NPE = the storm duration
a, b, and c = coefficients.

The three coefficients in Equation 8.11 were each related to a series of watershed characteristics which ultimately permitted each model parameter to vary from storm to storm and this relationship to vary among watersheds. On a regeneration test using the derived relationships, the model explained on the average 88 % of the variance, with an average absolute error on the peak discharge of 22 % and on the time to peak of 4 %. Simulation tests with the model were also successful.

The principle of variable response areas has been incorporated into some continuous flow models in a variety of ways. The Stanford Watershed Model (Crawford and Linsley, 1966), previously mentioned, uses a parametric linear cumulative-frequency distribution of infiltration capacity, conceptually shown in Figure 8.3, to allocate moisture for a storm-to-surface detention, interflow detention, and infiltration. This assumption implies that in a catchment there will be a range of infiltration capacities from zero to some maximum value (a parameter) and as the moisture supply increases, the infiltration capacity of increasingly larger areas will be exceeded.

Hollis (1970) developed a daily-flow model to study the effect of urbanization on Canons Brook in Essex. Initial runs of a rudimentary version of the model predicted that no runoff would result from even moderate summer rains that occurred during periods with large soil moisture deficits. However, streamflow was observed to increase following even small rains. He therefore included a feature in the model to allow for a saturated riparian area from which storm runoff would always occur. This riparian area was estimated to be 7 % of the Canons Brook catchment.

Betson (TVA, 1972) developed a parametric continuous daily-streamflow model designed to be used in water resource planning applications. This model incorporates the concept that the yield of storm runoff is proportional to the amount of moisture stored in the system. The algorithm used relates the moisture stored in the soil-moisture and groundwater reservoirs to surface runoff with an adaptation of an empirical relationship developed by Betson, Tucker and Haller (1969). On calibration runs, the model has been found to be capable of explaining some 85 % of the daily-streamflow variance. It has also been used to simulate streamflow under a variety of conditions and at ungaged locations (Betson, 1973; 1976).

The US Department of Agriculture Hydrograph Laboratory Model USDA HL-74 (Holtan and Lopez, 1975) is probably one of the more rational of the continuous-flow models available. A pedohypsograph, conceptually shown in Figure 8.10, is used as a basis for associating soils, topography, and land-use in a watershed. Soils are grouped by land-capacity classes to form hydrologic response zones for computing infiltration, evapotranspiration, and overland flow. The model thus approaches a three-dimensional capability.

The three zones, upland, hillslope, and bottom land, typically are 'likely to have characteristics that distinguish its hydrologic response from other zones within a watershed in several ways: (1) elevation affects the water balance of the site through its influence on evapotranspiration and drainage, (2) soil physical properties affect the disposition of storm rainfall, and (3) distance from the stream channel is an important factor in flood routing' (England and Onstad, 1968).

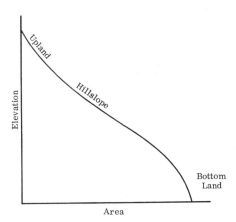

Figure 8.10 Watershed pedohypso-graph (after England and Onstad, 1968)

Infiltration in the USDA HL-74 model is determined by an exhaustion phenomenon equation presented by Holtan (1965).

$$f = aS_a^{1.4} + f_c \qquad (8.12)$$

where

f = the infiltration rate (in/hr)
a = the infiltration capacity (in/hr per inch of available storage)
S_a = the available storage in the surface layer (in)
f_c = the constant rate of infiltration after prolonged wetting (in/hr).

Infiltration and rainfall excess are computed for each soil zone by comparing observed rainfall to the infiltration capacities computed in Equation 8.12. Rainfall in excess of infiltration is then routed across each soil zone and cascaded downslope, subject to further infiltration, across designated subsequent soil zones en route to the channel (Holtan and Lopez, 1975). In tests on four watersheds across the United States ranging in size from 16·8 to 255 km², the correlation coefficient between observed and simulated monthly flows averaged about 0·92.

To avoid some of the damping effects in large basins, relatively large basins can be sub-divided into smaller, relatively homogeneous sub-watersheds. Each component area can then be treated as an entity. To complete the simulations the components can then be combined using a channel routing procedure. The program HYMO (Williams and Hann, 1972) is one approach. A technique such as this should allow the use of more detailed conceptual models in studies on larger basins. This approach, for example, would be particularly attractive in studies of the effect of land-use changes upon streamflow where the changes occur in one or more of the sub-watersheds.

Finally, a complete review of the available basin models and how each may or may not incorporate the concepts of hillslope hydrology is beyond the scope of this chapter. Brandstetter (1974) evaluated 18 current models and McPherson (1975) reported on 16 models and storm hydrograph prediction techniques. Both focused on urban models although many of the basin models could be applied in rural areas as well. There are many additional basin models developed in the United States and other countries. The number will probably continue to grow as the needs and requirements for these models change and as our ability to pragmatically incorporate hillslope hydrology concepts into these models improves.

In conclusion, the examples presented show that the dynamics of partial area runoff can be incorporated into both hillslope and basin models in a variety of ways. Predictions from these models, even in their present elementary state, may not always be superior to those obtained with some conventional lumped-parameter models. Nevertheless these models represent an improvement over traditional approaches. As the questions that models are designed to answer become increasingly complex and site-specific, models incorporating hillslope hydrology concepts will be used to meet these needs.

REFERENCES

Amerman, C. R., 1965, 'The use of unit-source watershed data for runoff prediction', *Water Res. Res.*, (1), 499–507.

Betson, R. P., 1964, 'What is watershed runoff?', *J. Geophys. Res.*, **69**, 1541–1552.

Betson, R. P., 1973, 'Agricultural watershed hydrology and modelling', presented at the *Ann. Am. Soc. Agronomy Meeting, Las Vegas, Nevada, November 1973*.

Betson, R. P., 1976, *Urban Hydrology—A Systems Study in Knoxville, Tennessee*, Tennessee Valley Authority, Knoxville, TN. 37902.

Betson, R. P. and Green, R. F., 1967, 'DIFCOR—A program to solve nonlinear equations', *Tennessee Valley Authority Research*, Paper No. 6, Knoxville, Tennessee.

Betson, R. P. and Green, R. F., 1968, 'Analytically derived unit graph and runoff,' Proc. Am. Soc. Civil Engrs. Hydraulics Div., **94**(HY6), 1489–1505.

Betson, R. P. and Marius, J. B., 1969, 'Source areas of storm runoff', *Water Res. Res.*, **5**(3), 574–581.

Betson, R. P., Tucker, R. L. and Haller, F. M., 1969, 'Using analytic methods to develop a surface-runoff model', Water Res. Res., **5**(1), 103–111.

Brandstetter, A., 1974, 'Comparative analysis of urban, stormwater models'. Presented at Short Course, 'Application of Storm Water Management Models' held at University of

Massachusetts, August 1974, Pacific Northwest Laboratories P.O. Box 999, Richlands, Washington 99352.

Burgess, R. L. and Swank, W. T., 1973, 'Analysis of ecosystems in the eastern deciduous forest biome, US International Biological Program,' *Proc. 7th World Forestry Congress, Buenos Aires, Argentina.*

Chow, V. T. (Ed.), 1964, *Handbook of Applied Hydrology*, McGraw-Hill, New York, N.Y.

Committee on Surface-Water Hydrology, 1965, 'Parametric and stochastic hydrology,' *Jour. Am. Soc. Civil Engrs.*, **91**(HY6), 119–122.

Crawford, N. H. and Linsley, R. K., 1966, 'Digital simulation in hydrology—Stanford watershed model IV,' *Stanford Univ. Dept. Civil Engr.*, Tech. Rept. No. 39.

Dawdy, D. R., 1969, 'Considerations involved in evaluating mathematical modelling of urban hydrologic systems,' *US Geolog. Surv. Water Supply Paper 1591-D.*

Dawdy, D. R. and Thompson, T. H., 1967, 'Digital computer simulation in hydrology', *J. Am. Water Works Assoc.*, **59**, 685–688.

Dunne, T. and Black, R. D., 1970, 'Partial area contributions to storm runoff in a small New England watershed', *Water Res. Res.*, **6**, 1296–1311.

England, C. B. and Onstad, C. A., 1968, 'Isolation and characterization of hydrologic response units within agricultural watersheds', *Water Res. Res.*, **4**(1), 73–77.

Engman, E. T., Gburek, W. J., Parmele, L. H. and Urban, J. B., 1971, 'Scale problems in interdisciplinary water resources investigation'. *Jr. Am. Water Res. Assoc. Water Res. Bull.*, June 1971. 495–505.

Engman, E. T. and Rogowski, A. S., 1974, 'A partial area model for stormflow synthesis', *Water Res. Res.*, **10**(3), 464–472.

Freeze, R. A., 1971, 'Three-dimensional, transient, saturated–unsaturated flow in a groundwater basin', *Water Res. Res.*, **7**(2), 346–366.

Goldstein, R. A. and Mankin, J. B., 1972, 'Prosper: A model of atmosphere–Soil–Plant–Waterflow', *Proc. Summer Computer Simulation Conference, San Diego, California,* June 1972.

Goldstein, R. A., Mankin, J. B., and Luxmoore, R. J., 1974, 'Documentation of PROSPER—a model of atmosphere–soil–plant water flow' Oak Ridge Nat. Lab. EDFB-IBP 73-9-UC48, Oak Ridge, TN 37830.

Grace, R. G. and Eagleson, P. S., 1965, 'Similarly criteria in the surface runoff process', *MIT Dept. of Civil Engr. Hydrodynamics Lab.*, Rept. No. 77.

Green, R. F., 1970, 'Optimization by the pattern search method', *Tennessee Valley Authority Res. Paper No. 7.*

Haan, C. T., 1972, 'Characterization of water movement into and through soils during and after rainstorms', *Univ. Kentucky Res. Rept. No. 56.*

Harley, M. B., 1967, *Linear Routing in Linear Channels*, M.Eng.Sci. thesis, National University of Ireland, Dept. Civil Engr.

Hewlett, J. D. and Nutter, W. L., 1970, 'The varying source area of streamflow from upland basins', *Proceedings of a Symposium on Interdisciplinary Aspects of Watershed Management*, American Society of Civil Engineers.

Hickok, R. B., Keppel, R. V. and Rafferty, B. R., 1959, 'Hydrograph synthesis', *Agric. Engr.*, October 1959.

Hickok, R. B. and Osborn, H. B., 1969, 'Some limitations on estimates of infiltration as a basis for predicting watershed runoff', *Am. Soc. Agric. Engrs.*, **12**, 738–803.

Hollis, G. E., 1970, 'The estimation of the hydrologic impact of urbanization: an example of the use of digital simulation in hydrology', occasional paper of the *Dept. Geography, Univ. College, London.*

Holtan, H. N., 1965, 'A model for computing watershed retention from soils', J. Soil and Water Conserv., **20**(3), 91–94.

Holtan, H. N. and Lopez, N. C., 1975, 'Usdahl-74 Revised model of watershed hydrology', *US Dept. Agric., Agric. Res. Ser.*, Tech. Bull. No. 1518.

Hursh, C. R., 1944, 'Reports on hydrology—1944—Appendix B—report of subcommittee on subsurface flow', *Trans. Am. Geophys. Union*, 743–746.

Kharchenko, S. I. and Roo, S. S., 1963, 'Experimental investigations of infiltration capacity of watersheds and the prospect of accounting for the variability in losses of rainwater in computing flood runoff', *Soviet Hydrology: Selected Papers, Am. Geophys. Union*, No. 6, 537–554.

Kirkby, M. J. and Chorley, R. J., 1967, 'Throughflow, overland flow and erosion'. *Bull. Internat. Assoc. Sci. Hydrology*, **12**, 5–21.

Kunkle, S. H., 1970, 'Sources and transport of bacterial indicators in rural streams', *Proceedings of a Symposium on Interdisciplinary Aspects of Watershed Management*, August 1970, American Society of Civil Engineers.

Lee, M. T. and Delleur, J. W., 1973, 'A Program for estimating runoff from Indiana watersheds—Part III Analysis of geomorphologic data and a dynamic contributing area model for runoff estimation', *Purdue University Water Res. Res. Centre Tech. Rept.* No. 24.

McPherson, M. B., 1975, 'Urban mathematical modelling and catchment research in the U.S.A.', *Am. Soc. Civil Engrs., Technical Memorandum No. IHP-1*.

Merva, G. E., Brazee, R. D., Schwab, G. O. and Curry, R. B., 1969, 'A proposed mechanics for the investigation of surface runoff from small watersheds. 1. Development', *Water Res. Res.*, **5**(1), 76–83.

Minshall, N. E. and Jamison, V. C., 1965, 'Interflow in claypan soils', *Water Res. Res.*, **1**(3), 381–390.

Munro, J. K., Luxmore, R. J., Begovitch, C. L., Dixon, K. R., Watson, A. P., Patterson, M. R., Jackson, D. R., 1976, 'Application of the unified transport model to the movement of Pb, Cd, Zn, Cu, and S through Crooked Creek watershed', Oak Ridge Nat. Lab. ORNL/NSF/EATC-28-UC-32. Oak Ridge, TN 37830.

Nash, J. E., 1967, 'The rôle of parametric hydrology', *J. Inst. Water Engrs.*, **21**, 435–474.

O'Connell, P. E., Nash, J. E. and Farrell, J. P., 1970, 'River flow forecasting through conceptual models. Part II. The Brosna Catchment at Farbane', *J. Hydrology*, **10**, 317–329.

Parlange, J. Y., 1972, 'Analytical theory of water movement in soils', *Joint Symposium on Fundamentals of Transport Phenomena in Power Media*, Univ. Guelph, Guelph, Ontario, Canada, August 1972.

Patterson, M. R., Munro, J. K., Fields, D. E., Ellison, R. D., Brooks, A. A. and Huff, D. D., 1974, 'A users manual for the Fortran IV version of the Wisconsin Hydrologic Transport Model' Oak Ridge Nat. Lab. ORNL-NSF-EATC-7/EDFB-IBP-74-9, Oak Ridge, TN 37830.

Philips, J. R., 1969, 'Theory of infiltration', *Adv. Hydrosci.*, (Ed.) Chow, V. T., Academic Press, New York.

Rawitz, E., Engman, E. T. and Cline, G. D., 1970, 'Use of mass balance method for examining the rôle of soils in controlling watershed performance', *Water Res. Res.* **6**(4), 1115–1123.

Rogowski, A. S., 1971, 'Watershed physics: model of the soil moisture characteristics', *Water Res. Res.* **7**(6), 1575–1582.

Rogowski, A. S., 1972a, 'Two and three point models of the soil moisture characteristic and hydraulic conductivity for field use', *Joint Symposium on Fundamentals of Transport Phenomena in Porous Media*, Univ. Guelph, Ontario, Canada, August 1972.

Rogowski, A. S., 1972b, 'Soil variability criteria', *Water Res. Res.*, **8**(4), 1015–1023.

Rogowski, A. S., 1972c, 'Variability of soil water flow parameters and their effect on the computation of rainfall excess and runoff', *Proceedings of an International Symposium on Uncertainties in Hydrologic and Water Resource Systems* (I), Tucson, Arizona, December 1972.

Rogowski, A. S., 1972d, 'Estimation of the soil moisture characteristic and hydraulic conductivity: comparison of models', *Soil Sci.*, **114**(6), 423–429.

Smith, R. E. and Woolhiser, D. A., 1971, 'Mathematical simulation of infiltrating watersheds', *Colorado State Univ. Hydrology Paper No. 47.*

Soil Conservation Service, 1966, 'Hydrology, Part 1. Watershed planning, in *National Engineering Handbook*, Washington, DC.

Tennessee Valley Authority, 1965, 'Area-stream factor correlation–A pilot study in the Elk River Basin', *Bull Internat. Assoc. Sci. Hydrology*, **10**(2), 22–37.

Tennessee Valley Authority and North Carolina State University, 1970, 'Watershed research in Western North Carolina,' Knoxville, Tennessee.

Tennessee Valley Authority, 1972, 'Upper Bear Creek project—A continuous daily-streamflow model', *Res. Paper No. 8*, Knoxville, Tennessee.

Tennessee Valley Authority, 1973a, 'Summary report on the Upper Bear Creek experimental project', Knoxville, Tennessee.

Tennessee Valley Authority, 1973b, 'Storm hydrographs using a double-triangle model', *Res. Paper No. 9*, Knoxville, Tennessee.

Viessman, W., Jr., 1968, 'Runoff estimation for very small drainage areas', *Water Res. Res.*, **4**(1), 87–93.

Weyman, D. R., 1973, 'Measurements of the downslope flow of water in a soil', *J. Hydrol.*, **20**(3), 267–287.

Wilde, D. J., 1964, *Optimum Seeking Methods*, Prentice-Hall, Englewood Cliffs, N.J.

Williams, J. R. and Hann, R. W., 1972, 'Hymo—A problem-oriented computer language for building hydrologic models', *Water Res. Res.*, **8**(1), 79–86

Woolhiser, D. A., 1971, 'Deterministic approach to watershed modelling', *Nordic Hydrology II*, 146–166.

Yevjevich, V., 1968, 'Misconceptions in hydrology and their consequences', *Water Res. Res.*, **4**(2), 225–232.

CHAPTER 9

Implications for sediment transport

M. J. Kirkby

School of Geography, University of Leeds, UK

9.1 INTRODUCTION

As we have studied the detailed flow of surface and subsurface water flow on hillslopes, it has become clear that existing models of sediment processes are based on hydrological assumptions which may be false, especially in humid areas. Little or no empirical work has been done on this topic to date. This final chapter therefore attempts to examine the implications of hillslope hydrological work for sediment transport from an analytical standpoint, and so suggest potentially-important topics for empirical work. In making this analysis, most attention will be paid to the process of soil wash, because it is the process most directly related to the quantity of overland flow, but work on hillslope hydrology also has potential importance for rates of removal in solution and the incidence of shallow landslides, which are also briefly analysed below.

Models of hillslope hydrology, particularly those described in Chapter 6, are beginning to provide excellent predictors of the detailed time and space distribution of surface and subsurface flow. To achieve their accuracy however, they require considerable detail of soil properties at a point, together with their spatial distribution. In this chapter a much simplified model will be used as the basis for the analysis of sediment transport. In it, the sequential simulation will be condensed into a temporal frequency distribution, mainly by ignoring the auto-correlation between successive rainfall events. Soils will also be assumed to be spatially uniform, so that the spatial distribution of water and sediment flow reflects only the influence of catchment topography. The influence of landforms on hydrological and sediment processes, and the formation of landforms by sediment transport are thus examined in their simplest form. This approach is not intended to deny that soil properties may vary consistently with topography, forming soil catenas, although it is clearly implied that the author believes that such influences, though important, are less significant than the primary dialogue between form and process with which this chapter is concerned.

⌐A discussion of the basis for a simple hydrological model, which is able to take account of the concepts of hillslope hydrology, is therefore critical to the sediment analysis below, and is not intended to compete with more complex, and more exact models already described in previous chapters. The model described here is able to predict–as observation shows that it must–a non-uniform production of overland flow over a hillslope or drainage basin. It thus differs from the assumption made by most existing sediment models (Bennett, 1974, for example), that overland flow is produced uniformly at all points.

From this point of departure, the implications for soil erosion are pursued at increasing time scales. In the short term of a few years, a sediment model attempts to predict the space and time distribution of sediment transport, and hence erosion and deposition, on a surface topography which is considered to be essentially fixed. This time scale is related to the practical problem of assessing the risk of damage on a given slope, and of designing remedial measures, such as drains and channelways, to reduce potential damage. On a longer time scale, of perhaps decades, the slope topography begins to evolve in response to sediment transport over it. The related practical problem is concerned with the design of an artificial slope to minimize erosion, and particularly with the elimination of areas in which concentrated erosion tends to form a growing gully. On a still longer time scale, the landscape erodes, crucially through the action of running water, into the familiar pattern of slopes and valleys. The scale of this pattern, measured most commonly as drainage density (km. of channel per km² of drainage area) is also determined by the distribution of sediment transport, and so is fundamental to fluvial geomorphology, although the time scale is too long to be of much practical importance.

For removal in ionic solution and in colloidal form, which are usually combined in measurements of 'dissolved load' in rivers, the implications of work on hillslope hydrology are less significant. They relate to higher moisture contents consistently occurring downslope, and consequently to the higher rates of actual evapotranspiration. In combination with increased overland flow, there is a distinct reduction in the amount of sub surface-flow production, and a consequent expected reduction in removal in solution. Shallow landslides have been shown (Skempton and Delory, 1957) to depend critically on piezometric levels in the soil which are once more related to the hillslope hydrology via subsurface flow. These implications are also pursued analytically below.

Not all the implications for sediment transport described above have been fully pursued previously, even for the case of uniform flow production. This case is therefore analysed fully beside the non-uniform case. This approach helps to make explicit the differences between the implications of uniform and non-uniform flow production, and hence the real contribution which work on hillslope hydrology can make to sediment studies.

9.2 INFILTRATION MODELS AND STORAGE MODELS

If hillslope hydrological models and empirical results are examined in previous chapters, certain generalizations can be made which contribute to a simplified model, which will form the basis of a sediment model below. These generalizations are as follows:

(i) Lateral downslope flow is most important close to the surface, especially over the surface and in near-surface soil layers, notably the A horizon. It is commonly dominated by flow in a saturated or near-saturated state. In designing a model, it is therefore reasonable to assume a constant lateral permeability (the saturated value) and to ignore hysteresis effects.

(ii) During a storm the lateral subsurface flow is a negligible component of the flood hydrograph, but subsurface flow over longer periods (i.e. between storms) is significant in establishing the areal distribution of soil-moisture storage at the beginning of a storm.

(iii) Overland flow travels at a rate much higher than subsurface flow. If a short time unit is used (< 2 hr, say), then overland flow should be routed using Manning's formula or an equivalent; but for longer time units, including the one-day unit used below, it may be considered to reach the slope base instantaneously.

(iv) Downward flow from the surface to a saturated level involves movement of water in the unsaturated phase. At the same time, an analysis of the output hydrograph form (Kirkby, 1975) shows that the flow hydrograph is not very sensitive to the exact details of the downward-flow process, provided that the overall quantity and average timing can be reliably estimated.

This section is concerned with comparing two simple models of downward flow which form two extremes on a continuum, both of which are good approximations under particular climatic conditions and which are reasonable approximations over a wide range of conditions. These models are the infiltration model and the storage model of surface runoff.

The infiltration mode of surface runoff (Horton, 1933) predicts the maximum *intensity* at which rainfall can enter the soil at a particular time. Rainfall in excess of this critical intensity (the infiltration capacity) runs off as overland flow. As water infiltrates into the soil, the infiltration capacity is reduced. It is commonly expressed as a function of time elapsed, or of total amount infiltrated during the storm. The former expression is analytically simpler, but is less adaptable to conditions of variable rainfall intensity than the latter type of expression. In a sequence of two storms, closely following one another, the higher initial moisture content at the start of the second storm allows a lower initial infiltration capacity to be predicted.

The storage model of surface runoff does not limit rainfall intensity, but only the available volume for cumulative infiltration. Infiltration of all rainfall is predicted until this storage is filled. Subsequent rainfall, even at low intensities, is predicted to produce overland flow. Storage may be depleted by downward percolation and by lateral downslope flow.

These two models identify two distinct modes for overland flow (OF) production, which can be called 'infiltration overland flow' and 'saturation overland flow'. In combination they model the near-surface soil as like a leaky bottle, which may not accept water poured into it either because the neck is too narrow (infiltration OF) or because the bottle is already full (saturation OF). If such

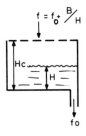

Figure 9.1   Near-surface soil model

a model is to incorporate lateral subsurface flows, it is argued that infiltration rates are best expressed in terms of near-surface storage, so that the best of both extremes may be combined in a very simple format, defined by three parameters. In Figure 9.1, the store has capacity H_c (measured as equivalent depth of water), leaks at constant rate f_0, and allows infiltration capacity at intensity

$$f = f_0 + B/H \qquad (9.1)$$

where

H = the current depth of storage
B = a soil constant

Equation 9.1 is very similar to the empirical Green and Ampt (1911) equation, and if it is solved for rainfall at the infiltration capacity as it varies from $H = 0$ at time $t = 0$, it leads to the Philip (1957) equation

$$f = f_0 + \left(\frac{B}{2}\right)^{\frac{1}{2}} t^{-\frac{1}{2}} \qquad (9.2)$$

A better approximation to the conditions of an actual storm may be obtained by solving Equation 9.1 for the volume infiltrated, V, before overland flow occurs by the two mechanisms, during rainfall at a steady rate, i. In each case, up until overland flow commences

$$V = it = H + f_0 t \qquad (9.3)$$

Substituting the conditions for overland flow by infiltration OF:

$$V_0 = \frac{B_i}{(i - f_0)^2} \tag{9.4}$$

and for saturation OF:

$$V_s = \frac{H_c i}{(i - f_0)} \tag{9.5}$$

These two expressions are compared schematically in Figure 9.2. In all cases the two curves intersect at an intensity

$$i_c = f_0 + B/H_c \tag{9.6}$$

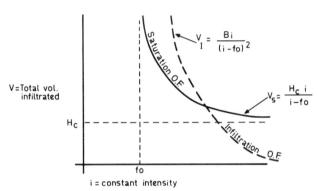

Figure 9.2 Volume infiltrated at constant

Where observed rainfall intensities commonly lie well above or below this critical value it is clearly appropriate to use a simple infiltration model or a simple storage model, respectively, although in general the combined model is preferable. Thus, a simple storage model is closest to reality under conditions where rainfall intensities are generally less than infiltration capacities; it is also most useful under conditions where rainfall is frequent. Great Britain might therefore be taken as a type-locality for its effective use. A simple infiltration model, on the other hand, is closest to reality under conditions where rainfall intensities commonly exceed infiltration capacities, and rainfall events are infrequent. It is worth noting however that an effective storage model has been developed for an area in Israel with 100–200 mm annual rainfall (Shanan, 1974). In the sections following, a daily storage model is developed, to show the influence of hillslope hydrology in the simplest way.

9.3 A DAY-BY-DAY STORAGE MODEL

The behaviour of a storage model can be understood most readily if storm rainfall events are treated as instantaneous and uncorrelated. The rainfall after allowance

N

for evapotranspiration and downward percolation, augments the subsurface storage everywhere, and may, in some parts of a slope, produce overland flow. The subsurface water travels slowly enough that its movement during a storm is more or less negligible, in agreement with many of Freeze's model results (e.g. Figures 6.17 and 6.18). Between storms, the slowly-moving water combines with soil water from previous storms, so that an approximation to average antecedent conditions can be obtained by assuming a suitably weighted rate, i, of soil-water accumulation. Weighting methods are discussed below.

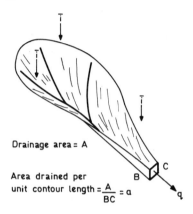

Drainage area = A

Area drained per
unit contour length = $\dfrac{A}{BC}$ = a

Figure 9.3   Discharge from a sub-catchment during indefinite (rainfall–evaporation–percolation) at rate i

In Figure 9.3, the area A, drains through a short section BC of a contour. The area drained per unit contour length (a) is defined as $a = \text{A}/\text{BC}$, and has the dimension of length (and is equal to distance from the divide if the contours are all straight and parallel). In the steady state thus established

$$q = ia = g(h) \cdot s \qquad\qquad (9.7)$$

where

q = discharge per unit contour length
h = depth of soil water flowing past BC
s = local hydraulic gradient at BC
$g(h)$ = a rating function which takes into account the varying hydraulic conductivity in different soil layers.

In a vertically-uniform soil, $g(h)$ is likely to be linear, with a constant equal to the saturated hydraulic conductivity. In general, $g(h)$ will be treated as an increasing, single-valued function, possibly of power or exponential form. The hydraulic gradients, s, will normally be almost the same as the topographic gradient, and will be assumed to be so.

Equation 9.7 can be inverted to give an expression for the steady-state value of h:

$$h = g^{-1}\left(\frac{ia}{s}\right) = r_0 \log_e\left[\phi\left(\frac{ia}{s}\right)\right] \qquad (9.8)$$

where

r_0 = a constant defined in Equation 9.11
g^{-1} = the inverse function of g
ϕ = a function defined by Equation 9.8, termed the 'hillslope-hydrology function' below.

Concentrating on the influence of topography alone, and therefore taking ϕ as areally uniform, the antecedent soil-water depth is seen to show a strong dependence on topography, as expressed by the term, a/s, the ratio of drainage area per unit contour length to slope. The variable a/s takes very high values in hollows,

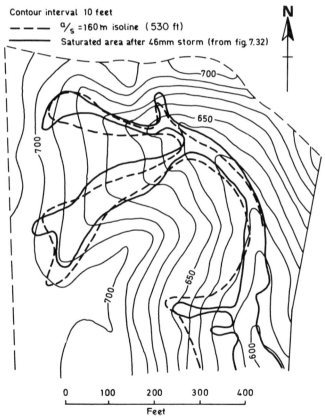

Figure 9.4 Distribution of area with $a/s > 100\,\text{m}$, calculated from contours shown, for basin WC-4 of Figures 7.28 and 7.32

because in them the area drained per unit contour length is high, and slopes tend to be low. Figure 9.4 gives an example of the distribution of high values of a/s along valley bottoms and extending up hollows, and compares it with an observed pattern of saturation.

If, again concentrating on the influence of topography, the storage capacity, h_c, of the soil is also assumed areally constant, then overland flow will be produced, for a storm of total rainfall, r, from those areas for which

$$r + h > h_c \tag{9.9}$$

so that one isoline for h or a/s should correspond to the areal limit of saturation during a particular storm, and result in the sort of agreement illustrated in Figure 9.4. The area defined in this way can, of course, be identified with the concept of a dynamic contributing area (Betson, 1964; TVA 1964; Weyman, 1974). Storm rainfall within this area will run off with little or no loss, and provide the main component of runoff, from the storm, particularly of the 'quick-flow' component (Hewlett, 1961). The prediction of the contributing area is therefore of major concern in the application of hillslope hydrology to the prediction of flood hydrographs, and on a smaller scale, to the prediction of the amount and areal extent of overland flow.

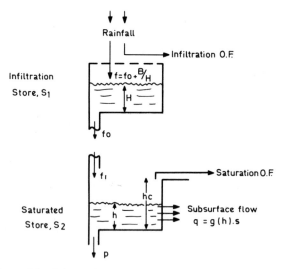

Figure 9.5   Schematic two-store soil hydrological model

If the value of i in Equation 9.8 is adjusted only seasonally, this simple storage model is sufficient to predict a dynamic contributing area which grows during storm in a realistic manner, and relies on only the hillslope-hydrology function, ϕ, which must be optimised or measured. Improvement in the prediction of contributing area can be obtained using a relatively simple method to update the value of i. One example of such a method involves the use of two non-linear stores to budget soil-water storage (Figure 9.5). The upper store, representing unsaturated water movement, is defined to behave according to Equation 9.1 above, but without a limitation on overall capacity. Inflow to the second store is from percolation down from the upper store, at a steady rate f_0, but with a variable delay, according to the level of the lower, saturated store (h). To allow for this possible complication, the inflow to the lower store is written in Figure 9.5 as f_1, but in many cases it is reasonable to take $f_1 = f_0$. In the lower store, the subsurface flow is generated, according to Equation 9.7 above, and saturation O.F. is generated when $h > h_c$. Where there is a distinct groundwater regime, it may be necessary to include a deep percolation leak from S_2 at rate p.

Using this type of model, a sequence of rainfalls can be converted to a current value of h, and, using Equation 9.7, this value of h can be converted back to an effective weighted mean-rainfall intensity, i. On the assumption of uniform soils, which is explicitly being made, Equation 9.8 allows an estimate of h at every point on a hillslope from its value at a particular point, h. Denoting properties at this point by asterisks throughout, Equation 9.8 gives

$$h = r_0 \log_e \left\{ \phi \left[\frac{g(h^*)a/s}{a^*/s^*} \right] \right\}$$
(9.10)

This method is discussed more fully in Kirkby (1975).

So far, the model described for the hillslope hydrology has not specified a particular unit time period. The model will now be made specific to a one-day unit, and the assumption of instantaneous overland flow made in consequence. The use of one-day rainfalls is justified by the widespread availability of such data, and their rather simple distribution. In an area without marked seasonal contrasts in rainfall, a total annual rainfall R, falling on N rain-days per year conforms closely to an exponential distribution. The number of days with a daily rainfall exceeding r is

$$N(r) = Ne^{-r/r_0}$$
(9.11)

where $r_0 = R/N$ is the mean rain per rain-day, and is the constant which appears in Equation 9.8. Under seasonal rainfall regimes, the parameters need only take distinct seasonal values.

It is now assumed that a part of each day's rainfall, drawn from this distribution, produces overland flow when it exceeds available storage h_c. The remainder, less-deep percolation and actual evapotranspiration, E, goes to produce a background soil storage level, h. Thus the mean infiltration intensity is obtained by subtracting these losses, and losses due to overland flow, from the total rainfall, and averaging over the year:

$$i = \frac{R - E - \bar{q}_{OF}}{365}$$
(9.12)

where $\bar{q}_{OF}$ is the overland flow contribution averaged over the area draining to the site considered. Since the overland-flow contribution generally increases downslope, this final term may be ignored until the area of perennial flow is approached. In this case, Equation 9.8 gives h, the depth of subsurface flow, and Equation 9.9 gives the total production of saturated overland flow:

$$q_{OF} = r + r_0 \log_e\left[\phi\left(\frac{ia}{s}\right)\right] - h_c \tag{9.13}$$

Summing over the rainfall distribution at a site each value of q_{OF} must be weighted by a frequency density which is obtained by differentiating Equation 9.11

$$n(r) = -\frac{d}{dr}\left[N(r)\right] = N/r_0 e^{-r/r_0} \tag{9.14}$$

The total annual overland flow contribution from a point on the hillside is (provided that the value is small) therefore:

$$q_{OF} = \int_{r=(h_c - r_0 \log \phi)}^{r=\infty} \left\{r + r_0 \log\left[\phi\left(\frac{ia}{s}\right)\right] - h_c\right\} \cdot N/r_0 e^{-r/r_0}\, dr$$

$$= Re^{-h_c/r_0} \cdot \phi(ia/s) \tag{9.15}$$

Similarly, the number of days a year with overland flow (N_{OF}) is given by

$$N_{OF} = Ne^{-h_c/r_0}\phi(ia/s) \tag{9.16}$$

These two expressions explain the form chosen in Equation 9.8 for defining the hillslope hydrology function, ϕ. It can be seen that a linear flow law in Equation 9.7, corresponding to a vertically-uniform soil, leads to an exponential form for ϕ, showing very great sensitivity of overland flow to topography. Higher powers of flow depth, h, in the rating function of Equation 9.7 lead to somewhat less-sensitive hillslope-hydrology functions.

Where the overland flow term in Equation 9.12 is too large to ignore, a correction must be applied. In the most extreme case, corresponding to constant a/s and hence constant overland flow production, $\bar{q}_{OF} = q_{OF}$ and

$$q_{OF} = Re^{-h_c/r_0}\phi\left[\frac{ia}{s}\left(1 - \frac{q_{OF}}{R - E}\right)\right] \tag{9.17}$$

Figure 9.6 shows a graphical solution to correct the simple approximation of Equation 9.15. In more complicated, and usual, cases where a/s increases downslope, the exact value of overland flow lies between the two curves shown in Figure 9.6, and becomes asymptotic to $q_{OF} = (R - E)$ at large values of a/s, that is, in hollows. The spatial implications of Equations 9.15–9.17 are taken up in the next section.

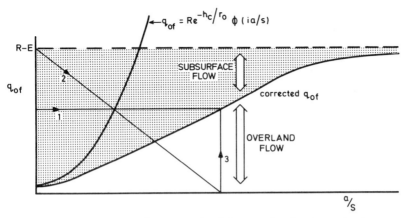

Figure 9.6   Graphical correction to overland-flow production in most extreme case
(constant overland-flow production)

9.4   THE SPATIAL DISTRIBUTION OF OVERLAND FLOW

Horton (1945) proposed that overland flow has to overcome a critical resistance
before appreciable erosion occurs. It is here argued that this threshold resistance is
very small, and that any appreciable overland flow will produce significant erosion,
as can be seen from the existence of rills within one metres of divides in semi-arid
badlands. The question of whether such erosion will develop into a valley is seen as
a different one, which is discussed below in the section on drainage density
(Section 9.7).

As a first approximation, then, appreciable overland flow and appreciable
erosion are considered to occur when N_{OF}, the overland flow frequency, exceeds
one day per year. This arbitrary value allows a comparison of different climates,
ignoring any soil differences. At this point, Equations 9.16 and 9.15 give
respectively

$$N_{OF} = 1 \quad \text{and} \quad q_{OF} = R/N = r_0 \qquad (9.18)$$

In order to obtain a better picture of what these expressions mean, four type-
climates are listed in Table 9.1, which gives values for the constants appearing in the
various equations for sites which are loosely associated with Britain (moderate and
high rainfall examples), Arizona and Washington DC. For illustrative purposes,
the flow law (Equation 9.7) is taken to be linear, with lateral saturated permeability,
$K = 25$ m/day in all climates. The constants in the last two rows will be introduced
at a later stage.

For the linear flow law, from Equation 9.7

$$q = ia = Khs, \qquad (9.7a)$$

$$\phi(z) = e^{z/Kr_0} \qquad (9.8a)$$

Table 9.1 Examples of constants in hillslope hydrology equations

Variable	Symbol	UK moderate rainfall	UK high rainfall	SE Arizona	Washington DC
Annual rainfall (mm)	R	700	2,500	300	1,000
Soil storage capacity (mm)	h_c	100	100	20	100
Mean rainfall per rain-day (mm)	$r_0 = R/N$	4	10	15	10
Average infiltration intensity over year (mm/day)	i	1	5	0	1
Number of raindays per year	N	175	250	20	100
Lateral saturated permeability (m/day)	K	25	25	25	25
'Throughflow critical distance' (m)	$Kr_0/i = u_T$	100	50	∞	250
'Overland flow critical distance' (m)	u_O	2.3×10^{10}	1,970	2·8	4,900

and, substituting in Equation 9.16 and taking natural logarithms

$$\log N_{OF} = (\log N - h_c/r_0) + \frac{i}{Kr_0} \cdot \frac{a}{s} \qquad (9.16a)$$

Figure 9.7 gives examples of this relationship, ignoring the correction of Equation 9.17, and hence valid only for N_{OF} less than about 10 days a year.

For the Arizona example. $h_c/r_0 < \log N$ and thus overland flow is expected at least once a year everywhere. In this case the effect of topography is negligible, and the analysis comes closest to Horton's (1945) model.

For the other three examples illustrated in Figure 9.7, $h_c/r_0 > \log N$ so that overland flow is only produced frequently enough to erode appreciably above some critical value of a/s. For the UK 2500-mm case, the critical value of a/s is about 200 m, a value which could readily be obtained on the side of a straight ridge. For the other two examples, the critical values of a/s are so high (> 1 km) that they can only be reached in definite hollows. It is these cases therefore that the three-dimensional topography, and its influence on hillslope hydrology, is of greatest importance, in that *only* hollows are eroded by wash erosion.

Examining the factors which contribute to these different cases, it can readily be seen that the slope of the lines in Figure 9.7 are determined mainly by the value of i,

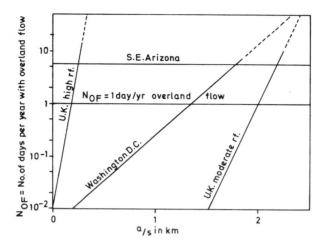

Figure 9.7 Examples of the relationship of equation 9.16. $N_{OF} = 1$ level corresponds to level of appreciable overland flow and hence wash erosion

the mean rate of water entry into the soil. High sensitivity to topography (i.e. a steep slope of the line in Figure 9.7) is thus favoured by high annual rainfall and relatively low evapotranspiration. In semi-arid areas, topography has no influence on the overland flow production, as is illustrated by the horizontal line for Arizona in Figure 9.7. In cool wet climates, such as the mountains of Britain, overland flow is very sensitive to topography, whereas most humid temperate areas show a rather low sensitivity, as is shown by the low slopes of the lines for UK moderate rainfall and Washington on the graph in Figure 9.7. The left-hand intercept on the graph, $(\log_e N - h_c/r_0)$, is most sensitive to the ratio h_c/r_0. The critical storage capacity has been shown (Carson and Kirkby, 1972) to depend strongly on vegetation cover and so, *under natural conditions*, on climate. h_c takes daily values of about 10 mm on bare ground, 40 mm under grass, and 100 mm under forest cover. Low h_c/r_0 is therefore typical for semi-arid areas with intense rains and sparse vegetation; and high values for h_c/r_0 are normal for densely-vegetated temperate areas with moderate average-rainfall intensities.

If occasional overland flow, as defined above, represents a minimum for effective wash erosion, a similar maximum can be estimated from the topographic position at which flow becomes perennial. Referring to Equation 9.15, an approximate criterion can be obtained from the position at which

$$q_{OF} = R \quad \text{or} \quad N_{OF} = N \tag{9.19}$$

The corresponding critical value of a/s may be obtained by extrapolation in Figure 9.7 or from Equation 9.16. Table 9.2 summarizes the result for the four type-examples of Table 9.1.

Table 9.2   Values of the ratio of drainage area per unit contour length to slope (a/s) depending on duration of overland flow, for the four locations in Table 9.1

Location	Critical a/s (km) for overland flow $\geqslant 1$ day/year	Critical a/s (km) for perennial overland flow
UK moderate rainfall	2·0	2·5
UK high rainfall	0·2	0·5
SE Arizona	0	∞
Washington DC	1·3	5·0

In between the two values of a/s (Table 9.2) is a region of ephemeral overland flow in which some erosion is taking place. These values may be seen as extreme possibilities for the definition of a 'stream head', although the Arizona example shows just how extreme either value may be. In Britain, on the other hand, the difference may be unimportant, and Ordnance Survey practice is to map streams if they normally flow continuously in winter (Harley, 1975). Some alternative criteria are considered in Section 9.7.

9.5   IMPLICATIONS FOR SOIL EROSION RATES

Commonly, models of soil erosion have expressed sediment transport rates as a function of distance from the divide multiplied by a power of slope gradient. It is clear, however, that distance is used as a substitute for overland flow. Where Q_{OF} is the accumulated overland-flow discharge summed over a year, then a good fit with available data can be obtained using an expression of the form

$$S \propto Q_{OF}^2 \cdot s \tag{9.20}$$

The fit remains good in experiments which compare different vegetation covers, the fifty-fold variation in overland flow runoff corresponding to a 2000-fold variation in erosion rates (Carson and Kirkby, 1972). To a first approximation such hydrological differences appear to outweigh any influences of soil resistance, although the latter are clearly important at a local scale.

In areas near divides, where there is least overland flow, creep and rainsplash processes tend to dominate wash processes, so that the transport equation 9.20 needs to be modified by the addition of a creep term, which is directly proportional to slope. Order of magnitude values from available data sources give

$$S = (0 \cdot 001 + 0 \cdot 02 \cdot Q_{OF}^2)s \tag{9.21}$$

where S and Q_{OF} are measured in m^2/year, and are considered to act down the line of greatest slope.

This transport model may be combined with the continuity equation for slope sediment

$$\mathbf{\nabla} \cdot \mathbf{S} + \frac{\partial y}{\partial t} = 0 \tag{9.22}$$

where

y = elevation
t = time elapsed
$\mathbf{V} \cdot \mathbf{S}$ = the vector divergence of the vector transport rate

Together, the transport law (Equation 9.21) and the continuity equation (9.22) provide the basis for simulating slope development from a given initial slope profile and given slope-base conditions. It is possible to proceed directly via a numerical computer simulation (e.g. Kirkby, 1976), but a partly analytical approach is preferred here for its generality.

If a slope-profile strip is outlined by two neighbouring lines of steepest slope, then the width, Ω, of the profile strip, is related to the area drained per unit contour length, at distance x from the divide, by

$$a = \frac{\int_0^x \Omega \, dx}{\Omega} \qquad (9.23)$$

and the continuity equation 9.22 may be written in the form

$$\frac{\partial(\Omega S)}{\partial x} + \frac{\partial(\Omega y)}{\partial t} = 0 \qquad (9.24)$$

To make the equations explicit, an expression for the total overland flow, Q_{OF}, must be obtained by summing the local overland flow production, q_{OF}, down the profile strip from the divide

$$Q_{OF} = \frac{\int_0^x \Omega q_{OF} \, dx}{\Omega} \qquad (9.25)$$

For the hillslope, the expression for overland flow production in Equation 9.15 is probably sufficiently precise for most purposes. Substituting it in Equation 9.25 gives

$$Q_{OF} = \frac{\int_0^x \Omega R e^{-h_c/r_0} \phi(ia/s) \, dx}{\Omega} \qquad (9.26)$$

Substituting in Equation 9.21 in its turn gives

$$S = \left\{ \cdot 001 + 2 \times 10^{-8} R^2 e^{-2h_c/r_0} \left[\frac{\int_0^x \Omega \phi(ia/s) \, dx}{\Omega} \right]^2 \right\} \cdot s \qquad (9.27)$$

By reverting to the definition of ϕ in Equation 9.8, it may be seen that the arid case ($i = 0$) corresponds to a zero depth, h, of subsurface flow, so that $\phi(0) = 1$, and for this simple case, the square-bracketed expression as a whole takes the value a.

Writing

$$u_O = \sqrt{\frac{\cdot 001}{2 \times 10^{-8} R^2 e^{-2h_c/r_o}}} = \frac{224}{R} \cdot e^{h_c/r_o} \text{ metres} \qquad (9.28)$$

where the mean annual rainfall, R, is measured in millimetres, Equation 9.27 becomes

$$S = \frac{\cdot 001}{u_O^2} \left\{ u_O^2 + \left[\frac{\int_0^x \Omega \phi(ia/s) \, dx}{\Omega} \right]^2 \right\} \cdot s \qquad (9.29)$$

in which the square-bracketed expression has the dimension of distance, and in the simple arid case, corresponding to uniform overland-flow production

$$S = \frac{\cdot 001}{u_O^2}(u_O^2 + a^2) \cdot s \qquad (9.30)$$

Equations 9.29 and 9.30 allow some predictions to be made about the spatial distribution of sediment transport, and hence erosion and deposition under conditions of non-uniform and uniform overland-flow production. To illustrate these patterns in a concrete form, the hillslope-hydrology function, ϕ, will again be assumed exponential in form (Equation 9.8a), and the additional assumption of parallel contours will be made ($\Omega = $ constant; $a = x$). Under these conditions

$$S = \frac{\cdot 001}{u_O^2} \left\{ u_O^2 + \left[\int_0^x e^{i/Kr_0 \cdot x/s} \, dx \right]^2 \right\} \cdot s \qquad (9.29a)$$

This equation contains two important length dimensions. The first is u_O, which has been seen above to be related to uniform overland flow production, and is here referred to as the 'overland flow critical distance'. The second is $u_T = Kr_0/i$, which is related to the subsurface flow term and is referred to as the 'throughflow critical distance'. Values for these distances have been calculated for the four type-climates and are listed in Table 9.1.

The implications are carried a stage further in Figure 9.8, where the increasing rates of surface lowering $(-\partial y/\partial t)$ downslope are estimated from Equation 9.29a for straight slopes at a gradient $\tan \beta = \frac{1}{2}$. It can be seen that erosion rates close to the divide reflect the values of the overland flow critical distance, u_O. Small values of u_O give high erosion rates near the divide and vice versa. At increasing distance downslope however, the erosion rate 'goes critical' at a distance from the divide which appears to reflect the values of throughflow critical distance listed in Table 9.1. In each case, the erosion rate changes from a negligible rate of less than one metre per million years to a catastrophic rate of more than one metre per year, over a relatively short distance. An initial indication that remedial measures are needed on a slope to prevent rapid wash erosion, can be obtained from an analysis similar

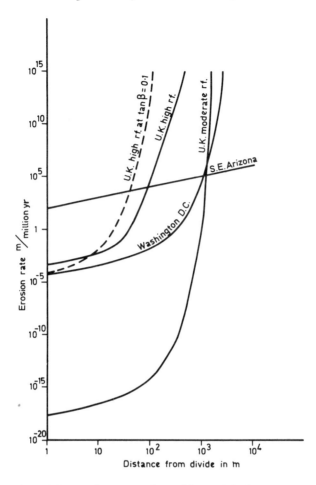

Figure 9.8   Erosion rates estimated for straight slopes at a
gradient of tan $\beta = 0.5$. Broken line shows one area for
tan $\beta = 0.1$.

to that of Figure 9.8. Although an additional approximation is involved, the
distances in the figure can be roughly interpreted as area drained per unit contour
length. Thus remedial measures might consist of modelling the slope into a spur
rather than a hollow, we well as solutions like the installation of resistant
channelways to reduce the maximum slope lengths. The effect of a lower slope
gradient is more complex, as is shown by the broken line in Figure 8.9. The erosion
rate is reduced close to the divide, and increased downslope. At the same time, the
distance at which the slope 'goes critical' is reduced.

The pattern of erosion shown diagrammatically in Figure 9.8, and implied by
Equation 9.29, is one which shows a dramatic response to the subsurface flow

regime in temperate or humid areas. The ultimately rapid increase in erosion and overland flow is one which is responding very sensitively to the soil hydrology. In the model presented, the response is via the slope hydrology function, ϕ, and through the storage capacity, h_c. Although these are to some extent artefacts of the model, it is thought that the sensitivity they predict is a real one. The sensitivity to slope gradient, though less marked, is also significant, and suggests that wash erosion can be controlled in part by design of suitable slope profile forms.

9.6   THE SLOPE PROFILES PRODUCED

The approach to a 'design slope' used here is to consider slopes which are in some sort of equilibrium with erosion processes. Two types of equilibrium are examined, and it will be shown that the appropriate slope profiles are similar to one another. It is further suggested that either forms a practical design slope on which erosion rates are effectively minimal. The first type of equilibrium slope considered is one on which the rate of lowering is constant at all points. The second type is one on which the rate of erosion is proportional to elevation above the basal point.

For the first type of equilibrium, the required slope is obtained by solving Equation 9.24 with $\dfrac{\partial y}{\partial t} = T$, a constant and Equation 9.29. Integrating the former over the drainage area above x gives

$$S = \frac{0 \cdot 001}{u_0^2}\left\{u_0^2 + \left[\frac{\int_0^x \Omega\phi(ia/s)\,dx}{\Omega}\right]^2\right\}s = aT \tag{9.31}$$

Rearranging gives

$$\int_0^x \Omega\phi\,(ia/s)\,dx = \Omega u_0(1000Ta/s - 1)^{\frac{1}{2}} \tag{9.32}$$

Now, put $ia/s = z$, a new variable, and differentiate with respect to x

$$\Omega\phi(z) = u_0\frac{d}{dx}\{(1000Tz/i - 1)^{\frac{1}{2}}\Omega\}$$

$$= 500Tu_0/i(1000Tz/i - 1)^{-\frac{1}{2}}\Omega\frac{dz}{dx}$$

$$+ u_0(1000Tz/i - 1)^{\frac{1}{2}}\frac{d\Omega}{dx} \tag{9.33}$$

This equation is not very tractable in general, but for parallel contours exactly, and approximately if the contours are only slightly curved:

$$\phi(z) = 500Tu_0/i(1000Tz/i - 1)^{-\frac{1}{2}}\frac{dz}{da} \tag{9.34}$$

This is an equation in z and a which can be solved by separation of variables. Thus:

$$\frac{500T}{i} \int \frac{(1000Tz/i - 1)^{-\frac{1}{4}} \, dz}{\phi(z)} = \int \frac{da}{u_O} \tag{9.35}$$

Let $(1000Tz/i - 1) = w^2$. $\tag{9.36}$

Then

$$a = u_O \int_0^{w = (1000Tz/i - 1)^{\frac{1}{2}}} \frac{dw}{\phi\left[\dfrac{1 + w^2}{1000T} \cdot i\right]} \tag{9.37}$$

From this general solution, particular solutions for particular slope hydrology functions, ϕ, can be derived. For example if, as before

$$\phi(z) = e^{z/Kr_0} \tag{9.8}$$

then, successively

$$a = u_O e^{-(i/1000TKr_0)} \int_0^w e^{-(w^2 i/1000TKr_0)} \, dw$$

$$= \left(\frac{\pi}{4}\right)^{\frac{1}{2}} \left(\frac{1000TKr_0}{i}\right)^{\frac{1}{2}} u_O e^{-(i/1000TKr_0)} \operatorname{erf}\left[w\left(\frac{i}{1000Tkr_0}\right)^{\frac{1}{2}}\right]$$

$$= \left(\frac{\pi}{4}\right)^{\frac{1}{2}} (1000Tu_T)^{\frac{1}{2}} u_O e^{-(1/1000Tu_T)} \operatorname{erf}\left[\left(\frac{z}{u_T} - \frac{1}{1000Tu_T}\right)^{\frac{1}{2}}\right] \tag{9.38}$$

Finally, putting

$$\frac{i}{1000Tu_T} = p^2 \tag{9.39}$$

this expression can be simplified to

$$\operatorname{erf}\left[\left(\frac{z}{u_T} - p^2\right)^{\frac{1}{2}}\right] = \left(\frac{4}{\pi}\right)^{\frac{1}{2}} p e^{p^2} \cdot a/u_O \tag{9.40}$$

This represents a dimensionless relationship between $z/u_T = ia/su_T$ and a/u_O, with a single parameter p. More usefully, it expresses a relationship between

$s \cdot \dfrac{u_T}{iu_O} = (a/u_O)/(ia/su_T)$ and x/u_O. This relationship is shown graphically in Figure 9.9, which shows a series of slope profiles corresponding to different values of the parameter p, and the locus of the points of inflexion on the slope profile. It can be

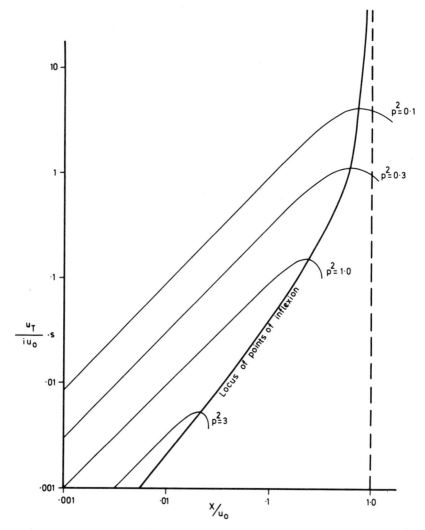

Figure 9.9   Equilibrium slope profiles developed under conditions of
constant down-cutting with a linear flow law (exponential slope hydrology
function)

seen that for low values of p, corresponding to high values of the constant
$u_T = Kr_0/i$, the slope form changes in elevation and gradient but not in *relative*
shape as the maximum slope is reduced. At higher values of p it can be seen that as
erosion reduces the maximum slope gradient, and its position moves towards the
divide. This difference is produced by the action of subsurface flow.

On a slope profile, the distance to the point of greatest slope (the point of
inflexion) is the most important single slope dimension. Where subsurface flow is

unimportant (p small and u_T large) the distance to the point of inflexion is very close to u_0, the overland flow critical distance. As u_T becomes small, the line joining points of inflexion tends towards a straight line at 45°, for which

$$\frac{u_T s}{u_0} \propto \frac{x}{u_0} \tag{9.41}$$

or

$$x \propto s \cdot u_T \tag{9.42}$$

at the point of inflexion. Thus, where subsurface flow becomes important, the dimension of the profile is determined by u_T, the throughflow critical distance, together with the slope gradient, s.

Near the divide, the value of $z = a/s$ is almost constant, at

$$z_0 = ia/s = p^2 u_T = \frac{i}{1000T} \tag{9.43}$$

This corresponds to the straight-line portion of the curves in Figure 9.9. The convexity of the divide is thus related in a simple way to the overall downcutting rate, T. Downslope, the value of z only slowly increases from its divide value of z_0, but, as far as the point of inflection, a linear approximation to $\phi(z)$, the slope hydrology function, is a reasonable one giving:

$$\phi(z) = \phi(z_0) + \lambda(z - z_0) \tag{9.44}$$

In this equation $\phi(z_0)$ is the value of the slope hydrology function, ϕ, at the divide, and λ is the rate of increase of ϕ with respect to z near $z = z_0$. Substituting into the more general solution of Equation 9.37:

$$a = u_0 \int_0^w \frac{dw}{\phi(z_0) + \dfrac{\lambda w^2 i}{1000T}} \tag{9.45}$$

Substituting $w = \left(\dfrac{1000T}{i} \cdot \dfrac{\phi(z_0)}{\lambda}\right)^{\frac{1}{2}} \tan\theta$ and integrating with respect to θ

$$a = u_0 \left[\frac{1}{z_0 \lambda \phi(z_0)}\right]^{\frac{1}{2}} \tan^{-1}\left[\frac{\lambda(z - z_0)}{\phi(z_0)}\right]^{\frac{1}{2}} \tag{9.46}$$

or, re-arranging to express z as a function of a:

$$z = z_0 + \frac{\phi(z_0)}{\lambda} \tan^2\left\{\frac{a[z_0 \lambda \phi(z_0)]^{\frac{1}{2}}}{u_0}\right\} \tag{9.47}$$

Recalling that $z = ia/s$, and re-arranging to obtain an expression for slope gradient, s:

$$s = \frac{ia}{z_0 + \dfrac{\phi(z_0)}{\lambda} \tan^2 \left\{ \dfrac{a[z_0 \lambda \phi(z_0)]^{\frac{1}{2}}}{u_0} \right\}} \tag{9.48}$$

Finally, setting

$$p^2 = \frac{\lambda z_0}{\phi(z_0)} \tag{9.49}$$

which is consistent with its definition in Equation 9.39; and defining

$$\gamma = p\phi(z_0) = [z_0 \lambda \phi(z_0)]^{\frac{1}{2}} \tag{9.50}$$

the expression for slope gradient may be written as

$$\frac{\gamma z_0}{iu_0} \cdot s = \frac{p^2(\gamma a/u_0)}{p^2 + \tan^2(\gamma a/u_0)} \tag{9.51}$$

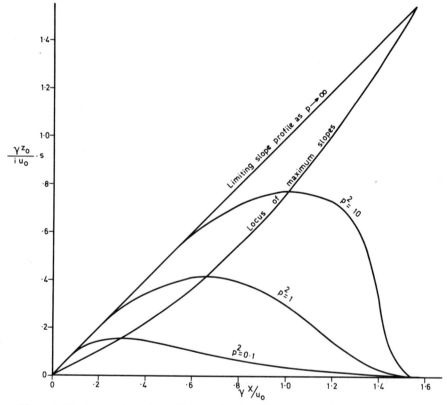

Figure 9.10 Approximate equilibrium slope profiles developed under conditions of constant down-cutting for a general slope hydrology function

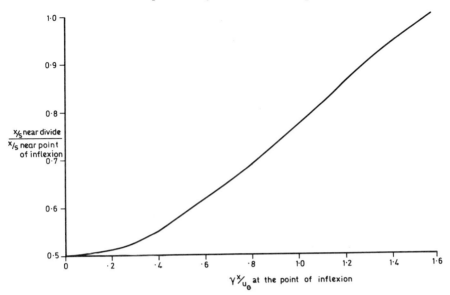

Figure 9.11   Graphical derivation of an approximate slope hydrology function ϕ from measured slope profiles. The derived variable γ is related to ϕ by equation 9.53

The maximum slope on these profiles occurs on the locus:

$$\frac{\gamma z_0}{i u_0} \cdot s = \gamma x/u_0 - \tfrac{1}{4}\sin\left(2\gamma x/u_0\right) \tag{9.52}$$

This locus, and examples of slope profile solutions, are shown in Figure 9.10. The right-hand portion of each profile, beyond the point of greatest slope, is limited in accuracy, insofar as the linear approximation of Equation 9.44 becomes inapplicable, but shows in a qualitative way the change in the character of the concavity as p becomes larger in more humid areas and on lower slopes.

The approximate profiles of Figure 9.10 may be used, it is suggested, to evaluate the soil hydrology function, ϕ, empirically from a series of measured slope profiles for the same soil and climate, but differing in divide convexity. Using Figure 9.11, which is derived from Figure 9.10, the change in a/s between the divide and the point of inflexion may be used to estimate the value of γ corresponding to each value of divide convexity (and hence $z_0 = ia/s$). From a series of profiles, γ can be built up as an empirical function of z_0. By noting that

$$\gamma^2(z_0) = z_0 \lambda \phi(z_0) = z_0 \phi(z_0)\frac{d\phi}{dz}, \tag{9.53}$$

evaluated at z_0, and by separating variables, the slope hydrology function is obtained as:

$$\phi^2(z) = \int_0^z 2\frac{\gamma^2(z_0)}{z_0}\,dz_0 \tag{9.54}$$

The design slopes produced above are characteristically convexo-concave in profile, with a point of inflexion at a distance of a few metres from the divide for semi-arid slopes (see Table 9.1), and at a distance of some tens or hundreds of metres in humid temperate areas. If the log–log plots of Figure 9.9 are viewed as a sequence of profiles as the overall lowering rate is gradually reduced, it may be seen that the evolution suggested is very close to one in which the slope profile is translated vertically downwards on the logarithmic scale, corresponding to a proportional loss of height in real terms; that is, to a state in which the rate of lowering is proportional to elevation at each point. This approximation is seen to be a good one for the upper profiles (small p, large u_T), and a reasonable one throughout. It can thus be seen graphically that the constant rates of lowering assumed above give profiles which are surprisingly close in form to the second type of design slope considered, for which it is assumed that the rate of lowering is proportional to elevation above the basal point. Such a slope has been defined as a 'characteristic form' (Kirkby, 1971) because it is characteristic of the assemblage of slope processes acting upon it. Its advantage as a practical design slope is that it recognises a fixed basal point, which is a common requirement of many designs, for example for a road-cutting.

As before, the basic equations are the continuity Equation (9.24), this time with

$$\frac{\partial y}{\partial t} = -m \cdot y \tag{9.55}$$

which is the basic assumption of the characteristic form together with Equation 9.29. The most practical ways of solving these equations are either numerically, or approximating analytically by iteration. In adopting the latter approach, the first approximation can be obtained by setting the elevation y equal to a constant, in which case the solution is identical to that already obtained, reinforcing the idea that the design procedures lead to broadly similar slopes. As a very simple example, the case of a straight slope on which only creep processes operate ($\phi \equiv 0$) may be cited. The first approximation to a slope with a divide at $x = 0$ at elevation y_0; and a basal point at $y = 0$, $x = x_1$ is

$$y = y_0[1 - (x/x_1)^2] \tag{9.56}$$

The exact solution for the characteristic form in this case is

$$y = y_0 \cos\left(\frac{\pi x}{2x_1}\right) \tag{9.57}$$

The two solutions differ by a maximum of $0.06y_0$, and such agreement is typical of the closeness of the two types of design profile to one another.

As a second example of a characteristic form, the case in which subsurface flow may be neglected is taken. This corresponds, as has been pointed out above, to

$\phi \equiv 1$. For a parallel-contour slope ($a = x$), the first approximation (constant down-cutting solution) gives, for the same boundary conditions as before:

$$y = y_0\left[1 - \frac{\log(1 + x^2/u_0^2)}{\log(1 + x_1^2/u_0^2)}\right] \tag{9.58}$$

This is, as might be expected from the previous analysis, a convexo–concave slope with a point of inflexion at $x = u_0$. If Equation 9.58 is substituted into Equation 9.55, a second approximation to the characteristic form is obtained

$$y = y_0\left[1 - \frac{\begin{array}{c}\frac{1}{2}\log(1 + x_1^2/u_0^2)\log(1 + x^2/u_0^2) - \frac{1}{4}\log^2(1 + x^2/u_0^2) \\ + \log(1 + x^2/u_0^2) - (\tan^{-1}x/u_0)^2\end{array}}{\frac{1}{4}\log^2(1 + x_1^2/u_0^2) + \log(1 + x_1^2/u_0^2) - (\tan^{-1}x/u_0)^2}\right] \tag{9.59}$$

The main qualitative difference between Equations 9.58 and 9.59 concerns the position of the greatest slope (point of inflexion). In Equation 9.59 its position depends on slope length, x_1, as well as on the dimension u_0. This relationship is plotted in Figure 9.12. It may be seen that the distance to the point of inflexion increases towards u_0 for very long slopes.

The characteristic form assumptions are no longer strictly true when the slope

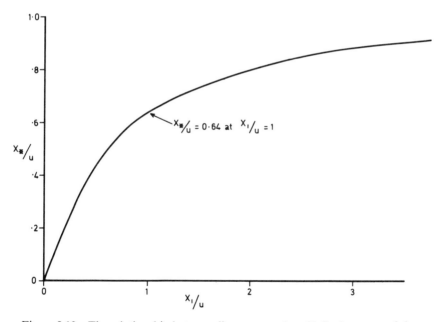

Figure 9.12 The relationship between distance to point of inflexion, x_*, and slope length x_1 for a slope with uniform overland-flow production. For equations 9.58 and 9.59, the value of u in this curve $= u_0$. In comparing with Figure 9.10, the value of u should be taken as the distance to the point of inflexion, obtained from Figure 9.10

hydrology function, ϕ, is not a constant, but only an approximation, valid over a range of slope gradients. Furthermore, even a second approximation to the characteristic form cannot be obtained analytically. Nevertheless, the form of the profiles shown in Figures 9.9 and 9.10 do not differ greatly from the form of Equation 9.58 in the upper part of the slope, down to the point of inflexion. It is therefore argued that Figure 9.12 shows approximately what happens in changing from a constant down-cutting profile to a characteristic-form profile. If the distance to maximum slope is read from Figure 9.9 or Figure 9.10 and this value is used for u in Figure 9.12, then for short slopes ($x \leqslant u$), Figure 9.12 shows roughly how much of the slope will be convex, and where the greatest slope will occur, in a characteristic form profile.

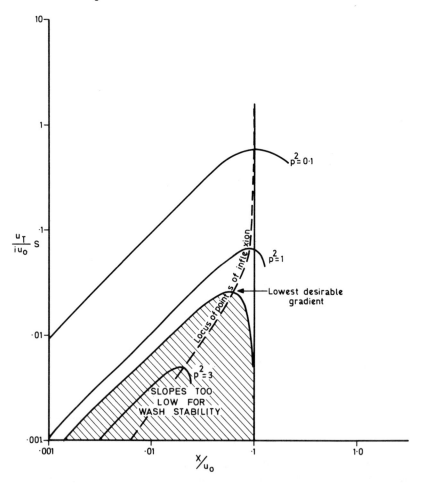

Figure 9.13   Example of the range of design slopes, stable to wash processes and of total length $0.1\ u_0$, for a *fixed* basal elevation. Slopes in the shaded area will tend to develop gullies

As a practical design procedure, the following guidelines are suggested. The parameter u_0 may be calculated from climatic considerations together with a soil-storage measure, h_c, which depends mainly on vegetation cover and hence land-use. The hillslope hydrology is represented by the parameters which can be calculated by either u_T (Figure 9.9) or γ (Figure 9.10). The required slope dimensions determine the divide convexity and hence the parameter p. Together these define a 'constant down-cutting' design slope. For reasons which will be discussed below, it is not advisable to include a concave section in this design. If the slope needs to be designed for zero basal erosion, then it should be modified to a characteristic-form design, on the basis of Figure 9.12. The proportionate shift in distance to the point of inflexion should be applied to re-scale both slope and distance axes to obtain a new design slope of the required length. A series of trials, starting from different constant down-cutting profiles, may be required to find a characteristic form profile of the correct total height. In Figure 9.13, such a series of transformed curves have been derived from Figure 9.9 for a slope of total length $x = 0 \cdot 1 u_0$. The point for which the untransformed distance to the point of inflexion is equal to the design slope length should be taken as the lowest desirable gradient for the overall profile. Lower-angle slopes will tend to become unstable, through the formation of gullies as described in the next section.

9.7 IMPLICATIONS FOR DRAINAGE DENSITY

Turning to the much longer time scale in which significant valleys are formed, the principal interest of a sediment model lies in predicting the scale of the landscape; or, in other words, its drainage density. An additional important consideration is the three-dimensional form towards which drainage basins tend, which can be treated as an independent 'shape' factor. Both of these aspects of the drainage basin are thought to respond significantly to the hillslope hydrology, and some progress may be made towards the prediction of drainage density.

Drainage density is a variable which is difficult to define in practise. For long-term studies, it must be defined in terms which are relatively permanent, so that dynamic definitions, for example, definitions in terms of instantaneous wet channel length (Gregory and Walling, 1968), do not seem appropriate. Four criteria which may be related to the empirical problem and to a sediment model are:

(i) the point at which water erosion first becomes effective
(ii) the point at which water flow becomes perennial
(iii) the point of inflexion at which profiles first become concave
(iv) the point at which valley formation first becomes important.

All but the last of these criteria have already been discussed in the sections above. If N_{OF} is set greater or equal to 1 in Equation 9.16 and the expression for u_0 (Equation 9.28) inserted, the distribution of effective erosion is given by

(i)
$$\phi\left(\frac{ia}{s}\right) \geqslant \frac{u_0 r_0}{224} \tag{9.60}$$

Similarly, perennial water flow occurs roughly (Equation 9.19) when $q_{OF} = R$, so that

(ii)
$$\phi\left(\frac{ia}{s}\right) \geqslant \frac{u_0 R}{224} \qquad (9.61)$$

The point of inflexion on a profile is given by a more complex expression, exemplified by Figures 9.9 or 9.10. In practise the criterion, for the constant down-cutting case, occurs at some intermediate position between these above. In semi-arid areas where subsurface flow is not important, it tends to be closer to the position given by criterion (ii). Where down-cutting at the base of the slope is less than the average rate, a case which includes the characteristic form, then the point of inflexion is shifted a short distance upslope, as has been described in the previous section.

The fourth criterion, concerning the formation of valleys, is the most difficult to apply in theory and practise, but is most clearly related to the long-term evolution of the landscape. It has been shown (Carson and Kirkby, 1972; Smith and Bretherton, 1972) that small-slope irregularities will tend to grow if, and only if,

$$a\frac{\partial S}{\partial a} > S \qquad (9.62)$$

where the partial differential indicates differentiation keeping slope gradient constant. Applying this criterion to a more general case, Smith and Bretherton have further shown that the critical instability condition is equivalent to the condition for the constant down-cutting slope profile to change from convex to concave. It is also a *sufficient* condition for the characteristic form profile to be concave. Thus, for a characteristic-form profile, the distance from the divide to the onset of wash instability is equal to the distance to the point of inflexion on the constant down-cutting profile, but the characteristic form itself is actually *concave* at this point. For a given slope profile, the maximum length it can attain before being subject to gully or valley formation is given by this criterion for wash stability. It clearly sets a lower limit on drainage density, on the assumption that all slopes reach their maximum size. If a profile has a maximum stable area-drained-per-unit-contour-length of a_c, and emerges into the valley axis of its drainage basin at a width Ω the minimum drainage density is obtained by summing along both sides of the valley network. Thus the total basin area, A, in a valley with total valley network length, L, is:

$$A = \sum_{2L} \Omega \cdot a_c = 2L\bar{a}_c \qquad (9.63)$$

where $\bar{a}_c$ is the mean value of a_c, so that the minimum drainage density,

$$D = \frac{L}{A} = \frac{1}{2\bar{a}_c} \qquad (9.64)$$

The stability distance, $\bar{a}_c$, is thus equivalent to the maximum 'distance of overland flow', in the terminology of Horton (1945) and others.

Substituting the wash stability criterion of Equation 9.62 into the general sediment transport, Equation 9.29, a slope profile can be shown to be stable if and only if

$$\frac{2al\left[\phi\left(\frac{ia}{s}\right) + I/\rho\right]}{u_0^2 + I^2} \leqslant 1 \qquad (9.65)$$

where I is written for the integral

$$\frac{1}{\Omega}\int_0^x \Omega\phi\left(\frac{ia}{s}\right)dx$$

Turning first to the simplest forms as an illustration of the meaning of the stability criterion, the case of soil creep ($\phi \equiv I \equiv 0$) on a straight slope indicates stability under all circumstances. For wash ($\phi \equiv 1; I = a$), the stability criterion is simply that $a \leqslant u_0$, whatever the form of the topography. For semi-arid areas therefore, in which this approximation is a fair one, the drainage density is estimated as $D = 1/2u_0$. Referring to the values shown in Table 9.1, a value of $180\,\text{km}/\text{km}^2$ is predicted for South-east Arizona. This value is certainly of the correct order of magnitude, but values for the other areas are clearly not appropriate.

Where subsurface flow is important, the analysis for wash stability is most easily done by using the predicted profiles under conditions of constant down-cutting, such as those shown in Figures 9.9 or 9.10. The stability distance is given by the maximum slope angle, but is clearly dependent on the value of that angle. In practise the maximum slope angles differ only slightly between constant down-cutting profile and characteristic-form profile, so that the values may be used interchangeably. This means that observed maximum valley side-slopes may be related to stability distances, and so to predicted minimum drainage densities. Using the values in Table 9.1 and the curves of Figure 9.9 (extended to lower values of x and s), the values obtained for maximum slopes of $\tan^{-1}\left(\frac{1}{2}\right)$ are shown in Table 9.3.

Table 9.3 Maximum stable distances and predicted minimum drainage densities for the four locations in Table 9.1

Location	Maximum stable distance (m)	Predicted minimum drainage density (km/km²)
UK moderate rainfall	1,500	0·3
UK high rainfall	90	5·6
SE Arizona	2·8	180
Washington DC	360	1·4

The values in Table 9.3 appear to be of the correct order of magnitude, and suggest that a knowledge of hillslope hydrology may help in the understanding of differences in drainage density in *humid* areas, a topic which has hitherto proved intractable. Inspection of Figure 9.9 indicates, for instance, that drainage density should be an increasing (and roughly-linear) function of i/s in an area of uniform climate and broadly-similar soils. Thus, areas of permeable bedrock (low i) and high slopes may be expected to have relatively low drainage density. Areas of permeable bedrock might also be expected to show a large increase in average subsurface intensity, i, under former permafrost conditions, so that the ratio of current to maximum former drainage density would be less than in impermeable areas. Such predictions are broadly valid for southern England, and show that much can be done.

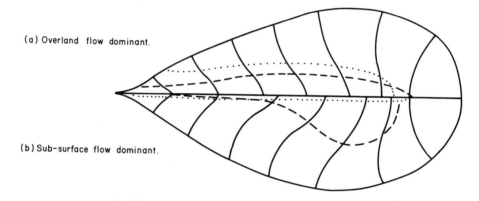

(a) Overland flow dominant.

(b) Sub-surface flow dominant.

Figure 9.14   Idealised first-order basins produced by dominant flow regimes (dotted line = limit of concave profiles; dashed line = limit of concave contours)

Characteristic-form profiles can be shown to be close to eventual profiles produced, irrespective of initial slope forms. If this argument is applied to the assemblage of profiles in a drainage basin, it can be argued that the entire three-dimensional basin form tends towards some characteristic form in the same way. Although no formal analysis has been completed, the examination of some simulated examples and measurement of real basins shows that areas with subsurface flow rather than overland flow or vice versa tend to show a trade-off between profile concavity and contour concavity. Where overland flow is dominant, profiles tend to have an appreciable basal concavity without showing convergence of flow down the profiles except very close to the valley axis. There thus tends to be a greater area of concave profiles (25%, say) than of concave contours (20% say) in a typical first-order basin (Figure 9.14(a)). Where subsurface flow is dominant (Figure 9.14(b)), the profile concavity is commonly exceedingly narrow (10% of the area), but a spoon-shaped hollow is common in the stream head area, giving a total of 30% or more of the basin with concave contours. It can be argued that this distinction reflects the most efficient form under each flow regime. For overland-flow regimes, hollows within the stable area tend to produce flow interference rather than reinforcement, and so tend to reduce the area of effective erosion. Where subsurface flow is important however, the hollows are needed to generate saturated conditions and extend the area of erosion towards the valley head. Increased subsurface flow produced by more frequent rain and a high soil-water capacity in near-surface layers (mainly in organic layers), is thus able to influence the areal distribution of surface wash erosion, and so mould the landscape both in terms of the overall scale or drainage density and through the three-dimensional shape of drainage basins.

9.8 IMPLICATIONS FOR REMOVAL OF MATERIAL IN SOLUTION

Runoff waters also carry dissolved and colloidal material from hillslopes, and, in humid areas, the combined total is commonly the major component of the sediment budget. Overland flow is significant in picking up highly soluble materials which are derived from surface organic litter, fertiliser and rainfall. Most soluble rock material however, comes into equilibrium with water in a matter of days rather than minutes. A first approximation to the rate of net solute removal (including colloids finer than $2\,\mu$m) can therefore be obtained by assuming that overland flow picks up no solutes, and that subsurface flow comes into some equilibrium with soil minerals.

In a humid area therefore, the rate of chemical removal is roughly proportional to the subsurface flow, which in turn depends on the hillslope hydrology. As has been described above, the overland flow increases downslope, leaving a smaller proportion of the rainfall to add to the downslope subsurface flow. As a result, the rate of solute removal should be somewhat reduced at downslope sites and in hollows. In a humid area, the increase in evaporation in hollows is usually unimportant, though it tends to accentuate the reduction in subsurface flows. In arid and semi-arid areas the contribution of solutes to the sediment budget is generally slight compared to mechanical sediment removal, so that only the humid case need be considered here. In hollows, where the effects of reduced solute removal are most important, the correction to overland flow of Equation 9.17 needs to be included:

$$q_{OF} = R\,e^{-h_c/r_0}\phi\left[\frac{ia}{s}\left(1 - \frac{q_{OF}}{R - E}\right)\right] \qquad (9.17)$$

For large a/s, the exact form of the flow law ϕ is unimportant, and the left-hand side of Equation 9.17 becomes more or less constant. Then

$$a/s \simeq \frac{B}{1 - \dfrac{q_{OF}}{R - E}} \qquad (9.66)$$

for some constant B, and

$$q_{OF} \simeq R - E - \frac{B}{a/s} \qquad (9.67)$$

and the subsurface flow contribution

$$q_{SF} = R - E - q_{OF} \simeq \frac{B}{a/s} \qquad (9.68)$$

The rate of solute removal has previously been seen to be proportional to the subsurface-flow contribution, and so is therefore roughly inversely proportional to a/s. This approximation is valid in the areas of high a/s, which are the only areas in which the solute removal rate is appreciably reduced in this way.

As hollows have been argued to be areas of appreciable overland flow and wash erosion in humid areas, the reduction in solute removal is probably not significant in determining the form of the landscape. The deduction of reduced solute removal in hollows does, however, reinforce the inference that hollows are a product of mechanical erosion by wash processes rather than loci of accelerated soil formation. Observations of greater soil depth in hollows are therefore interpreted

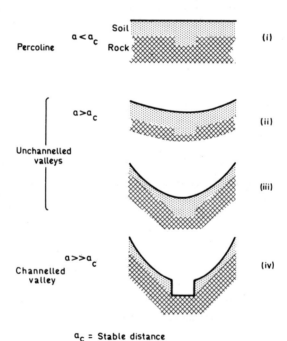

Figure 9.15 Hypothetical spatial sequence of valley-bottom cross-sections in a downstream direction

in general as showing an alternation of channel extension in extreme storms, and infilling by slope processes in more moderate storms. This alternation may, in the long run, lead either to the formation of a normally dry valley-head if the distance from the divide exceeds the stable value, or to a 'percoline' feature (Bunting, 1961) if the distance is less than the stable value, in which case the former channel need leave no trace on the surface topography. A hypothetical spatial sequence of sections is shown in Figure 9.15. At (i) a percoline is shown, with no tendency for the occasional major storm to form a permanent valley. At (ii), the stable distance has been exceeded, so that the occasional storm channels cumulatively produce a valley, even though infilling occurs most of the time. At (iii) and (iv) a surface channel becomes more frequent, although there may be occasional periods of infilling (iii) until the perennial channel stage (iv) is reached.

9.9 IMPLICATIONS FOR SHALLOW LANDSLIDES

Almost planar landslides have been shown to occur on clay slopes (Skempton, 1964; Chandler, 1971), and their existence inferred for residual soils on sandstones and limestones (Carson and Petley, 1970). In all cases, a good approximation to the failure conditions can be obtained if it is assumed that the slides are infinite in extent, and that the effective cohesion has been reduced to zero. Figure 9.16 shows a

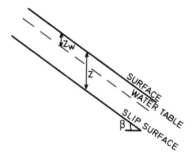

Figure 9.16   Notation for analysis of shallow slides

slope at angle β, with a hydrostatic water table at depth z_w below the surface. For a potential slide surface at depth z, the safety factor (SF) is:

$$SF = \frac{\tan\psi\left[\left(1 - \frac{\gamma_w}{\gamma}\right) + \frac{\gamma_w}{\gamma}\cdot\frac{z_w}{z}\right]}{\tan\beta} \qquad (9.69)$$

where

γ = unit of weight of soil
γ_w = unit weight of water
ψ = angle of friction for soil

The maximum stable slope angle α, is obtained when the safety factor $= 1$, when $\beta = \alpha$ and

$$\tan \alpha = \tan \psi \left[\left(1 - \frac{\gamma_w}{\gamma} \right) + \frac{\gamma_w}{\gamma} \frac{z_w}{z} \right] \qquad (9.70)$$

The ratio γ_w/γ is constant, and commonly has a value of approximately 0.5, so that

$$\tan \alpha \simeq \tfrac{1}{2} \tan \psi \left(1 + \frac{z_w}{z} \right) \qquad (9.71)$$

For a constant value of ψ with depth, Equation 9.71 suggests failure at a large depth. In a typical real soil, however, ψ is thought to be a minimum in the clay-rich B-horizon, and rises to large values below the weathering front, as interlocking

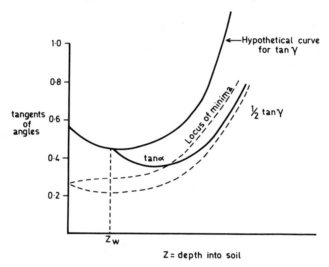

Figure 9.17   Hypothetical stability analysis for shallow slides in a soil with a 'normal' variation in properties with depth

improves towards sound bedrock. Equation 9.71 is illustrated in Figure 9.17, showing

$$\tan \alpha = \tan \psi \quad \text{for} \quad z < z_w$$

(water table below the slip surface), and an asymptotic approach to $\tan \alpha = \tfrac{1}{2} \tan \psi$ at large depths (below the water table). Failure can be expected if the slope is greater than the minimum on the $\tan \alpha$ curve, at a depth greater than z_w. Over time, as the water table fluctuates seasonally, failure will occur if the slope is greater than the minimum point attained by the locus of these minima, which is thought to correspond to the minimum depth to water table attained in practice.

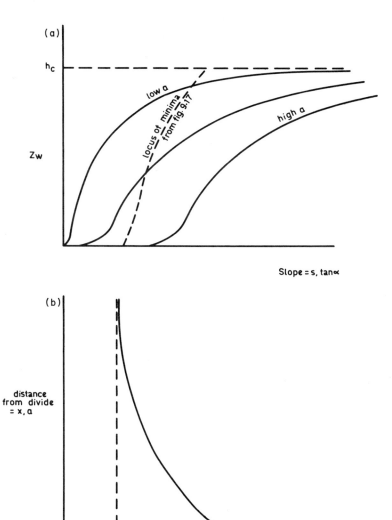

Figure 9.18:(a) The influence of hillslope hydrology on critical angles for slope stability, on the basis of Figure 9.17 and Equation 9.71: (b) Hypothetical variation of stable angle with distance from the divide

As piezometric pressures are only transmitted slowly through soil pores, the relevant moisture levels are thought to be subsurface levels, which persist for at least several days. As a guide to the wettest likely conditions, a value of 2 to 5 times the average value of daily effective rainfall is thought appropriate to an exceptional year in the wet season. The relationship between z_w and position on the slope depends on the hillslope hydrology, since for an effective rainfall i, the depth to the water table is

$$z_w = h_c - h = h_c - r_0 \log_e [\phi(ia/s)] \qquad (9.72)$$

Figure 9.18(a) combines this family of curves, as 'a' varies, with the locus of minima from Figure 9.16, to produce a relationship shown diagrammatically in Figure 9.18(b) between distance from the divide and maximum stable slope. At long distances from the divide, failures are close to the surface at $\tan \alpha = \frac{1}{2} \tan \psi_s$ where ψ_s is the angle of friction at the soil surface. Closer to the divide, the stable angle is thought to increase, allowing either an overall concave slope, or steeper straight slopes where the total length is less. The argument illustrated in Figures 9.17 and 9.18 shows that lower-angle slopes, although they have higher moisture levels, normally still tend to be more stable.

As slides occur on a hillside, more weathered material tends to accumulate downslope, so that there is also a tendency for angles of friction to decrease systematically downslope. This effect will tend to enhance the influence of the rising piezometric levels, and therefore enhance the tendency to overall concavity. Non-zero values for cohesion have a similar effect which is probably dominant at least in the short term. It is however recognized that landslides occur as large discrete units. As a result, any simple model which treats the hillslope as a continuously-varying system can, at most, describe the gross morphology of a hillside, while ignoring the considerable heterogeneity produced by individual slide blocks.

9.10 CONCLUSIONS

The application of field results and modelling methods for hillslope hydrology appears to open up important avenues for research into sediment processes, which have been explored above from a theoretical point of view. They indicate potentially important directions for sediment modelling, for slope design and for the spatial pattern of hillslope-sediment production.

The work of Bennett (1974) and Foster and Meyer (1972) represents the present state-of-the-art in hillslope sediment modelling. One of its significant assumptions is that overland flow is produced uniformly over a hillslope. The analysis above shows that the assumption is a reasonable one for the semi-arid areas where wash erosion is most important, but breaks down severely in humid and temperate areas. The day-by-day rainfall model put forward shows how non-uniform overland-flow production can be estimated. It is argued above that the resulting pattern of wash erosion is significant enough in amount to have a major effect on hillslope and valley forms in the long term. It is therefore important that it should be accurately modelled.

In modelling subsurface flow, the increasing discharge downslope is very sensitive to the exact form of the rating function relating subsurface discharge to the soil-moisture level. The linear model which has mainly been used above, corresponds to the simplest form of Darcy's law. It leads to a very rapid build-up of soil-moisture level downslope, and a resulting high estimate for overland flow. Higher power laws or an exponential rating curve predict a somewhat less-rapid build-up. Possible errors from this source have a rather slight influence unless slopes are concave, but are then crucial.

In the short term, patterns of subsurface and overland flow may have important implications for localizing the area from which significant amounts of sediment are removed. In the longer term, patterns of erosion mould the landscape and control its scale and three-dimensional form. The scale is commonly measured as drainage density (km of channel per km^2 of drainage area), and an attempt has been made above to predict its variation for both arid and humid areas. The success of such predictions underlines the importance of modelling hillslope flows.

From a series of criteria for stable lengths, the criterion for small hollows to enlarge into valleys (Smith and Bretherton, 1972) is suggested not only as the key control on drainage density, but as a useful design criterion for a slope which will

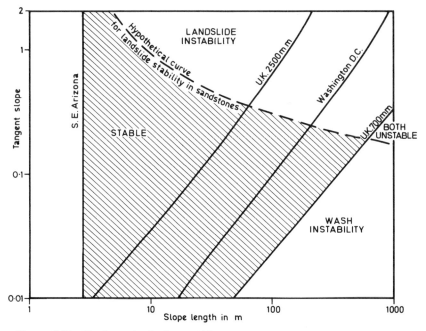

Figure 9.19 Design criteria for stability of sandstone slopes under shallow landslides and wash. Stable designs are to *left* of curves

o

not tend to gully severely. Curves such as those of Figure 9.19 can be constructed for particular areas to indicate stability with regard to wash erosion. When combined with a curve for landslide stability (Fig. 9.18), which also involves a consideration of hillslope hydrology, it is suggested that slope-stability domains may be established for a given climate and lithology, as shown diagrammatically in Figure 9.19. Together, they perhaps put another limitation on predicted drainage density, at the value of stable slope length where the two curves intersect (indicating, for example, in Figure 9.18 a drainage density of $0.8 \, km/km^2$ for lowland Britain).

The implications for sediment research are very plain. Predictions above, of the pattern of wash erosion, urgently require field observations to back them up. Appropriate forms for the rating curve relating subsurface discharge to subsurface moisture level are also required, although at first approximations may be obtained from hydrograph recession curves at one scale, and from actual slope profiles at another scale. These are perhaps the most immediate questions raised, but it is clear that as we absorb the implications of hillslope hydrological measurements and models, a new generation of sediment process studies will be born.

REFERENCES

Bennett, J. P., 1974, 'Concepts of mathematical modelling of sediment yield'. *Water Res. Res.*, **10**(3), 485–492.

Betson, R. P., 1964 'What is watershed runoff?', *J. Geophys. Res.*, **69**(8), 1541–1552.

Bunting, B. T., 1961, 'The rôle of seepage in soil formation, slope development and stream initiation', *Am. J. Sci.*, **259**, 503–518.

Carson, M. A. and Kirkby, M. J., 1972, *Hillslope Form and Process*, Cambridge University Press, pp. 212 and 394.

Carson, M. A. and Petley, D. J., 1970, 'The existence of threshold hillslopes in the denudation of the landscape', *Trans. Institute of British Geographers*, **49**, 71–95.

Chandler, R. J., 1971, 'Landsliding on the Jurassic escarpment near Rockingham, Northamptonshire' *Inst. Brit. Geographers, Special Publication* No. 3, 111–128.

Foster, G. R. and Meyer, L. D., 1972, 'A closed form soil erosion equation for upland areas', in *Sedimentation Symposium to Honour Professor Hans Albert Einstein*, Shen, H. W. (Ed.), pp. 12.1–12.19, Colorado State Univ., Fort Collins.

Green, W. H. and Ampt, G. A., 1911, 'Studies on soil physics. 1. The flow of air and water through soils', *J. Agric. Sci.*, **4**(1) p 1–24.

Gregory, K. J. and Walling, D. E., 1968, 'The variation of drainage density within a catchment', *Internat. Assoc. Sci. Hydrology*, XIII Annee, No. 2, 61–68.

Harley, J. B., 1975, *Ordnance Survey Maps—A Descriptive Manual*, Ordnance Survey, Southampton.

Hewlett, J. D., 1961, 'Soil moisture as a source of base flow from steep mountain watersheds', *US Dept Agric. Forest Ser., Southeastern Forest Experimental Station, Asheville, North Carolina* Station paper 132.

Horton, R. E., 1933, 'The rôle of infiltration in the hydrologic cycle', *Transactions of the Amenial Geophysical Union*, **14**, 446–460.

Horton, R. E. 1945, 'Erosional development of streams and their drainage basins: hydrophysical approach to quantitative morphology', *Bull. Geolog. Soc. Am.*, **56**, 275–370.

Kirkby, M. J., 1971, 'Hillslope process–response models based on the continuity equation', *Inst. Brit. Geographers, Special Publication* No. 3, 15–30.

Kirkby, M. J., 1975, 'Hydrograph modelling strategies', in *Process in Physical and Human Geography*, Peel, R. F., Chisholm, M. D. and Haggett, P. (Eds.), Heinemann, pp. 69–90.

Kirkby, M. J., 1976, 'Hydrological slope models—the influence of climate', in *Geomorphology and Climate*, Derbyshire, E. (Ed.), John Wiley, London.

Philip, J. R., 1957/1958, 'The theory of infiltration', *Soil Sci.*, **83**, 345–57 & 435–48; **84**, 163–77, 257–64 & 329–39; **85**, 278–86 & 333–37.

Shanan, L. T., 1974, personal communication.

Skempton, A. E., 1964, 'Long term stability of clay slopes', *Geotechnique: the 4th Rankine lecture*, **14**, 77–102.

Skempton, A. W. and Delory, F. A., 1957, 'Stability of natural slopes in London clay', *Proc. 4th Internat. Conf. Soil Mechs. Foundation Engr., London.*

Smith, T. R. and Bretherton, F. P., 1972, 'Stability and the conservation of mass in drainage basin evolution', *Water Res. Res.*, **8**, 1506–24.

Tennessee Valley Authority, Office of tributary development, Knoxville, Tennessee, 1963: *Res. Paper No. 2*, 'A water yield model for analysis of monthly runoff data'. April 1964: *Res. Paper No. 4*, 'Bradshaw creek—Elk River. A pilot study in area–stream factor correlation'. May 1964: *Synopsis of Bradshaw Creek–Elk River project.*

Weyman, D. R., 1974, 'Runoff processes, contributing area and streamflow in a small upland catchment', *Inst. Brit. Geographers, Special Publication* No. 6, 33–43.

Glossary of terms

R. J. Chorley

Department of Geography, University of Cambridge, UK

(References are included in the references for Chapter 1.)

Absorption
The entry of water into the soil or rock by all natural processes, including infiltration and gravity flow from streams into valley alluvium (Langbein and Iseri, 1960).

Antecedent moisture
Soil-moisture content preceding a given storm.

Baseflow
Stream discharge derived from effluent groundwater seepage (Butler, 1957, p. 337). A time-based definition relating to sustained or fair-weather runoff, largely composed of groundwater effluent (Langbein and Iseri, 1960). Outflow from extensive groundwater aquifers which are recharged by water percolating down through the soil mantle to the water table (Tischendorf, 1969, p. 4).

Canopy
The layer, or layers, of foliage in a tree or bush cover.

Canopy interception loss
Rainfall retained on standing vegetation and evaporated without dripping off or running down the stems or trunks (Helvey and Patric, 1965, p. 194).

Capillary conductivity
The constant term in the Darcy equation expressing the ability of a soil or rock to transmit water under a given gradient of hydraulic head (Philip, 1957–8, p. 163; Hillel, 1971, p. 109; Carson and Kirkby, 1972, p. 42; Wind, 1972, p. 229). For saturated soils it depends on their texture and geometry, for unsaturated soils it is inversely related to soil-moisture content and may be orders of magnitude less than under saturated conditions.

Capillary forces
Those acting on soil water at the grain contacts, and attributable to molecular forces between soil particles and the water (Baver, 1937, p. 433). When less than atmospheric pressure (i.e. producing tension) they are directly related to the surface tension of water and inversely related to the acceleration of gravity and the size of the capillary films.

Capillary fringe
The unsaturated zone containing water in direct hydraulic contact with the water table, and held above the water table by capillary forces (Butler, 1957, p. 337).

Capillary potential
A concept introduced by Buckingham in 1907 to express the specific potential energy of soil water relative to that of water in a standard reference state (i.e. a hypothetical reservoir of pure and free water at atmospheric pressure, at the same temperature as the soil water and at a given and constant elevation). A negative pressure potential resulting from the capillary and adsorptive forces due to the soil matrix (Hillel, 1971, pp. 50 and 57).

Conductivity flow
The flow of soil moisture under gravity (Carson and Kirkby, 1972, pp. 44–46).

Contributing area
The area of a catchment contributing to storm runoff (Betson, 1964).

Critical distance
The distance from a divide at which the depth of overland flow is sufficient to develop a tractive force great enough to entrain surface particles (Horton, 1945).

Darcy's law
An experimentally-derived relationship stating that velocity of fluid flow through a permeable medium is directly proportional to the hydraulic gradient and to the hydraulic conductivity. It is valid only for low velocities well within the laminar range. Being originally stated for saturated flow, it was extended by Richards in 1931 to embrace unsaturated flow. (Swatzendruber, 1960, p. 4037; Hillel, 1971, p. 109; Wind, 1972, p. 229.)

Deep seepage
Infiltration which reaches the water table (Butler, 1957, p. 338).

Delayed flow
That part of measured streamflow which lies below the arbitrarily-chosen line of separation of the hydrograph, and conventionally identified with baseflow (Tischendorf, 1969, p. 5).

Depression storage
The volume of water, forming part of surface detention, which is contained in small natural depressions in the land surface during or shortly after rainfall, none of which runs off (Horton, 1933, p. 447; Horton, 1935; Langbein and Iseri, 1960; Tischendorf, 1969, p. 5).

Detention storage
(*See* Surface detention).

Diffusion component
The term in Philip's (1957–8) infiltration equation representing flow in very discrete steps, from one pore space to the next, in a more or less random fashion (Kirkby, 1969, p. 216).

Diffusivity flow
The slow flow of soil moisture under capillary forces (Carson and Kirkby, 1972, pp. 44–46).

Diffusivity function
The ratio of the hydraulic conductivity to the specific water capacity (Hillel, 1971, p. 111).

Direct runoff
A time-based definition of runoff entering stream channels promptly after rainfall or snow melt; superimposed on baseflow, it forms the bulk of the flood hydrograph (Langbein and Iseri, 1960). The sum of channel precipitation, overland flow and subsurface stormflow (Tischendorf, 1969, p. 5).

Dynamic watershed model
A conceptual model based on the assumption that the area of a catchment contributing to storm runoff is dynamic and can vary in size between storms and during the course of a single storm (TVA, 1965; Hewlett and Hibbert, 1967).

Effective rainfall
  (effective precipitation)
That part of rainfall (or precipitation) that produces runoff (Langbein and Iseri, 1960). Rainfall minus evaporation and interception storage.

Effective surface storage
The depth of water that would have run off into stream channels were it not for the existence of surface storage. It is equal to depression storage plus that water infiltrated out in transit (Cook, 1946, p. 727).

Evaporation
The process by which moisture passes into the atmosphere as vapour (Linsley, Kohler and Paulhus, 1949, p. 154).

Evapotranspiration
Water withdrawn from a land area by evaporation from water surfaces and moist soil, and by plant transpiration (Langbein and Iseri, 1960).

Exfiltration
A term coined by J. R. Philip to define the removal of water from the soil at the ground surface, together with the associated unsaturated upward flow (Freeze, 1974, p. 628).

Exit point
The point on a slope profile separating the zone of infiltration from that of exfiltration (Freeze, 1974).

Field moisture capacity
The amount of water which can be permanently held in the soil in opposition to the downward pull of gravity after the excess water has drained away under gravity and

the rate of downward movement of water has materially decreased. (Horton, 1933, p. 447; Langbein and Iseri, 1960; Liakopoulos, 1965, p. 64.)

Field moisture deficiency
The amount of water which would be required to restore the soil-moisture content to field moisture capacity (Horton, 1933, p. 447). (*See* Retention storage.)

Footslope
The transitional area between the valley sideslope and the channelway (Hack and Goodlett, 1960, p. 6). Sometimes termed a 'cove'.

Gravitational potential
That part of the total potential of soil water due to gravity (Hillel, 1971, p. 52).

Gross rainfall
Rainfall per storm, measured in the open or above the vegetation canopy (Helvey and Patric, 1965, p. 194).

Ground rainfall
That part of the rainfall which actually reaches the ground. It is equal to the total rainfall minus that intercepted by vegetation (Horton, 1933, p. 447).

Groundwater flow
That part of streamflow that has passed into the ground, has become groundwater (i.e. has passed into the zone of saturation), and has been discharged into a stream channel as spring or seepage water (Langbein and Iseri, 1960).

Horton overland flow
Direct surface runoff across saturated soils where rainfall intensity exceeds infiltration plus depression storage (Knapp, 1970b, p. 40; Carson and Kirkby, 1972, pp. 50–51).

Hydraulic conductivity
(*See* Capillary conductivity.)

Hydraulic gradient
The gradient of capillary suction in the soil profile, expressing the head drop per unit distance in the direction of flow (Hillel, 1971, pp. 75 and 83).

Hydraulic force
That force acting on soil water which is produced by a pressure or tension gradient (Carson and Kirkby, 1972).

Hydraulic head
The equivalent height of a liquid column corresponding to a given pressure (Hillel, 1971, p. 59).

Hydraulic potential
(*See* Total potential.)

Hydraulic resistance
The ratio of the thickness of the soil layer to the capillary conductivity of the soil (Swartzendruber, 1960; Hillel, 1971, p. 90).

Hydraulic resistivity
The reciprocal of the hydraulic (capillary) conductivity.

Infiltration
The passage of water through the surface of the soil, via pores or small openings, into the soil mass (Horton, 1933 and 1942; Cook, 1946, p. 727).

Infiltration capacity
The maximum rate at which a given soil can absorb falling rain (or melting snow), when it is in a specified condition (Horton, 1933, p. 447; Horton, 1941, p. 399).

Infiltration rate
The volume rate of infiltration (Cook, 1946, p.727).

Interception
Water retained for some period, however short, after rain has struck the vegetal material above the mineral soil surface (Tischendorf, 1969, p. 6).

Interception loss
The amount of water evaporated or subliminated from rain or snow caught by living or dead plant material. This may be made up both of crown and forest floor litter interception losses (Tischendorf, 1969, p. 6).

Interflow
An intermediate component of runoff, between overland flow and groundwater flow, made up of subsurface flow which returns to form surface runoff without reaching the water table before arriving at the watershed outlet (Amerman, 1965, pp. 503–504).

Limiting infiltration rate
The relatively low and steady infiltration rate of a soil surface, free from sun cracks and biologic structures, which has been wetted at capacity rate for a period long enough to permit full swelling of colloids and all the other associated adjustments of the soil structure to a stable field condition (Horton, 1939b, p. 702; Butler, 1957, p. 344).

Litter
The surface layer of fallen, dead vegetation lying on the mineral soil.

Litter flow
Downslope flow of water in the litter layer (Lamson, 1967; Tischendorf, 1969, p. 89).

Litter interception loss
Rainfall retained in the litter layer and evaporated without adding to the moisture in the underlying mineral soil (Helvey and Patric, 1965, p. 194).

Matric potential
(*See* Capillary potential.)

Net rainfall
Rainfall which enters the mineral soil (Helvey and Patric, 1965, p. 194).

Osmotic potential
The change in potential energy of soil water due to the presence of solutes, which lower the vapour pressure of water and affect its uptake by plants (Hillel, 1971, p. 59).

Overland flow
That part of streamflow which originates from rain which fails to infiltrate the mineral soil surface at any point as it flows over the land surface to channels (Langbein and Iseri, 1960; Tischendorf, 1969, p. 7; Hewlett and Nutter, 1970, pp. 65–66).

Partial area runoff
Storm runoff generated by only a part of the surface of a catchment (Betson, 1964).

Penman formulae
Formulae for the calculation of the energy utilized in evapotranspiration (Penman, 1948 and 1963).

Percolation
The passage of water under hydrostatic pressure through the interstices of a soil or rock, excluding the movement through large openings (Meinzer, 1923; Horton, 1933, p. 446; Langbein and Iseri, 1960).

Percolines
Orthogonal zones of relatively deep soil of distinctive phase on a slope along which downslope movement of moisture becomes concentrated (Bunting, 1961, p. 503).

Percolines
Orthogonal zones of relatively deep soil of distinctive phase on a slope along which downslope movement of moisture becomes concentrated (Bunting, 1961, p. 503).

Permeability
The ability of a soil or rock to transmit water.

Pipe flow
Concentrated subsurface flow of water in natural pipes (Knapp, 1970b, p. 40; Jones, 1971).

Piping
The formation of natural pipes in soil or other unconsolidated deposits by eluviation or other processes of differential subsurface erosion (Jones, 1971).

Porosity
The volume of voids or pore spaces in a soil or rock expressed as a fraction of the bulk volume (Butler, 1957, p. 342).

Potential evapotranspiration
The total amount of evaporation which would take place from a small area in the midst of a large, unbroken, completely-vegetated surface, where moisture amount does not limit the evaporation process (Tischendorf, 1969, p. 5).

Pressure potential
(*See* Capillary potential.)

Quickflow
That portion of the total runoff which is immediately attributable to a precipitation event (Woodruff and Hewlett, 1971, p. 1). (*See* Direct runoff.)

Quick return flow
(Jamison and Peters, 1967, p. 471.) (*See* Quickflow.)

Rainfall excess
Rainfall delivered at the surface in excess of the instantaneous infiltration rate, which forms depression storage and surface detention (Horton, 1933, p. 447; Langbein and Iseri, 1960).

Rainfall intensity
The rate at which rain water reaches the soil or vegetation surface at any instant (Cook, 1946, p. 727).

Rate of surface runoff
The rate at which surface runoff is passing from a given area at any instant (Cook, 1946, p. 727).

Recession flow
Discharge contributing to the recession limb of the storm hydrograph.

Retention storage
That amount of soil moisture by which it is less than the field moisture capacity (Hoover, 1962, p. 40). (*See* Field moisture deficiency.)

Return flow
Infiltrated water which returns to the land surface having flowed for a short distance in the upper soil horizon (Dunne and Black, 1970, p. 483).

Saturated (saturation) interflow
Interflow occurring under saturated conditions (Betson, Marius and Joyce, 1968, p. 603).

Saturated (saturation) overland flow
Surface runoff occurring where the soil is saturated, even though the infiltration capacity has not been exceeded (Carson and Kirkby, 1972, pp. 50–51).

Saturated (saturation) throughflow
Lateral flow in the soil under saturated conditions (Calver, Kirkby and Weyman, 1972).

Secondary baseflow
(Barnes, 1939.) (*See* Interflow)

Seepage lines
(*See* Percolines.)

Seep zone
Saturated soil layer at the slope base from which water enters the stream through its banks (Hewlett and Hibbert, 1963).

Sheet flow
A thin film or sheet of water running over the surface when rainfall intensity exceeds infiltration capacity and surface storage has been filled (Horton, Leach and Van Vliet, 1934).

Soil detention storage
That portion of soil water which is in excess of field moisture capacity and slowly drains under gravity through large soil pores (Fletcher, 1952, p. 359; Hoover, 1962, p. 40).

Soil moisture
All water which is stored in the weathered soil mantle, be it stored only for short periods, or be it depleted by slow drainage which sustains baseflow (Tischendorf, 1969, pp. 7–8).

Soil retention storage
That portion of soil water which is less than field moisture capacity and is retained in the soil, or evaporated or transpired into the atmosphere.

Soil-water detention
Soil water moving under gravity that reaches the stream channel in sufficient time to contribute to the storm hydrograph (Hursh and Fletcher, 1942, p. 486). (*See* Soil detention storage.)

Specific water capacity
The change of water content per unit change of capillary potential (Hillel, 1971, p. 65).

Stemflow
That portion of the gross rainfall which is caught on the vegetation canopy and reaches the litter or mineral soil by running down the stems or trunks (Helvey and Patric, 1965, p. 194).

Stormflow
(*See* Direct runoff.)

Storm hydrograph
The graph of stream discharge against time for a given storm event.

Storm interval
The interval of time from the beginning of a rain, through the rain period and the subsequent dry or rainless period to the beginning of the next subsequent rain period (Horton, 1933, p. 447).

Storm runoff
(*See* Direct runoff.)

Storm seepage
That portion of storm precipitation which infiltrates the surface and moves laterally through the surface soil or through colluvial stream bank material as ephemeral, shallow, perched groundwater above the main groundwater level. It may form part of direct runoff (Hursh, 1936; Barnes, 1939; Hursh and Brater, 1941, p. 870; Langbein and Iseri, 1960).

Stream-head hollow
Concave-valley head where runoff converges (Hack and Goodlett, 1960, p. 6).

Subsurface runoff
The movement of subsurface storm water within the soil layers to stream channels at a rate more rapid than the usual groundwater flow (Hursh, 1936, pp. 301–302). (*See* Interflow and Storm seepage.)

Subsurface Stormflow
That part of streamflow which derives from the lateral subsurface flow of water in saturated soil zones above water-impeding layers, especially in basal hillslope soils, and which discharges into the stream channel without entering the groundwater zone, arriving at the gauging station so quickly as to become part of the stormflow associated directly with a given rainstorm (Hursh, 1936; Whipkey, 1965b, p. 84; Whipkey, 1967b, p. 255; Tischendorf, 1969, p. 8).

Surface detention
That portion of rainwater, other than depression storage, which remains in temporary storage on the land surface as it moves downslope by overland flow and either runs off, is evaporated or is infiltrated after the rain ends (Horton, 1933, p. 447; Horton, 1937a, p. 403; Butler, 1957, p. 338; Ven Te Chow, 1964, p. 20–5).

Surface runoff
That part of runoff which travels over the soil surface to the nearest stream channel, having not passed beneath the surface since precipitation (Langbein and Iseri, 1960).

Surface storage
Surface detention plus depression storage.

Surface stormflow
That portion of rainwater (or snowmelt) which fails to infiltrate the mineral soil at any point along its path of travel to the gauging stations in a perennial stream (Hewlett, 1974, p. 606).

Tension of soil moisture
Soil moisture pressures which are less than atmospheric pressure (Carson and Kirkby, 1972, pp. 43–44).

Throughfall
That portion of gross rainfall which directly reaches the forest litter through spaces in the vegetative canopy and as drip from leaves, twigs and stems (Helvey and Patric, 1965, p. 194).

Throughflow
Downslope flow of water occurring physically within the soil profile, usually under unsaturated conditions except close to flowing streams, occurring where permeability decreases with depth (Kirkby and Chorley, 1967).

Total potential of soil water
The sum of the gravitational head, the pressure (capillary) potential and the osmotic potential of the soil water (Hillel, 1971, p. 52).

Translatory flow
The lateral throughflow of 'old' rainwater stored in the soil and released to channel flow by a process of displacement by new rain (Hewlett and Hibbert, 1967, p. 279).

Transmission component
The constant in Philip's (1957–58) infiltration equation representing the steady unimpeded downward laminar flow of water through a continuous network of large soil pores (Kirkby, 1969, p. 216; Gregory, and Walling, 1973, p. 285).

Transmission zone
Zone at the surface of the soil of essentially constant hydraulic conductivity and high saturation developing during a rainstorm (Hansen, 1955, p. 104).

Transpiration
The process by which water in plants is transferred to the atmosphere as water vapour (Linsley, Kohler and Paulhus, 1949, p. 169).

Total interception loss
Rainfall per storm retained by the canopy and litter, and evaporated without adding to the moisture in the mineral soil (Helvey and Patric, 1965, p. 194).

Transmission of soil moisture
The steady movement of water through the soil under gravity (Philip, 1957–58).

Unit hydrograph
The hydrograph of direct surface runoff (or quickflow) resulting from a storm of unit amount and duration (Sherman, 1932).

Vadose water
Water in the zone of aeration (i.e. that volume of soil and bedrock above the water table).

Valley side-slope
The straight portion of the valley side (Hack and Goodlett, 1960, p. 6).

Variable source model
Alternative name for Dynamic watershed model.

Wetting front
The lower limit of the wetting zone of the soil (Bodman and Coleman, 1943; Hansen, 1955; Philip, 1957–58, pp. 171–173).

Wetting zone
Zone of low hydraulic conductivity and saturation lying below the soil-transmission zone and limited below by the wetting front (Hansen, 1955, p. 104).

Author index

The *italic* page numbers refer to entries in the lists of references at the ends of chapters.

Subject index